C0-CEA-022

DATE DUE

DEC 1 5 2007

DEC 0 3

NOV 2 5 2016

Library Store #47-0108 Peel Off Pressure Sensitive

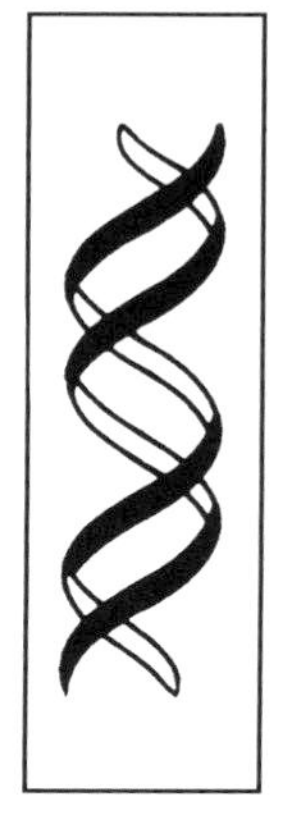

Genetics of Sexual Differentiation and Sexually Dimorphic Behaviors

Advances in Genetics, Volume 59

Serial Editors

Jeffery C. Hall
Waltham, Massachusetts

Jay C. Dunlap
Hanover, New Hampshire

Theodore Friedmann
La Jolla, California

Veronica van Heyningen
Edinburgh, United Kingdom

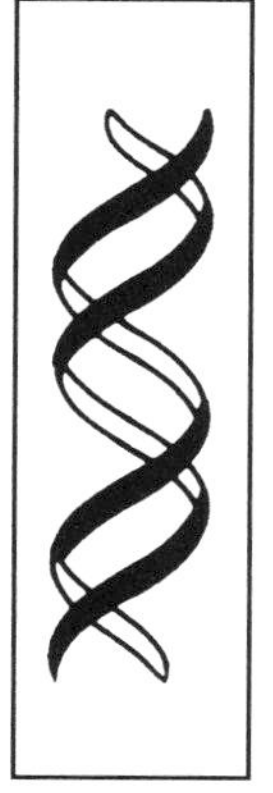

Genetics of Sexual Differentiation and Sexually Dimorphic Behaviors

Edited by
Daisuke Yamamoto
Division of Neurogenetics
Graduate School of Life Sciences
Tohoku University
Sendai, Japan

WITHDRAWN
FAIRFIELD UNIVERSITY
LIBRARY

ELSEVIER

AMSTERDAM • BOSTON • HEIDELBERG • LONDON
NEW YORK • OXFORD • PARIS • SAN DIEGO
SAN FRANCISCO • SINGAPORE • SYDNEY • TOKYO
Academic Press is an imprint of Elsevier

Academic Press is an imprint of Elsevier
84 Theobald's Road, London WC1X 8RR, UK
Radarweg 29, PO Box 211, 1000 AE Amsterdam, The Netherlands
Linacre House, Jordan Hill, Oxford OX2 8DP, UK
30 Corporate Drive, Suite 400, Burlington, MA 01803, USA
525 B Street, Suite 1900, San Diego, CA 92101-4495, USA

First edition 2007

Copyright © 2007 Elsevier Inc. All rights reserved

No part of this publication may be reproduced, stored in a retrieval system
or transmitted in any form or by any means electronic, mechanical, photocopying,
recording or otherwise without the prior written permission of the publisher

Permissions may be sought directly from Elsevier's Science & Technology Rights
Department in Oxford, UK: phone (+44) (0) 1865 843830; fax (+44) (0) 1865 853333;
email: permissions@elsevier.com. Alternatively you can submit your request online
by visiting the Elsevier web site at http://elsevier.com/locate/permissions, and selecting
Obtaining permission to use Elsevier material

Notice
No responsibility is assumed by the publisher for any injury and/or damage to persons
or property as a matter of products liability, negligence or otherwise, or from any use
or operation of any methods, products, instructions or ideas contained in the material
herein. Because of rapid advances in the medical sciences, in particular, independent
verification of diagnoses and drug dosages should be made

ISBN: 978-0-12-017660-1
ISSN: 0065-2660

For information on all Academic Press publications
visit our website at books.elsevier.com

Printed and bound in USA
07 08 09 10 11 10 9 8 7 6 5 4 3 2 1

Working together to grow
libraries in developing countries

www.elsevier.com | www.bookaid.org | www.sabre.org

ELSEVIER BOOK AID International Sabre Foundation

Contents

CONTRIBUTORS

Numbers in parentheses indicate the pages on which the authors' contributions begin.

Gary K. Beauchamp (129) Monell Chemical Senses Center, Philadelphia, Pennsylvania 19104

Sven Bocklandt (245) Laboratory of Sexual Medicine, Department of Urology, David Geffen School of Medicine at UCLA, Gonda Center, Los Angeles, California 90095

Manfred Gahr (67) Max Planck Institute for Ornithology, Seewiesen, Germany

Elizabeth A. D. Hammock (107) Department of Pharmacology, Vanderbilt Kennedy Center for Research on Human Development, Vanderbilt University, Nashville, Tennessee 37232

E. B. Keverne (217) Sub-Department of Animal Behaviour, University of Cambridge, Madingley, Cambridge, CB3 8AA, United Kingdom

Michael J. Meaney (173) Developmental Neuroendocrinology Laboratory, Douglas Hospital Research Centre, McGill University, Montreal, QC, Canada H4H 1R3

Douglas S. Portman (1) Department of Biomedical Genetics and, Center for Aging and Developmental Biology, University of Rochester, Rochester, New York 14642

Kazushige Touhara (147) Department of Integrated Biosciences, The University of Tokyo, Chiba 277–8562, Japan

Eric Vilain (245) Laboratory of Sexual Medicine, Department of Urology, David Geffen School of Medicine at UCLA, Gonda Center, Los Angeles, California 90095

Daisuke Yamamoto (39) Division of Neurogenetics, Graduate School of Life Sciences, Tohoku University, 6-3 Aoba, Aramaki, Aoba-ku, Sendai, Miyagi 980-8578, Japan

Kunio Yamazaki (129) Monell Chemical Senses Center, Philadelphia, Pennsylvania 19104

Preface

There has been a long-standing debate on whether gender differences in our "behaviors" are learned or innate in principle. Even though this "nature-versus-nurture issue" in behavioral gender differences will remain a controversial subject for many years to come, everyone seems to agree that there exist behavioral gender differences indeed. That is, phenotypic differences between sexes are evident at the behavioral level. This observation was sufficient to prompt geneticists to initiate genetic analysis of and neuroscientists to commence their search for the neural basis of behavioral gender differences.

As a result of enormous efforts to elucidate the sex differences in brain anatomy and physiology in several model organisms, the neural correlates to behavioral sexual dimorphism are now emerging. In studies using animals amenable to genetic analysis, candidate genes responsible for neural sexual differences underlying behavioral sexual dimorphism have been identified.

Comparative studies of neural sex differentiation revealed a marked divergence in underlying molecular mechanisms. However, scientists found clues to evolutionary conservation in such apparently diverged sex differentiation. The best example of this is the discovery of a protein family with the DM domain in many different animals including insects and vertebrates, which represent the two "extreme" phylogenetic groups that had been considered to share nothing in common in their sex determination mechanisms.

Vertebrates and insects have been regarded to situate at opposite poles in terms of the logic in sex determination, as the former relies on a non-cell autonomous mechanism mediated by circulating hormones, whereas the latter utilizes an autonomous mechanism based exclusively on the genetic code. Even this dogma has been challenged by the observations that gynandromorphic birds have brains in which the female part and the male part are closely situated but separated by a distinct border. How are such insect-like gynandromorphic brains formed in birds if one postulates hormonal signaling as the sole mechanism operating in brain sex determination?

Yet another breakthrough was provided by the findings that genetic systems are typically controlled by epigenetic mechanisms that induce long-lasting chromatin alteration as triggered by environmental stimuli provided at a critical developmental stage. This type of genomic change may explain why certain key stimuli "predispose" organisms to particular behavioral patterns often for lifetime. Imprinted behaviors and developmental plasticity of brain functions may all be ascribed to this epigenetic regulation of the genome.

Thus, the conventional framework for understanding animal behavior as a patchwork of a bit of "innate" and a bit of "learned" is now to be replaced by a more dynamic view that genes and the environment interact, resulting in specific genomic conditions under which individual organisms develop behavioral patterns most appropriate for a particular situation they encounter.

In this volume, cutting-edge research on the genetic and epigenetic mechanisms that shape the brain gender are reviewed and discussed, with the aim of developing a new paradigm for understanding the nature-versus-nurture issue in animal and human behavior.

Daisuke Yamamoto

1

Genetic Control of Sex Differences in *C. elegans* Neurobiology and Behavior

Douglas S. Portman

Department of Biomedical Genetics and Center for Aging and Developmental Biology, University of Rochester, Rochester, New York 14642

ABSTRACT

As a well-characterized, genetically tractable animal, the nematode *Caenorhabditis elegans* is an ideal model to explore the connections between genes and the sexual regulation of the nervous system and behavior. The two sexes of *C. elegans*, males and hermaphrodites, have precisely defined differences in neuroanatomy: superimposed onto a "core" nervous system of exactly 294 neurons, hermaphrodites

Advances in Genetics, Vol. 59
Copyright 2007, Elsevier Inc. All rights reserved.
0065-2660/07 $35.00
DOI: 10.1016/S0065-2660(07)59001-2

and males have 8 and 89 sex-specific neurons, respectively. These sex-specific neurons are essential for cognate sex-specific behaviors, including hermaphrodite egg-laying and male mating. In addition, regulated sex differences in the core nervous system itself may provide additional, though poorly understood, controls on behavior. These differences in the nervous system and behavior, like all known sex differences in the *C. elegans* soma, are controlled by the master regulator of *C. elegans* sex determination, *tra-1*. Downstream of *tra-1* lie specific effectors of sex determination, including genes controlling sex-specific cell death and a family of regulators, the DM-domain genes, related to *Drosophila doublesex* and the vertebrate DMRT genes. There is no central (i.e., gonadal) regulator of sexual phenotype in the *C. elegans* nervous system; instead, *tra-1* acts cell-autonomously in nearly all sexually dimorphic somatic cells. However, recent results suggest that the status of the gonad can be communicated to the nervous system to modulate sex-specific behaviors. Continuing research into the genetic control of neural sex differences in *C. elegans* is likely to yield insight into conserved mechanisms of cell-autonomous cross talk between cell fate patterning and sexual differentiation pathways. © 2007, Elsevier Inc.

I. INTRODUCTION

A. *Caenorhabditis elegans* as a model for sex differences in nervous system structure and function

Coordinating the development and function of the multiple attributes that distinguish the sexes of a given species is a significant challenge to the genetic programs that govern metazoan development. Organisms must make a decision, usually early in development, about which sex to become; the outcome of this decision then instructs the development of multiple sex-specific structures and guides the sex-specific modification of common structures. This process is particularly complex in the nervous system, where sexual information must interface with the intricate genetic mechanisms that pattern the nervous system and regulate the properties of neural circuits.

In vertebrates, the sexual differentiation of the nervous system and the control of sex-specific behavior have been thought to rely completely on the action of gonadal steroid hormones (reviewed by Morris *et al.*, 2004). However, a series of recent findings has demonstrated that this model is oversimplified: some sex differences are cell-intrinsic, depending on sexual karyotype in the nervous system itself rather than on the influence of gonadal cues (reviewed by Arnold, 2004 and herein). However, in neither the hormonal nor the cell-intrinsic pathways are the links between sex determination, neural differentiation, and behavior well understood. It is in this regard that invertebrate model systems, such as the nematode *C. elegans* and the fruit fly *Drosophila*, can make their most significant contributions, bringing about fundamental new insights into the genetic mechanisms that regulate complex biological processes like sex determination and differentiation.

Since Sydney Brenner's "taming" of the *C. elegans* as a genetic model system in the 1960s, there has been great interest in exploiting the unique experimental tractability of this system to dissect the genetic underpinnings of neural development and behavior (Ankeny, 2001; Brenner, 1974; Brown, 2003). Indeed, the most common class of mutants isolated in Brenner's original screens were the so-called *unc* (*unc*oordinated) mutants that exhibited defects in locomotion, the most obvious *C. elegans* behavior (Brenner, 1974). Subsequent studies of these mutants have led to a series of fundamental insights into neurobiology: the identification of the *unc-6*/Netrin family of axon guidance cues and their receptors, *unc-40*/DCC and *unc-5* (Chan *et al.*, 1996; Hamelin *et al.*, 1993; Hedgecock *et al.*, 1990; Ishii *et al.*, 1992); characterization of *unc-86*, a founding member of the POU-HD family of transcriptional regulators of neural fate (Baumeister *et al.*, 1996; Finney and Ruvkun, 1990; Finney *et al.*, 1988); identification of *unc-13*, a conserved regulator of synaptic vesicle fusion (Lackner *et al.*, 1999; Maruyama and Brenner, 1991; Richmond *et al.*, 1999); and many others.

At the same time, sex determination and sexual differentiation were also the subject of intensive early genetic analysis in *C. elegans* (Doniach and Hodgkin, 1984; Hodgkin and Brenner, 1977; Kimble *et al.*, 1984; Klass *et al.*, 1976; Madl and Herman, 1979; Nelson *et al.*, 1978). Beginning with the isolation and study of many mutants with completely or partially sex-reversed phenotypes, including the *tra* (transformer), *her* (hermaphroditization), and *fem* (feminization) mutants, a robust genetic pathway has been elucidated that links differential chromosome content (see below) to dosage compensation and sexual fate in the germ line and soma (for reviews see Ellis and Schedl, 2006; Meyer, 2005; Zarkower, 2006). Interestingly, several genes with sequence similarity to *D. doublesex* (the DM-domain genes *mab-3*, *mab-23*, and *dmd-3)* have been found to be terminal effectors of the *C. elegans* sex determination pathway, indicating that some components of sex determination are likely to be shared across metazoa (Lints and Emmons, 2002; Raymond *et al.*, 1998; Shen and Hodgkin, 1988; Yi *et al.*, 2000; D. A. Mason and D. S. P., unpublished data). As we describe here, understanding the relationships between these two complex regulatory networks—patterning and regulation of neural circuitry on one hand, sex determination and differentiation on the other—is the focus of active current investigation that is likely to lead to critical insights into the mechanisms that control sex-specificity in neural development and behavior throughout the animal kingdom.

II. MAIN TEXT

A. Sex determination and differentiation in *C. elegans*

C. elegans is an androdioecious species. As shown in Fig. 1.1A, its two sexes are hermaphrodite and male. The hermaphrodite is essentially a modified female that produces and stores some sperm that can be used to self-fertilize its own oocytes. Animals of this sex lack male genital structures; thus, *C. elegans* hermaphrodites are unable to cross-fertilize each other. In contrast, the male produces only sperm, and males can reproduce only by cross-fertilizing a hermaphrodite. Hermaphroditism is a recent evolutionary innovation in *C. elegans*, as its nearest phylogenetic neighbors are gonochoristic (i.e., male–female) species (Kiontke *et al.*, 2004), indicating that the hermaphrodite is generated from minor modification of an ancestral female developmental program. In self-fertilizing hermaphrodite populations, males arise very infrequently (<0.3%). Despite the relative rarity of the male, its developmental program is maintained in the genome, indicating that males may be required to provide outcrossing. However, the rate of outcrossing in wild populations seems to be very low, and over evolutionary time the role of males in the species may be dwindling (Barriere and Felix, 2005; Chasnov and Chow, 2002; Stewart and Phillips, 2002). Owing to its ability to self-fertilize, the hermaphrodite offers experimental advantages that have led to a much more thorough characterization of its neuroanatomy and development; in several regards, the biology of the *C. elegans* male is much less well understood than that of the hermaphrodite.

As in most animals, sex determination in *C. elegans* is chromosomal: embryos with two sex chromosomes (XX) develop as hermaphrodites; those with only one (X0) become males (Brenner, 1974; Nigon and Dougherty, 1949). There is no heteromorphic (Y) sex chromosome in *C. elegans*. Interestingly, the number of X chromosomes per se is not the primary sex-determining cue; rather, the sex chromosome to autosome ratio X:A is assessed according to the relative copy numbers of specific autosomal and sex chromosome "signal element" genes (Carmi *et al.*, 1998; Madl and Herman, 1979; Powell *et al.*, 2005). Downstream of these signal element genes lies a complex regulatory hierarchy that independently controls both dosage compensation (the reduction of gene expression from each hermaphrodite X chromosome by half) (reviewed by Meyer, 2005) and sexual differentiation (Fig. 1.1B). This latter pathway relies on successive repressive interactions to control the activity of the gene *tra-1*, the master sexual regulator in *C. elegans*. XX animals carrying a null mutation in *tra-1* develop as somatic males, whereas X0 animals are essentially unaffected, indicating that *tra-1* functions in hermaphrodites to promote female and/or repress male fates (Hodgkin, 1987; Hodgkin and Brenner, 1977). *tra-1* also has a role in the development of the gonad

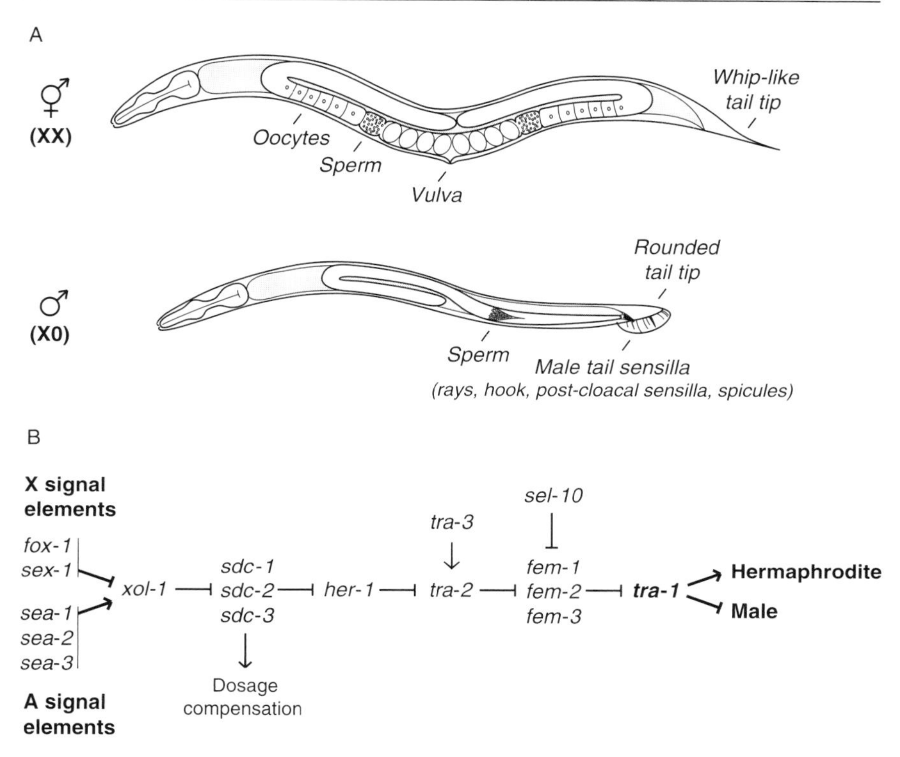

Figure 1.1. Sex determination and differentiation in C. *elegans*. (A) The primary sexual dimorphisms in C. *elegans*. Adult hermaphrodites (above) are XX and have a tapered, whiplike tail, a vulva, and a bilobed gonad that produces both oocytes and sperm. Adult males (below) are X0, slightly smaller in size, and have a rounded tail tip, several classes of tail sensilla, and a single-armed gonad that makes sperm. Modified from Zarkower (2006). (B) The somatic sex determination pathway. The X:A ratio is read by signal-element genes on the X chromosomes and autosomes; these converge onto the regulator *xol-1*, which is active only in X0 animals (Rhind *et al*., 1995). Downstream of the *sdc* genes, dosage compensation (not shown) reduces gene expression from the X chromosome by half and is controlled independently from the differentiation of somatic characteristics (Meyer, 2005). Essentially all somatic characteristics are controlled through a series of repressive interactions that regulate the master sex-determining gene *tra-1*. In XX animals, *tra-1* is ON, repressing male fates and perhaps promoting hermaphrodite fates. In X0 animals, *tra-1* is OFF, allowing male development to proceed.

in both sexes; this is thought to be an ancestral function that is separate from the role of *tra-1* in sex determination (Hodgkin, 1987; Mathies *et al*., 2004). Interestingly, neither *tra-1* nor its upstream regulators have conserved functions in sex determination outside nematodes, consistent with the idea that upstream events in sex determination evolve rapidly (Wilkins, 1995; Zarkower, 2001).

Through alternative splicing, *tra-1* encodes two Zn-finger proteins with sequence similarity to Ci/GLI transcription factors, indicating that a *hedgehog*-like pathway may have been co-opted in the worm to carry out a sex determination function (Zarkower and Hodgkin, 1992). The longer of these forms, called TRA-1A, is thought to be the active form, as the shorter is unable to bind DNA *in vitro* (Zarkower and Hodgkin, 1993). TRA-1A has been shown to act as a transcriptional repressor to block the expression of male-specific genes in XX animals (Conradt and Horvitz, 1999; Yi *et al.*, 2000); TRA-1A may also act to promote hermaphrodite gene expression. Genetic analysis has suggested that the sex-specific regulation of *tra-1* activity comes about posttranslationally (de Bono *et al.*, 1995). Consistent with this, it has been found that sex-specific proteolysis may be critical in generating a hermaphrodite-specific active form of TRA-1A (Schvarzstein and Spence, 2006). The mechanisms by which the *fem* genes, the most downstream genetic regulators of *tra-1*, control sex-specific TRA-1A activity remain an area of active investigation.

Elegant genetic studies have shown that *tra-1* acts cell-autonomously in the specification of nearly all sexually dimorphic cell fates in the *C. elegans* soma (Hunter and Wood, 1990). This stands in stark contrast to vertebrate sex determination mechanisms, in which sex determination events in the early gonad conscript peripheral tissues to adopt sex-specific characteristics through the influence of gonadal steroids (Morris *et al.*, 2004). In *C. elegans*, most sex-specific extragonadal characteristics do not depend on the gonad, and gonadal precursor cells can be completely removed by laser ablation very early in development with essentially no effect on sex-specific somatic development (Kimble, 1981; Klass *et al.*, 1976). Instead of relying on a gonadal cue, the sexual fate of somatic cells in *C. elegans* begins with cell-autonomous establishment of the X:A ratio early in embryonic development (Rhind *et al.*, 1995). A series of repressive interactions reinforces this decision, leading to local non-cell-autonomous activity of the secreted protein HER-1, which acts through the FEM proteins to regulate TRA-1A activity (Hodgkin, 1986; Hunter and Wood, 1992). Thus, the status of TRA-1A activity in any given somatic cell is sufficient to determine whether that cell adopts a male (*tra-1* OFF) or hermaphrodite (*tra-1* ON) fate. The only known exception to this is the hermaphrodite vulva, whose development relies on an inductive signal from the hermaphrodite somatic gonad (Hunter and Wood, 1992; reviewed by Sternberg, 2005).

Therefore, the genes repressed or activated by TRA-1A are thought to wholly account for the somatic characteristics that distinguish hermaphrodites from males, although recent evidence has suggested that there may be some relatively minor *tra-1*-independent functions that are necessary for the full complement of sex-specific differences (Grote and Conradt, 2006; van den Berg *et al.*, 2006). While some genes downstream of *tra-1* are known, the genetic mechanisms that link *tra-1* to sex-specific differentiated characteristics—particularly in the nervous system—remain incompletely described. As described below, two genes controlled by *tra-1*, *egl-1* and *ceh-30*, regulate the sexual phenotype of the nervous system by controlling sex-specific programmed cell death. The Hox genes *mab-5* and *egl-5* are also regulated at least indirectly by *tra-1* and have important roles in shaping male-specific cell lineages and neural cell fates (Chisholm, 1991; Chow and Emmons, 1994; Costa *et al.*, 1988; Kenyon, 1986). Finally, three related genes, *mab-3*, *mab-23*, and *dmd-3*, control partially overlapping subsets of male-specific development and function in the nervous system and elsewhere, though their relationship to *tra-1* is complex and not fully understood (Lints and Emmons, 2002; Shen and Hodgkin, 1988; D. A. Mason, K. H. Lee, R. M. Miller, and D. S. P., unpublished data). As mentioned above, these genes encode proteins harboring one or more DM (*doublesex/mab-3*) domains, the only known conserved molecular link between effectors of sex determination in metazoans (reviewed in Zarkower, 2001). The importance of these factors in the sexual differentiation of the *C. elegans* nervous system has led to speculation about the potential for the vertebrate relatives of these genes, the DMRT family, to be regulators of sexual dimorphism in the central nervous system (CNS).

B. The *C. elegans* nervous system

The *C. elegans* nervous system simultaneously exhibits a minimalist simplicity and an astonishing complexity. The adult hermaphrodite nervous system comprises exactly 302 neurons, compared to 383 in the adult male. The identity and developmental history of each of these neurons have been completely described and are essentially identical between individuals. Moreover, through the painstaking work of John White and colleagues, the complete neuroanatomy of the adult hermaphrodite, including patterns of synaptic connectivity, has been reconstructed from serial electron micrographs (White *et al.*, 1986). This "wiring diagram" for the *C. elegans* nervous system provides a substrate for understanding the neural control of behavior unrivaled in any system.

The worm nervous system is organized into several major ganglia in the head and tail (Hall *et al.*, 2006; White *et al.*, 1986). Sensory information is received through several classes of ciliated sensory neurons, mostly in the head, that synapse onto integrating interneurons. Most neural processing is carried out in the nerve ring, a circumferential neuropil that surrounds the isthmus of the pharynx. Locomotory behavior is controlled by the activity of a small set of command interneurons that regulate the function of several classes of motor neurons in the ventral nerve cord.

In the laboratory, *C. elegans* exhibits a variety of behaviors, including regulated forward and reverse locomotion, response to touch, temperature, food availability, and a large variety of chemosensory cues. As discussed below, hermaphrodite egg-laying and male mating are the primary sex-specific behaviors in this species. None of these behaviors is essential for laboratory viability, and all have been subjected to genetic analysis. Indeed, it is this tractability across the molecular, cellular, circuit, and systems levels that makes C. *elegans* such an attractive model for the genetic studies of behavior (reviewed by Rankin, 2002; Whittaker and Sternberg, 2004). The worm nervous system has also been shown to be capable of several forms of plasticity, including habituation to mechanical stimuli, adaptation to the presence of chemoattractants, and associative learning of chemosensory cues coupled to food availability and quality (Kuhara and Mori, 2006; Tomioka *et al.*, 2006; Zhang *et al.*, 2005; reviewed by Giles *et al.*, 2006; Tsalik and Hobert, 2003). The study of non-sex-specific behavior in this organism has made many important contributions to understand the mechanisms by which genes control neural development, circuit function, and behavior; more information may be found in the excellent reviews on this topic (Bargmann, 2006; Goodman, 2006; O'Hagan and Chalfie, 2006; Sengupta, 2007; Von Stetina *et al.*, 2006).

C. Sexual dimorphism in the *C. elegans* nervous system

The sex differences exhibited by the C. *elegans* nervous system are of two general forms. First, and more conspicuous, is the presence of sex-specific neurons in both hermaphrodites and males (Fig. 1.2A; Table 1.1). This sex-specific component of the nervous system comprises 8 neurons in hermaphrodites and 89 in males, all of which are thought to function in circuits dedicated to sexually dimorphic behavior. In both sexes, the sex-specific nervous system is overlaid onto the 294 neurons common to both sexes, referred to here as the "core" nervous system. Owing to the experimental accessibility of C. *elegans*, the cellular composition of the sex-specific nervous system of both sexes is precisely known (Sulston and Horvitz, 1977; Sulston *et al.*, 1980, 1983; White *et al.*, 1986). In contrast, a second and much less well-understood dimension of sexual dimorphism in the nervous system is sex differences in the core nervous system itself (Table 1.1). In principle, these differences could encompass alterations in morphology, gene expression, neurophysiology, synaptic connectivity, and synaptic strength in shared neurons.

Interestingly, the existence of these two forms of sexual dimorphism is also characteristic of more complex organisms. As in C. *elegans*, vertebrate species have sex-specific neural structures, such as those that innervate the genitalia. In contrast, sex differences in the C. *elegans* core nervous system may be more analogous to subtler sexual dimorphisms in the vertebrate CNS, such as the difference in size of the hypothalamic medial preoptic area in mammals (Raisman and Field, 1971) and the high-vocal center in songbirds (Nottebohm and Arnold, 1976). Though, unlike C. *elegans* core nervous system differences, these structures likely have sex differences in cell number, in both cases these can be seen as modulations in the properties of neural structures common to both sexes.

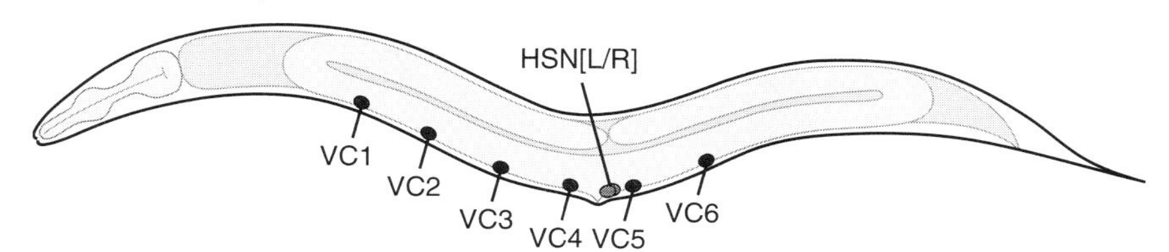

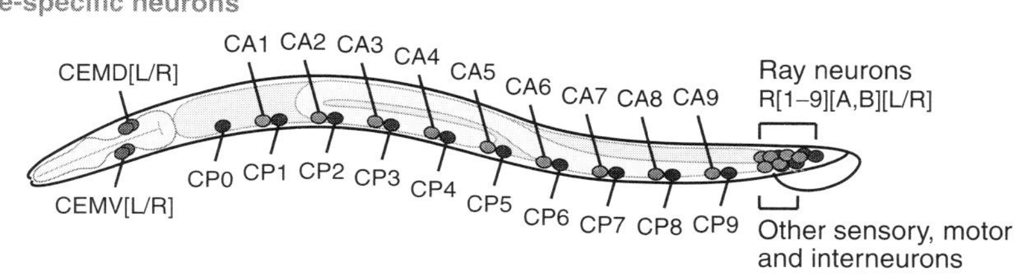

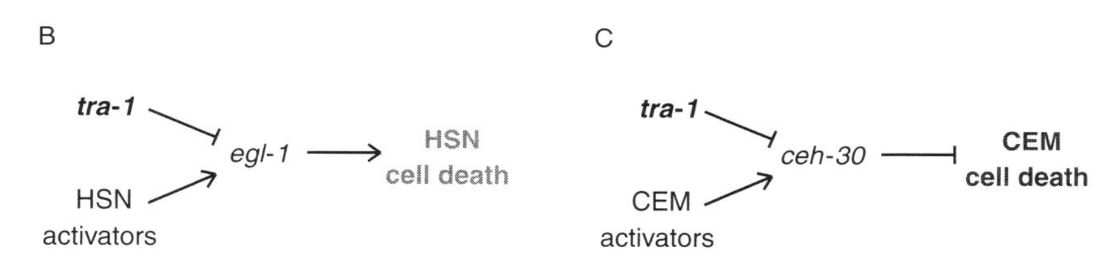

Figure 1.2. Sexual dimorphism in the C. *elegans* nervous system. (A) The adult hermaphrodite (above) has eight sex-specific neurons: six VC neurons in the ventral cord and two HSN motor neurons. These neurons regulate hermaphrodite egg-laying behavior. The adult male (below) has 89 sex-specific neurons. Four of these, the CEMs, are sensory neurons in the head and are thought to have a role in detecting hermaphrodite pheromone cues (see text). The male ventral cord contains the CA and CP motor neurons; both of these are implicated in specific steps of male mating behavior (Loer and Kenyon, 1993; Schindelman *et al.*, 2006). The male tail contains a variety of sensory, motor, and interneurons; the largest class of these is the RnA and RnB ray sensory neurons. (B) Sexual dimorphism in the presence of the HSN is regulated by differential programmed cell death. In XX animals, *tra-1* is active and represses *egl-1* expression in the HSNs. In X0 animals, unknown activators promote *egl-1* expression in the HSNs, leading to cell death (Conradt and Horvitz, 1999). (C) Differential cell death also underlies the sex difference in the CEM neurons. In XX animals, *tra-1* prevents the expression of the survival factor *ceh-30* in the CEM neurons, leading to their death. In X0 animals, *ceh-30* prevents HSN death; whether this occurs through regulation of *egl-1* is unknown (see text for details). (See Color Insert.)

Table 1.1. Sex Differences in C. *elegans* Neuroanatomy

Cells	No.	Description
Hermaphrodite-specific neurons(8)		
VC[1–6]	6	Cholinergic motor neurons
HSN[L,R]	2	Serotonergic motor neurons
Male-specific neurons (89)		
CA[1–7]	7	Ventral cord motor neurons
CA[8,9]	2	Neuron-like cells; lack obvious synapses
CEM[D,V][L,R]	4	Cephalic sensory neurons
CP[0–7]	8	Ventral cord motor neurons; express serotonin
CP[8,9]	2	Tail interneurons
DV[E,F]	2	Tail interneurons
DX[1–4]	4	Tail interneurons
EF[1–4]	4	Tail interneurons
HO[A,B]	2	Hook sensory neurons
PC[A–C][L,R]	6	Postcloacal sensilla sensory motor neurons
PDC	1	Tail interneuron
PGA	1	Tail interneuron
PV[V,Z]	2	Tail motor neuron
PV[X,Y]	2	Tail interneuron
R[1–9][A,B][L,R]	36	Ray sensory neurons
SPC[L,R]	2	Spicule sensory motor neurons
SP[D,V][L,R]	4	Spicule sensory neurons
Sexually dimorphic core neurons		
ADF[L,R]	2	Amphid sensory neuron; expresses *srd-1* in males only
AIM[L,R]	2	Ring interneuron; expresses *srj-54* in males only
PDA	1	Motor neuron (hermaphrodite Y, male Y.a); sexually dimorphic connectivity
PDB	1	Motor neuron; sexually dimorphic connectivity
PHC[L,R]	2	Tail sensory neuron (striated rootlet in male); sexually dimorphic connectivity
PVP[L,R]	2	Called PVU and PVS in male; sexually dimorphic connectivity

As such, the genetic mechanisms that impart sex differences to these structures could share common elements, especially given recent evidence that downstream effectors of sexual differentiation may be conserved in animals (reviewed in Bratus and Slota, 2006; Zarkower, 2001).

1. The hermaphrodite-specific nervous system

The *C. elegans* hermaphrodite-specific nervous system comprises just eight neurons: one pair of *h*ermaphrodite-*s*pecific *n*eurons (HSNs) and six VC motor neurons (Table 1.1). These two neuron classes regulate egg-laying, the primary hermaphrodite-specific behavior (reviewed in Schafer, 2006). The developmental basis for HSN sex-specificity has been elucidated through a series of elegant genetic studies; in contrast, the mechanisms underlying hermaphrodite-specificity of the VC neurons are not completely clear.

a. The HSNs

The HSNs (so-called even though they are not the only neurons specific to this sex) are a bilateral pair of serotonergic motor neurons that innervate the vm2 vulval muscles to stimulate egg-laying. The HSNs are essential for normal egg-laying behavior: genetic or microsurgical ablation of these cells renders animals unable to lay eggs (Trent *et al.*, 1983). As a result, many genes important for HSN development and function have been identified in genetic screens for egg-laying defective (Egl) mutants (Desai and Horvitz, 1989; Desai *et al.*, 1988; Trent *et al.*, 1983).

The sexual dimorphism in the presence of the HSNs occurs through differential cell death. In embryos of both sexes, the HSNs are born in the tail region and subsequently migrate to their final position in the midbody (Sulston *et al.*, 1983). In hermaphrodites, the HSNs remain quiescent until L4. Presumably in response to an extrinsic cue, they then extend short branched processes to innervate vulval muscles and long anterior processes that form synapses in the nerve ring. In contrast, in males the HSNs undergo programmed cell death during migration to the midbody (Sulston and Horvitz, 1977; Sulston *et al.*, 1980), accounting for the sex-specificity of this cell type in adults.

HSNs born in males undergo programmed cell death through a canonical mechanism that is shared with nearly all other developmental cell deaths in the *C. elegans* soma: activation of the BH3-only factor EGL-1 blocks CED-9/Bcl-2 to activate CED-4/Apaf-1 and the caspase CED-3, triggering apoptosis (Conradt and Horvitz, 1998; Desai *et al.*, 1988; reviewed in Conradt and Xue, 2005). Interestingly, a series of experiments have demonstrated that the male-specificity of this cell death results from the sex-specific activation of *egl-1* itself. In wild-type males, *egl-1* is activated in the embryonic HSNs, triggering their death; in contrast, *egl-1* is repressed in hermaphrodites, allowing their survival and eventual differentiation (Conradt and Horvitz, 1998). Important insights into the mechanism of differential *egl-1* activation arose from the study of an unusual dominant *egl-1* allele, *egl-1(n1084)*, in which HSNs undergo inappropriate (i.e., male-like) apoptosis in hermaphrodites (Desai *et al.*, 1988; Trent *et al.*, 1983), suggesting that *egl-1* is inappropriately activated in hermaphrodite HSNs by the *egl-1(n1084)* mutation. Molecular analysis showed that *n1084* is a single-nucleotide change in a conserved *cis*-acting regulatory element 5.6 kb downstream of *egl-1*-coding sequence. Using a combination of genetics and biochemistry, Conradt and colleagues demonstrated that this element is in fact a binding site for TRA-1A (Conradt and Horvitz, 1999). In wild-type hermaphrodites, TRA-1A binds to this regulatory element to repress *egl-1* expression in HSNs; in contrast, the lack of TRA-1A activity in males allows the activation of *egl-1* expression in the HSNs through unknown factors (Fig. 1.2B). The disruption of this element in *n1084* prevents the repression of *egl-1* by TRA-1A in hermaphrodite HSNs, causing them to adopt the male-like fate of programmed cell death. Thus in this case, *tra-1* acts to directly repress male-specific gene expression in the hermaphrodite. This mechanism, therefore, provides a striking example of a well-understood genetic pathway that regulates sexual dimorphism in the *C. elegans* nervous system.

b. The VC neurons

The VC neurons are cholinergic motor neurons that have synaptic connections to the HSNs and the vm2 vulval muscles (White *et al.*, 1986). The VCs have a more subtle effect on egg-laying behavior, acting to negatively regulate vulval muscle contraction. This function is mediated by their release of acetylcholine, which may inhibit the HSN-mediated stimulation of vulval muscle contraction (Bany *et al.*, 2003).

Rather than sex-specific cell death, it is a cell fate change that underlies the sexually dimorphic presence of the VCs, and the genetic pathway mediating this switch is not well understood. As shown in Fig. 1.3, the six VCs are derived postembryonically from a set of ventral precursor cells, the Pn.a cells, that also produce several classes of core motor neurons (VA, VB, AS, and VD). The VCs themselves arise from the Pn.aap branch of the lineage; in males, the Pn.aap branch instead produces the male-specific CA and CP motor neurons (see below). (In *C. elegans* lineage nomenclature, *a* and *p* refer to anterior or posterior daughter cells; thus, Pn.aap is the posterior daughter of the anterior daughter of the anterior daughter of the Pn cell.) Thus, this sexual dimorphism—the production of VCs in hermaphrodites and CAs and CPs in males—arises from a change in the fate of the Pn.aap cells.

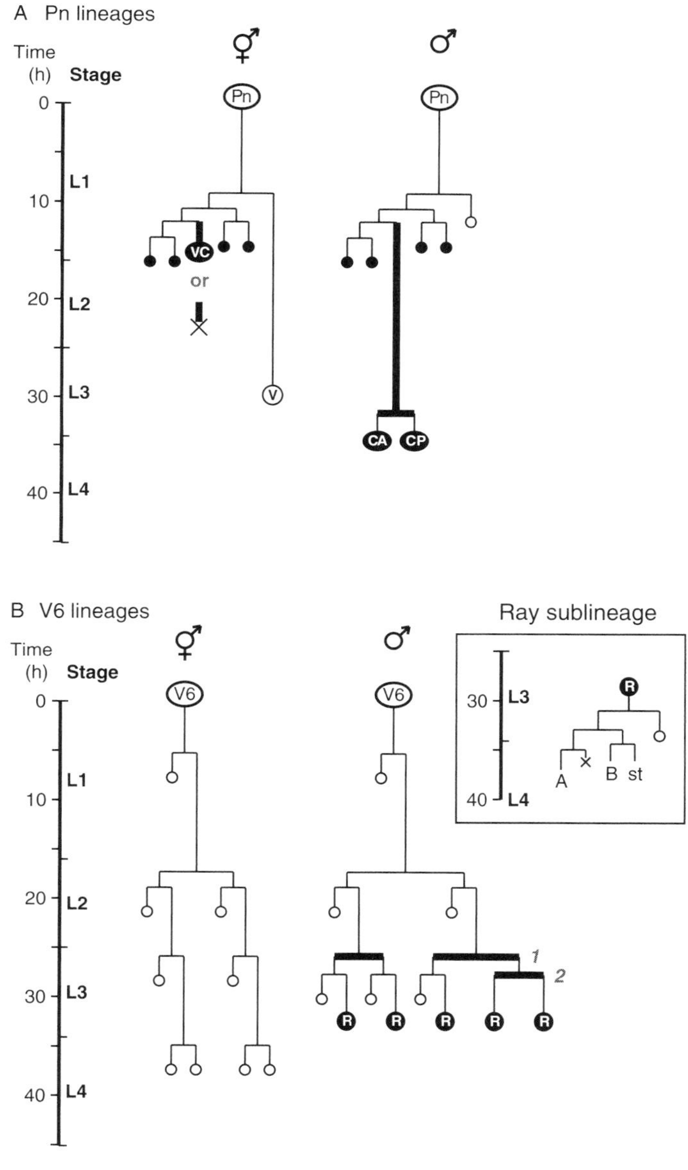

Figure 1.3. Sex differences in cell lineage give rise to differences in ventral cord and tail neurons. Developmental time is on the vertical axis, reading down; cell divisions are shown on the horizontal axis with anterior to the left. Open circles represent hypodermal cells, closed circles, neurons. Bold lines indicate differences between hermaphrodites (left) and males (right). On the left is shown larval stage and postembryonic developmental

In both sexes, the specification of neural fates in Pn.aap requires the midbody Hox cluster gene *lin-39*. In *lin-39* mutants, neither the hermaphrodite VC nor the male CA/CP neurons are generated (Clark *et al.*, 1993; Ellis, 1985; Loer and Kenyon, 1993). However, the fate of the Pn.aap cells themselves is still sexually dimorphic in *lin-39* mutants. In *lin-39* hermaphrodites, all of the Pn.aap cells that would become VC neurons undergo programmed cell death; in *lin-39* males, some of the Pn.aap cells die, some survive with small neuronal-like nuclei, and some divide but fail to give rise to fully differentiated motor neurons (Clark *et al.*, 1993; Ellis, 1985; Loer and Kenyon, 1993). Thus, it seems that the generation of sexual dimorphism in the Pn.aap lineage involves a mechanism of at least two steps: a *tra-1*-mediated propensity for cell death in hermaphrodites (perhaps through differential regulation of *egl-1)* and a sexually dimorphic response to *lin-39*. Efforts to understand the genetic basis for sex differences in Pn.aap-derived neurons are ongoing and should lead to interesting insights into the cross talk between sex determination and axial patterning pathways.

2. The male-specific nervous system

As mentioned above, the sex-specific nervous system of the *C. elegans* male is significantly more complex than that of the hermaphrodite, comprising a set of 89 sensory-, motor-, and interneurons that are connected to the core nervous system (Table 1.1; Fig. 1.2). However, unlike that of hermaphrodites, the ultrastructure of the male nervous system has not been fully determined so that a high-resolution integration of most of these cells into the worm "wiring diagram" is not yet possible. The primary obstacle to this project has been the complexity of circuitry in the male tail, thought to rival the complexity of the nerve ring, the main site of neural processing in the worm. However, with the help of recent

time in hours. V indicates a vulval precursor cell and R a ray precursor cell. (A) Both the hermaphrodite VC and the male CA and CP neurons arise from the Pn ventral cord blast cells. In XX animals, the Pn.aap branch of this lineage either differentiates into a VC neuron (in the P3-P8 lineages) or undergoes cell death (X in P1, P2, and P9-P12). In contrast, the Pn.aap cell in males remains undifferentiated until late L3, when it divides to give rise to one CA and one CP neuron. At the extreme ends of the P cell domain, there are exceptions to this (not shown) (Sulston and Horvitz, 1977; Sulston *et al.*, 1980). In P2, the lineage is altered such that Pn.aap does not divide and gives rise only to the CP0 neuron (there is no CA0). In P12, the Pn.aap cell becomes PVX, a male-specific tail interneuron. (B) The V6 lineage gives rise to rays 2–6 in males; in hermaphrodites, V6 produces tail hypodermal cells. Two male-specific doubling divisions (indicated by the bold horizontal lines labeled ***1***) require *mab-5* (Kenyon, 1986; Salser and Kenyon, 1996) and account for the additional proliferation in V6. The division labeled **2** can also be considered a symmetrical doubling division; this requires the Hox gene *egl-5* (Chisholm, 1991). The inset shows the ray sublineage, the program by which a single Rn cell (R) gives rise to a hypodermal cell, an A-type and B-type neuron, a ray structural cell (st), and a cell death (X).

advances in computer-aided analysis, ultrastructural reconstruction of the male nervous system is now underway and has already led to interesting insights into male sensory circuits and the ways in which they are integrated into the core nervous system (S. W. Emmons, D. Hall, and M. Xu, personal communication).

Sensory neurons are the major constituent of the male-specific nervous system, accounting for 50 of the 89 male-specific neurons. This highlights the importance of sensory function in the male: unlike hermaphrodites, *C. elegans* males must locate potential mates in order to reproduce and must carry out a mating behavioral program that integrates multiple sensory cues. The majority of these sensory neurons are in the tail, innervating four classes of male-specific sensilla, the rays, hook, postcloacal sensilla, and spicules. Four additional sensory neurons, the CEMs, are present in the head of the male. The functions of many of these male-specific sensory neurons have been defined by laser ablation studies and the analysis of mutants (Barr and Sternberg, 1999; Liu and Sternberg, 1995); several excellent reviews address the roles of these neurons in mediating sensory behaviors (Barr and Garcia, 2006; Emmons, 2006).

The male also has sex-specific interneurons that are likely to be important for the integration of sensory information. All of these are located in the tail. Again, because the male nervous system is unreconstructed, potential connections between male sensory neurons, interneurons, and the core nervous system remain essentially unknown. Finally, the male tail and ventral cord contain several classes of sex-specific motor neurons. The CA neurons send processes to the dorsal nerve cord (White, 1988); they have also been implicated through an unknown mechanism in sperm-transfer behavior during mating (Schindelman *et al.*, 2006). The CP neurons are serotonergic and innervate the male-specific diagonal muscles that are important for controlling tail posture during mating (Loer and Kenyon, 1993). Some male-specific neurons (the PC [A,B,C] neurons of the postcloacal sensilla and the spicule SPC neurons) have both sensory- and motor functions (Sulston *et al.*, 1980).

For some of these male neurons, including the CEMs and the ray neurons, the mechanisms that impart sexual dimorphism are at least partly understood. For most others, however, essentially nothing is known about how *tra-1* effects alterations in cell lineage and fate to specify male neural development. Below we explore what is known for three classes of neurons present only in the male: the CEM head sensory neurons, the tail ray sensory neurons, and the CA/CP ventral cord motor neurons. For all other male-specific neurons, no specific link from *tra-1* to sexual dimorphism has yet been made.

a. The CEM neurons

The four CEM neurons, the so-called cephalic companion sensory neurons, are the only male-specific neurons in the *C. elegans* head. These cells have, by virtue of their location and morphology, been long thought to have a role in the response of males to secreted chemical cues (pheromones) produced by hermaphrodites (Sulston *et al.*, 1980). A series of recent findings indicate that the CEMs are indeed important for the ability of males to detect hermaphrodite-secreted chemical cues (see below).

Like the HSNs, sexually dimorphic programmed cell death establishes sex differences in the CEMs (Sulston *et al.*, 1983). In this case, however, the situation is reversed: these neurons die in the hermaphrodite embryo but survive in males. A critical event in this pathway is the sex-specific expression of the Bar-homeodomain transcription factor gene *ceh-30* (H. Schwartz and H. R. Horvitz, personal communication). This gene acts in the embryonic male CEMs to promote their survival: the CEMs die (i.e., they adopt a hermaphrodite-like fate) in males lacking *ceh-30* function. Interestingly, *ceh-30* seems to be a direct target of *tra-1*, such that TRA-1A activity represses *ceh-30* expression in the CEMs of hermaphrodites. In males, *ceh-30* expression (presumably via unknown CEM-specific activators) is thought to block the function of the canonical cell death pathway, though the mechanism through which this occurs is not known (Fig. 1.2C) (H. Schwartz and H. R. Horvitz, personal communication). Thus, the CEMs provide a second example of the regulation of sex-specific cell death in the nervous system by *tra-1*.

b. The CA and CP motor neurons

The only male-specific cells of the ventral nerve cord are the CA and CP motor neurons. Lineally, the CAs and CPs are analogous to the hermaphrodite VC neurons, as they arise from the division of the Pn.aap cells, which either become VC neurons or die in hermaphrodites (Fig. 1.3). In males, the Pn.aap cells do not die (this may be the default hermaphrodite fate), but instead remain undifferentiated until they divide during L3 to give rise to one CA and one CP. As discussed above, it is possible that both sexually dimorphic activation of the cell death program and differential responses to the activity of Hox genes may underlie the switch from VC to CA/CP fate.

c. The rays

The rays are sensory organs that protrude laterally from the male tail and are the most prominent among the several classes of sensory specializations with which the male is endowed. Each ray is an independent sensillum innervated by two distinct sensory neurons, called A-type and B-type (Sulston *et al.*, 1980). As adult *C. elegans* males have nine bilateral pairs of rays, the ray neurons account for 40% (36/89) of the male-specific nervous system. Though much is known about ray development (reviewed in Emmons, 2005), we do not yet have a full understanding of the genetic mechanisms that make ray formation male specific. We do know, however, that the mechanism is significantly more complex than those that underlie sex differences in the HSNs, CEMs, and ventral cord neurons, and involves multiple sexually dimorphic regulatory events.

Each ray forms through a self-contained developmental module, called the ray sublineage, that generates the three cell types of the differentiated ray (Sulston *et al.*, 1980). The ray sublineage is triggered in a set of nine bilateral pairs of ray precursor cells called the Rn cells, where *n* represents a number from 1 to 9, designating the precursor to a specific ray. The ray precursor cells themselves are lateral hypodermal cells that are part of the body "seam" that runs along each side of the animal from head to tail. During late larval development, each ray precursor cell executes the ray sublineage program to give rise to multiple differentiated cell types: two sensory neurons (called RnA and RnB, or simply A-type and B-type), an associated glial-like ray structural cell (Rnst), a cell that undergoes programmed cell death (Rn.aap), and a hypodermal cell (Rn.p). Despite their similar cellular composition, all rays are not identical: each has a unique identity that is defined by its position, morphology, patterns of gene expression, and perhaps neural connectivity. A gene that is critical for ray development is *lin-32*, an *atonal/MATH*-class bHLH gene that acts, together with its heterodimerization partner HLH-2, as the ray proneural gene (Portman and Emmons, 2000; Zhao and Emmons, 1995). Males null for *lin-32* function lack nearly all ray development (R. M. Miller and D. S. P., unpublished data) and ectopic *lin-32* expression can in some cases be sufficient to trigger ray development (Zhao and Emmons, 1995).

The sexual dimorphism in ray development does not come about through a simple mechanism (e.g., death of the ray precursor cells in hermaphrodites). Instead, the sex determination pathway impinges on male development in at least three ways to make the rays male specific. First, alterations in the lineage of the seam cell progenitors of the ray precursor cells occur in males to generate additional numbers of lateral hypodermal cells (Sulston *et al.*, 1980). Second, the expression of *lin-32* occurs in these tail hypodermal cells only in males (Zhao and Emmons, 1995). Third, the response of these cells to *lin-32* activation is sexually dimorphic, allowing *lin-32* to direct the formation of differentiated ray cells rather than other non-sex-specific cell types (Zhao and Emmons, 1995). In recent years, our understanding of the control of each of these three regulatory points by the sex determination pathway has become more complete, though several significant gaps remain.

The first of these sex-specific modifications is the alteration of seam cell lineages to produce nine ray precursor cells on each side of the larval male tail. The full complement of ray precursor cells arises only in males as a result of a set of male-specific proliferative divisions in the hypodermal V5, V6, and T seam cells lineages that give rise to the ray precursor cells (Sulston *et al.*, 1980). As a result, males have a row of 11 hypodermal seam cells in the L3 tail, 9 of which give rise to rays; hermaphrodites have only 4 such seam cells. In wild-type males, the ray precursor cell R1 arises from the V5 lineage, whereas R2 through R6 descend from V6, and R7 through R9 from T. As shown in Fig. 1.3, V6 undergoes an additional round of doubling divisions to generate extra progeny in the male; V5 and T undergo similar but less extensive changes in cell lineage (Sulston *et al.*, 1980). In V5 and V6, this additional round of division is controlled by the *C. elegans ftz* orthologue, the Hox gene *mab-5* (Aboobaker and Blaxter, 2003; Costa *et al.*, 1988; Kenyon, 1986; Salser and Kenyon, 1996). In *mab-5* mutant males, the V5 and V6 lineages are identical to those in the hermaphrodite. Interestingly, they are also identical to the lineages of the more anterior seam cells V1–V4 in both sexes, indicating that *mab-5* imparts both spatial and sexual patterning. *mab-5* is known to bring about the patterning of these lineages cell-autonomously (Kenyon, 1986). Additionally, MAB-5 protein is detectable in both sexes in the early stages of seam cell development (when these lineages are sexually equivalent), but disappears from hermaphrodites and remains on in males once seam cell lineages diverge sexually in early L3 (Salser and Kenyon, 1996). However, the downstream targets of *mab-5* that bring about this symmetric doubling division are unknown. One attractive model is that *tra-1* acts together with a temporal cue to repress *mab-5* expression in the V5 and V6 lineages of L3 and L4 hermaphrodites (Fig. 1.4A); this possibility has not yet been tested directly. In contrast to V5 and V6, the lineage alterations in T are apparently not regulated by Hox genes; the pathway that controls sex differences in the T lineage remains undefined.

A second important sex-specific step in ray development is the activation of *lin-32*. As *lin-32* is not expressed in the tail seam of L3 hermaphrodites, this is clearly a sexually dimorphic event, though *lin-32* does not seem to be a direct target of *tra-1*. Instead, *lin-32* is activated in males through the combined effects of two pathways, a Hox pathway that involves *mab-5* and the *Abd-B* orthologue *egl-5*, and a second pathway involving the DM-domain gene *mab-3*. A series of elegant experiments has shown that *mab-5* acts to promote ray development late in the V5 and V6 lineages—after it promotes the earlier doubling divisions—presumably by activating *lin-32* (Salser and Kenyon, 1996). In addition, *egl-5* acts downstream of *mab-5* to allow *lin-32* expression in V6.ppppa, such that the two daughters of V6.pppp both produce rays (Chisholm, 1991). Thus, the full extent of *lin-32* activation in V5 and V6 requires both *mab-5* and *egl-5*. Acting in parallel with Hox genes is *mab-3*, which is expressed non-sex-specifically in all seam cells (Yi *et al.*, 2000). *mab-3* indirectly activates *lin-32* by blocking expression of the Hes-class bHLH gene *ref-1*, a repressor of *lin-32* (Ross *et al.*, 2005). As *mab-3* is expressed throughout the seam in both sexes, it does not act as an instructive factor for *lin-32* expression in the male tail. Instead, it seems that the combined function of *mab-3* and the Hox genes is to limit *lin-32* expression to the ray precursor cells (Fig. 1.4B), presumably functioning together with an L3-specific temporal cue. The means by which *tra-1* acts to make these events male specific is unclear, though it is again likely to occur through sex-specific regulation of Hox genes. As above, the mechanisms acting in the T-derived rays are likely to be different, as their development is largely independent of Hox genes and *mab-3* (Chow and Emmons, 1994; Shen and Hodgkin, 1988).

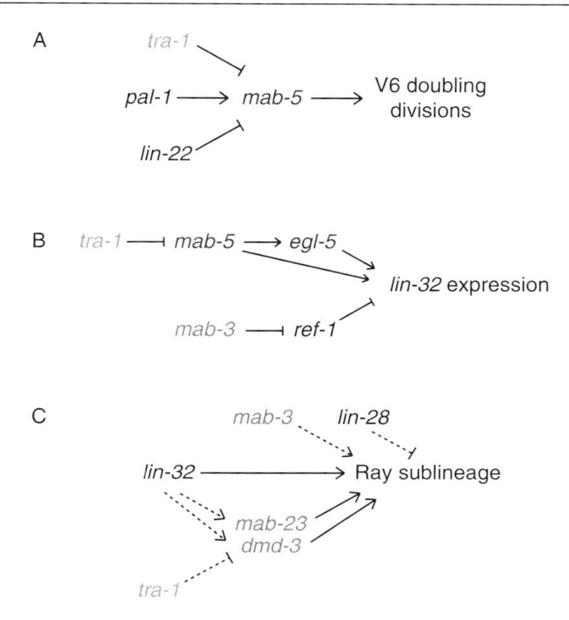

Figure 1.4. Sexually dimorphic steps in ray development. (A) The male-specific lineage alterations in V6 require the Hox gene *mab-5*. *mab-5* expression requires the *caudal* orthologue *pal-1* (Hunter *et al.*, 1999); its anterior boundary is limited by the bHLH gene *lin-22* (Kenyon, 1986; Wrischnik and Kenyon, 1997). The male-specific maintenance of *mab-5* in the V6 lineage is likely to require *tra-1*, though the mechanism is not known. (B) Expression of *lin-32*, the ray proneural gene, requires the Hox genes *mab-5* and *egl-5* (Yi *et al.*, 2000; Zhao, 1995). In addition, the DM-domain gene *mab-3* promotes *lin-32* expression by inhibiting the action of the bHLH gene *ref-1* (Ross *et al.*, 2005). (C) The mechanisms that generate male-specific cell types (ray cells) from the ray sublineage are not well understood. *mab-3* is necessary for *lin-32* to efficiently drive ray production (Yi *et al.*, 2000), and *mab-23* and *dmd-3* have a role in the differentiation of the RnA neuron (Lints and Emmons, 2002; R. Lints and D.S. Portman, unpublished data). Whether *mab-23* and *dmd-3* act downstream of or in parallel with *lin-32* is not known, nor is whether *tra-1* may block *dmd-3* expression in the hermaphrodite seam. *lin-28* may act to confer temporal specificity to the ray sublineage, preventing a ray from being made in the L2 postdeirid lineage (Ambros and Horvitz, 1984). (See Color Insert.)

A final layer of sex-specificity in ray development is the ray sublineage itself. Though this part of the mechanism is not well understood, several findings point to the importance of this step. First, *lin-32* has many other functions during *C. elegans* development, specifying the development of other neurons and neural structures in both embryos and larvae, including the touch cells and head sensory and interneurons (Chalfie and Au, 1989; Shaham and Bargmann, 2002; D. S. P., unpublished data). This indicates that some mechanism must give specificity to *lin-32* such that the ray sublineage is triggered when it is expressed in male seam cells. Though ectopic expression of *lin-32* in the hermaphrodite can cause seam cells to produce neural structures (Zhao and Emmons, 1995), it is not clear whether these are rays or other neural cell types. Second, overexpression of *lin-32* is not able to efficiently rescue the ray loss in *mab-3* mutants, suggesting that *mab-3* may act in the male seam to potentiate *lin-32* function (Yi *et al.*, 2000). Together, these findings indicate that the cellular context of the male seam, presumably comprising spatial, temporal, and sexual information, dictates the execution of the ray sublineage in response to *lin-32* expression.

Though no direct targets of *lin-32* in the rays are known, the four *lin-32*-dependent differentiated cells that descend from the ray precursor cell must require the activation of multiple, asymmetrically distributed cell type-specific regulatory factors (Portman and Emmons, 2000). The temporal component of the mechanism that gives ray specificity to *lin-32* may involve the heterochronic gene *lin-28*. In *lin-28* mutants of both sexes, the V5.pa seam cell generates a ray-like structure in place of the postdeirid, a non-sex-specific sensory structure that requires *lin-32* (Ambros and Horvitz, 1984). This indicates that *lin-28* may normally function to repress a male-specific *lin-32* cofactor in V5.pa. Hox genes and DM-domain genes are also likely to have important roles in this process. Both *mab-5* and *egl-5* have important roles in specifying ray identity, patterning the rays along the anterior–posterior axis. Additionally, the DM-domain genes *mab-23* and *dmd-3* have important roles in controlling cell fates in the ray sublineage itself, particularly in regulating differentiated properties in the RnA neurons (Lints and Emmons, 2002; R. Lints and D. S. P., unpublished data). One appealing model is that *mab-3*, *mab-23*, and/or *dmd-3* may act to "masculinize" the progeny of the ray precursor cell by specifying the development of male-specific neural types, rather than the non-sex-specific neural structures generated by *lin-32* function elsewhere (Fig. 1.4C).

3. Sex differences in the core nervous system

Relatively less attention has been given to the development and function of sex differences in the core nervous system itself, though this is likely to be of central importance to generating distinct male and hermaphrodite behaviors. Table 1.1 outlines our understanding of the extent of these differences, which are of two general types. First, some neurons that are lineally equivalent in both sexes show subtle differences in morphology and connectivity, such as the PDA pre-anal ganglion motor neuron. This may allow these cells to integrate information coming from the male-specific nervous system, or to allow the core nervous system to have output onto male-specific musculature. A second type of sex difference in sexually isomorphic neurons is at the level of changes in gene expression. As discussed below, these differences are only easily detectable through the use of fluorescent gene-expression reporter constructs, and only two examples of this sort are known.

a. Ultrastructural differences

Several examples of sex differences of the core nervous system were identified in the cell lineage and ultrastructural studies of the 1970s and 1980s. In one case, these differences are lineal: in the hermaphrodite, the tail cell called Y becomes the PDA neuron, a motor neuron in the pre-anal ganglion; in males, Y divides and its anterior daughter (Y.a) becomes PDA (Sulston *et al.*, 1980). Several core neurons also demonstrate differences in connectivity: this is the case for PDA and another pre-anal ganglion motor neuron, PDB, as it is for the PVP(R/L) interneurons (called PV(S/U) in males) and the PHC sensory neurons (Sulston *et al.*, 1980; S.W. Emmons, personal communication). The full extent of these synaptic differences awaits the completion of the male tail wiring project. The PHCs also differ ultrastructurally, having a striated rootlet only in the male (Sulston *et al.*, 1980). In general, the genetic mechanisms underlying these sex differences are unknown, as these differences are subtle and difficult or impossible to observe by light microscopy. Nevertheless, it is expected that these are in some way under the control of *tra-1*.

b. Molecular differences

In addition to these subtle structural differences, two examples of sex differences in gene expression in the core nervous system are known. First, the seven-transmembrane putative chemoreceptor gene *srd-1* has been found to be expressed in the ADF sensory neurons only in males, even though these cells are present and appear identical in both sexes (Troemel *et al.*, 1995; White *et al.*, 1986). A second serpentine receptor gene, *srj-54*, is expressed in the common AIM interneurons in males only (D. S. P., unpublished data). In neither of these cases is the functional significance of sexually dimorphic gene expression understood; however, it is noteworthy that both of these neurons are implicated in sensory behavior. The ADF neurons regulate entry into the long-lived dauer larva stage (Bargmann and Horvitz, 1991a), have secondary roles in sensing several classes of ions and small molecules (Bargmann and Horvitz, 1991b), and encode a learned aversion to pathogenic bacteria by expressing serotonin (Zhang *et al.*, 2005). Interestingly, males and hermaphrodites differ in their propensity to enter the dauer stage (Ailion and Thomas, 2000), suggesting that sex differences in ADF could have functional consequences. In contrast, no precise functions have been defined for the AIM neurons; however, they do synapse extensively onto the AIA interneurons, an important center for the integration of chemosensory information (Sengupta, 2007; White *et al.*, 1986). It is quite possible that these differences represent only a small glimpse of the true extent of sex differences in gene expression in the core nervous system, as *C. elegans* gene expression patterns are often examined only in hermaphrodites. Several microarray experiments designed to detect sex-specific gene expression in *C. elegans* have been carried out (Jiang *et al.*, 2001; Reinke *et al.*, 2004; Thoemke *et al.*, 2005), but these may not be sufficiently sensitive to reliably identify expression differences in a small number of neurons. Further investigation of this area should lead to significant insight into the sex-dependent regulation of behavior, especially as sexual dimorphism in sexually isomorphic circuitry is a common theme in the nervous systems of more complex animals.

D. Sex differences in *C. elegans* behavior

The full extent of sex differences in C. *elegans* behaviors is only beginning to be understood. The more obvious sex-specific behaviors—that is, hermaphrodite egg-laying and male mating—are well characterized and, to a first approximation, controlled by the sex-specific nervous system. However, it is becoming increasingly apparent that behaviors common to both sexes—locomotion, for example, and a variety of sensory responses—also differ between the sexes. In some cases, more complex behaviors (e.g., male food-leaving; see below) may arise through the modification of basic parameters of behaviors common to both sexes. Here we review the known sex differences in C. *elegans* behavior and, to the extent to which it has been determined, discuss the genetic and neural underpinnings of these differences.

1. Hermaphrodite behavior

a. Egg-laying behavior

Under favorable conditions, hermaphrodites incubate early newly fertilized eggs (produced either by self- or cross-fertilization) in the uterus for several hours before they are laid. Egg-laying is a result of regulated contraction of the vulval muscles: on contraction, the vulva opens transiently and a single egg is laid. The rate of egg-laying is regulated by several factors, particularly the availability of food.

The genetic and neural control of egg-laying has been the subject of an extensive recent review by Schafer (2006). As described above, the HSNs are thought to be the primary mediator of egg-laying, promoting vulval muscle contraction through serotonergic innervation of these muscles. In addition, the VC motor neurons provide cholinergic input to the vulval muscles and HSNs to inhibit muscle contraction (Bany *et al.*, 2003). Both the HSNs and VCs have extensive contacts with core nervous system neurons, which may allow the regulation of egg-laying behavior according to the state of the animal. The extent to which there may be anatomical and/or functional sex differences in the neurons that are pre- or postsynaptic partners of the HSNs and VCs is unknown.

b. Mating behavior

Historically, the *C. elegans* hermaphrodite has been considered to be a passive partner in mating behavior. However, recent evidence suggests that the hermaphrodite's response to male copulatory behavior may be regulated. Perhaps to promote the generation of self- rather than cross-progeny, young (i.e., self-fertile) hermaphrodites appear to express a locomotory "escape" response when a male attempts to copulate. In contrast, older animals—those that are no longer self-fertile owing to the depletion of self-sperm—may be more receptive to males (Kleemann, 2005). In addition, self-fertile hermaphrodites have been observed to actively expel sperm deposited by males, while those depleted of sperm show no such behavior (Kleemann, 2005). The genetic and neural mechanisms that regulate these behaviors are unknown. However, the dependence of hermaphrodite behavior on reproductive status suggests the existence of a pathway, perhaps endocrine, by which the state of the germ line and gonad can be communicated to the nervous system. It is quite possible that additional aspects of hermaphrodite behavior are regulated during mating; this remains relatively unexplored.

c. Avoidance of hermaphrodite-conditioned medium

It has recently been found that adult hermaphrodites may aversively respond to liquid medium conditioned by other hermaphrodites, possibly as a way to avoid competition for food or other resources (J. White and E. M. Jorgensen, personal communication). The neural and genetic bases for this behavior are unknown as is the identity of the secreted factor(s) to which animals respond. Interestingly, as described below, males display the opposite response to hermaphrodite-conditioned medium. This finding has interesting implications for the potential mechanisms that could bring about this sexually dimorphic response.

2. Male behavior

The behavior of *C. elegans* adult males differs from that of hermaphrodites in several respects. Males exhibit some behaviors, such as mating, that are completely absent in the hermaphrodite; males also have generally increased activity and have a propensity to leave a food source devoid of potential mates. As two excellent reviews have focused on male behavior (Barr and Garcia, 2006; Emmons, 2006), the emphasis here is on the neural and genetic mechanisms that underlie sex differences in the expression of these behaviors.

a. Male mating behavior

As a stepwise behavior that must integrate several sensory cues and produce several types of coordinated motor output, male mating is thought to be the most complex behavior encoded in the *C. elegans* nervous system. As these steps are likely hardwired into the male nervous system, male mating is considered to be an innate behavior. However, it is not unlikely that male mating is regulated by experience, though this remains unexplored.

Male mating behavior entails a series of distinct steps: response to hermaphrodite contact, turning, vulva location, spicule insertion, and sperm transfer (Liu and Sternberg, 1995). Each of these steps requires male-specific neuromuscular or gonadal structures. In the first step, response to contact, the male apposes the ventral surface of his tail against the hermaphrodite body and several coordinated changes in activity occur. The male tail assumes a characteristic "clenched" posture (a result of the contraction of tail diagonal muscles), allowing tail sensilla to efficiently detect cues on the hermaphrodite body. The male also ceases forward locomotion and head foraging movements, and initiates backward (tail-first) locomotion along the hermaphrodite body in search of the vulval opening. Response to contact can be triggered by several putative mechanosensory structures in the male tail, primarily the rays and hook, depending on where on the male's body contact is first made (Liu and Sternberg, 1995). Response is mediated by signaling through the polycystin-class TRP channels LOV-1 and PKD-2, which may act as mechanosensors (Barr and Sternberg, 1999), though other sensory pathways are likely to be involved (Barr *et al.*, 2001; Peden and Barr, 2005; R. M. Miller and D. S. P., unpublished data). The sexual dimorphism in the initiation of mating behavior can, therefore, largely be explained by its reliance on male-specific sensory organs and muscles, though backward locomotion and the cessation of feeding almost certainly involve circuitry of the core nervous system.

The goal of the male's reverse locomotion is to locate the hermaphrodite vulva. If the male has initiated response on the dorsal side of the hermaphrodite (or on the ventral side past the vulva), the male will need to turn at least once in order to search the other side of the hermaphrodite body. Turning involves continuous contact between the male tail and the hermaphrodite body and requires a tight curling of the tail mediated at least in part by the diagonal muscles. Serotonergic innervation of the diagonal muscles by the CP neurons is necessary for this step (Loer and Kenyon, 1993), as are the dopaminergic ray neurons, R5A, R7A, and R9A (Lints and Emmons, 1999; Liu and Sternberg, 1995). Ablation studies have shown that the most posterior rays (the so-called "T-rays," rays 7, 8, and 9) are essential for coordinating turning behavior (Liu and Sternberg, 1995), which is thought to be triggered by a change in the shape of the hermaphrodite body at the head or tail (Liu and Sternberg, 1995; P. W. Sternberg and A. Whittaker, personal communication).

The location-of-vulva step of mating behavior is mediated primarily by the hook, a sensory organ on the ventral side of the male tail (Liu and Sternberg, 1995). Unknown mechanical cues are first sensed by the hook, causing the male to stop in the vicinity of the vulva; like response, this step also requires polycystin signaling (Barr and Sternberg, 1999). High-resolution location of the vulval slit involves the postcloacal sensilla and spicules; the spicules rapidly prod the hermaphrodite body as the male slowly moves back and forth (Garcia *et al.*, 2001; Liu and Sternberg, 1995). Penetration of the vulva by the spicules occurs through cholinergic innervation of the spicule protractor muscles. This event is thought to be sensed by proprioceptive neurons in the spicules that regulate the transfer of sperm through the vas deferens and cloaca into the hermaphrodite uterus (Garcia *et al.*, 2001; Schindelman *et al.*, 2006). Pharyngeal pumping also slows dramatically when the spicules are inserted (Liu and Sternberg, 1995). The initiation of sperm transfer requires neurons in the male ventral cord, most likely the male-specific CA motor neurons (Schindelman *et al.*, 2006). Once sperm transfer is complete, the spicules are withdrawn and the mating behavioral program terminates.

As mentioned above, the execution of each of these steps depends critically on sex-specific components of the male nervous system. However, the neural and genetic control of these steps clearly also involves the core nervous system. The precise means by which these interactions are important for mating behavior are not clear, though some steps, such as response, must involve communication with the core nervous system at several levels to modulate locomotion and block pharyngeal pumping and foraging. It is also likely that regulatory events in the core nervous system modulate mating behavior in several ways. Males are reluctant to initiate mating behavior in the absence of food; in addition, feeding behavior is linked through the pharyngeal NSM neurons to the regulation of spicule protraction (Gruninger *et al.*, 2006). Recently it has been found that components of the core mechanosensory circuit are necessary for turning behavior (T. Liu, C. Li, and M. M. Barr, personal communication). Together these findings indicate that multiple connections between the male-specific nervous system and the core nervous system are likely to be important for the regulation and execution of mating behavior.

The sex-specificity of mating behavior has been thought to be completely under the control of *tra-1*. Though *tra-1* males have fertility defects, this was attributed to a role for *tra-1* in the development of the germ line, a function of *tra-1* that is thought to be non-sex-specific and separate from its role in sex determination (Hodgkin, 1987; Mathies *et al.*, 2004). However, evidence has indicated that *tra-1* XX pseudomales may also have defects in mating behavior, particularly in the response step of mating (Grote and Conradt, 2006; R. M. Miller and D. S. P., unpublished data). This may indicate that additional pathways act in parallel to *tra-1* to completely masculinize the male nervous system (Grote and Conradt, 2006; van den Berg *et al.*, 2006). Alternatively, it may be that existing alleles of *tra-1* are not truly null such that *tra-1* may still have residual feminizing activity in *tra-1* XX pseudomales. A third possibility is that the gonadal defects in *tra-1* animals could secondarily cause behavioral phenotypes, consistent with the speculation that the gonad may have some role in regulating neural function in the male (see below) (Lipton *et al.*, 2004).

Acting downstream of *tra-1*, the DM-domain gene *mab-23* has been shown to have an important role in mating behavior. *mab-23* mutant males have severe defects in turning behavior as a result of defects in the differentiation of diagonal muscles. The transfer of sperm is also blocked in *mab-23* males owing to defects in male-specific proctodeum or cloaca formation (Lints and Emmons, 2002). This is consistent with the idea that *mab-23* acts to masculinize both sex-specific and non-sex-specific tissues to bring about male behavior. Interestingly, two other DM-domain genes, *mab-3* and *dmd-3*, also have important roles in masculinizing neural circuits in the male-specific and/or core nervous system (see below). Understanding how these factors act downstream of *tra-1* to engender male-specific behavioral programs remains an important goal for the field.

b. Male food-leaving

In laboratory culture, males in the absence of hermaphrodites will leave a food source, often lethally. As this behavior is exhibited only by adult males, and is not expressed when a hermaphrodite is present, it has been interpreted as a male mate-searching or sex-drive behavior (Lipton *et al.*, 2004). While the specific signals that control this behavior have not been identified, one possibility is that the presence of a hermaphrodite inhibits a mate-searching drive state. Interestingly, the rate at which males leave food over time is constant, indicating that solitary males are constitutively in a mate-searching mode that can be expressed stochastically, possibly as result of random encounters with the edge of the food source. Recent evidence indicates that the ray sensory neurons are important for implementing food-leaving (A. Barrios and S. W. Emmons, personal communication), suggesting that tonic input from the male-specific nervous system regulates this behavioral state. Like the hermaphrodite escape response described above, food-leaving behavior in both sexes depends on the state of the gonad (Lipton *et al.*, 2004), consistent with the possibility of endocrine signals modulating the function of the nervous system. Recently, the steroid hormone receptor *daf-12* has been found to regulate food-leaving, suggesting the intriguing possibility that a steroid could be involved in the sex-specific expression of this behavior (G. Kleemann and S. W. Emmons, personal communication). Again, because expression of this behavior occurs ultimately through modulation of locomotory patterns, components of the core locomotory circuit (Gray *et al.*, 2005; Tsalik and Hobert, 2003; Wakabayashi *et al.*, 2004) may be actively regulated to bring about sex-specific outcomes. It is also known that the DM-domain genes *mab-3*, *mab-23*, and *dmd-3* are important for male food-leaving behavior (Yi *et al.*, 2000; L. Jia, G. Kleemann, and S. W. Emmons, personal communication; R. M. Miller and D. S. P., unpublished data), again indicating that this class of genes may have important roles in controlling male-specific behavior.

c. Response to hermaphrodite pheromones

The existence of hermaphrodite-produced pheromone(s) detected by males has long been postulated but has remained elusive. Recently, several groups have demonstrated that males can recognize culture medium conditioned by hermaphrodites (Simon and Sternberg, 2002; J. White and E. M. Jorgensen, personal communication; K. L. Chow, personal communication; A. Barrios and S. W. Emmons, personal communication) or by females from other *Caenorhabditis* species (Chasnov *et al.*, 2007). Depending on the assay conditions, males will linger in regions containing conditioned medium (Simon and Sternberg, 2002) or will display taxis toward pheromone signals. Interestingly, it has been found that both the sex-specific CEM sensory neurons as well as non-sex-specific sensory neurons of the head are required for males to be able to detect these signals, indicating that sex differences in the core nervous system, in addition to the male-specific nervous system, may generate male-specificity in the response to conditioned medium (J. White and E. M. Jorgensen, personal communication; W. K. So and K. L. Chow, personal communication). The genetic basis for these sex differences are unknown; however, recent evidence suggests that *dmd-3* males fail to recognize hermaphrodite-conditioned medium, indicating that *dmd-3* is necessary for some aspect of this sexually dimorphic behavior (R. Miller and D. S. P., unpublished data).

3. Sex differences in common behaviors

Some behaviors are characteristically expressed in both sexes and are known or thought to emerge through the functions of sexually isomorphic circuits and gene activity. However, as discussed above, this view is likely to be an oversimplification, and the establishment of sex differences in the core nervous system may have significant implications for *C. elegans* behaviors, both those that are wholly sex-specific (i.e., egg-laying and mating) and those that occur through differential regulation of behaviors common to both sexes (e.g., the hermaphrodite escape response and male food-leaving).

a. Locomotion

As for all nematodes, locomotion in *C. elegans* is driven by the propagation of dorsoventral sinusoidal waves along the length of the body. Animals modify their direction of movement by turning, often in conjunction with multiple brief reversals, in a maneuver termed a "pirouette" (Pierce-Shimomura *et al.*, 1999). Though male and hermaphrodite locomotion is qualitatively similar, males have been observed to be generally more active under standard culture conditions and can be considered to be hyperkinetic compared to hermaphrodites. Recent more quantitative analysis of sex differences in locomotion has indicated that males differ significantly from hermaphrodites in their locomotory waveform, body bend frequency, as well in the frequency and duration of spontaneous reversals (W. R. Mowrey and D. S. P., unpublished data). These differences could in principle reflect a modification of the core locomotory circuit or could result from the influence of male-specific circuitry on locomotory behavior. Males also display significant differences in their responses to drugs that interfere with cholinergic transmission (aldicarb, levamisole, and nicotine), suggesting specific differences in the characteristics of acetylcholine signaling at the neuromuscular junction (Matta *et al.*, 2007; W. R. Mowrey and D. S. P., unpublished data). As for other sex differences in behavior, these differences in locomotion are expected to be under the control of *tra-1*, but specific downstream effector genes that may act in this pathway have not yet been identified.

b. Olfaction

Chemosensory behavior, particularly olfaction, has been the subject of extensive analysis in the *C. elegans* adult hermaphrodite, revealing important insights into the genetic and neural control of sensory responses (for reviews see Bargmann, 2006; Sengupta, 2007). As described above, the putative chemoreceptor *srd-1* is male specifically expressed in the shared sensory neuron ADF, suggesting the possibility of sex differences in chemosensory behavior (Troemel *et al.*, 1995). Recently, it has been found that adult males are significantly less responsive than hermaphrodites to several attractive olfactory cues in standard olfaction assays. Males also display significantly different olfactory preferences when animals are simultaneously exposed to two odorants, indicating that the *C. elegans* olfactory system may be modified to generate characteristic sex-specific repertoires of sensory behavior (K. H. Lee and D. S. P., unpublished data). Again, these sex differences are downstream of *tra-1*, and may at least partially require *dmd-3* function, as *dmd-3* males display some partially feminized olfactory behavior characteristics (K. H. Lee and D. S. P., unpublished data). The specific cellular focus of *tra-1* and *dmd-3* action is not known; these genes may act autonomously in the olfactory system or indirectly either through modification of olfactory behavior by the male-specific nervous system or by influences from the gonad and/or germ line. Interestingly, at least some sex differences in *C. elegans* olfactory preference are developmentally regulated, suggesting that sex-specific maturation of the nervous system may underlie these changes (K. H. Lee and D. S. P., unpublished data). Understanding the genetic and neural control of sex differences in olfaction is likely to reveal important mechanisms by which sex modifies neural function in the worm.

c. Learning and memory

It has been recently reported that there are significant sex differences in the ability of *C. elegans* adults to associate a sodium chloride chemosensory cue with the lack of food. Under standard assay conditions, hermaphrodites form this associative memory quite efficiently, while males perform much more poorly

(Vellai *et al.*, 2006). Interestingly, insulin-like growth factor signaling (through the insulin receptor DAF-2 and its downstream target DAF-16/FOXO) is required for the formation of this memory, and mutations in this pathway abolish sex differences in this behavior, indicating that differential insulin-like signaling may underlie these sex differences (Vellai *et al.*, 2006). In a different paradigm, males and hermaphrodites have been shown to have similar nonassociative habituation responses to mechanical stimulation. However, male responses to tap stimuli were greater than those of hermaphrodites, perhaps because of the additional sex-specific mechanoreceptors present in the male tail (Mah and Rankin, 1992).

III. CONCLUSIONS

A. Genetics of sex-specific neural development

A number of important insights into genetic mechanisms have emerged from the study of sex differences in the *C. elegans* nervous system. During worm development, the master regulator *tra-1* acts cell-autonomously in many tissues to control sexual dimorphism. In the nervous system, one of the functions of *tra-1* is to effect two important sets of sex-specific cell deaths, those of the HSNs in males and the CEMs in hermaphrodites. Both of these occur through embryonic regulation of programmed cell death, such that sexual dimorphism in the nervous system is already present in newly hatched larvae. These pathways may provide paradigms for other systems in which the role of sex-specific cell death in shaping sex differences in the nervous system is not well known.

Other sex differences in the nervous system—for example, motor neurons in the ventral cord and sensory neurons in the male tail rays—arise through more complex developmental mechanisms. In the ventral cord, both cell death and Hox-dependent changes in cell fate are likely to be important. In the rays, Hox-dependent cell lineage alterations, differential expression of bHLH factors, and the specification of male-specific neural cell types are all ultimately under the control of *tra-1*. Though these pathways are not completely understood, two important principles are already apparent. First, as in *Drosophila*, the interface between Hox-dependent spatial patterning and sex determination is critical for the emergence of sex-specific cell types, highlighting the importance of understanding the interface between these pathways. Second, three genes of the DM-domain family are essential for male-specific cell lineage and fate in sensory ray development, again suggesting that these genes may be ancient regulators of sexual dimorphism and may have similar roles in other systems.

The most subtle sort of sex difference in the *C. elegans* nervous system, ultrastructural and molecular differences in the core nervous system, is just beginning to be investigated. Again, these dimorphisms are almost certainly regulated by *tra-1*, but nothing is known about the genetic mechanisms by which they occur. Given the emerging evidence that nonhormonal, cell-autonomous mechanisms shape neural properties in the vertebrate CNS, further investigation of the pathways regulating these processes in *C. elegans* may lead to important insights into conserved genetic pathways that control sex differences in neural character in other organisms.

B. Genetics of sex-specific behavior

Considerably less is known about the genetic mechanisms that engender behavioral differences between the sexes in *C. elegans*. Like the developmental mechanisms discussed above, these differences are controlled by *tra-1*, though recent evidence suggests that this gene may not completely account for the full extent of sexual dimorphism. Again, DM-domain genes are proving to be critical regulators of this process, as three genes of this family in *C. elegans* are required for some aspects of male-specific behavior. Additionally, recent results have led to the surprising possibility that the status of the gonad regulates neural function and sex-specific behavior through unknown mechanisms. Further investigation to understand how the sex-specific nervous systems interface with the core nervous system, the nature of the neural circuits that control sex-specific behaviors (particularly in the male), and the ways in which shared behaviors are differentially modified to produce higher-order sex-specific outcomes is certain to be fruitful.

C. Relevance for vertebrate systems

How can research in a simple invertebrate illuminate the mechanisms that shape sex differences in the vertebrate nervous system? Though *C. elegans* has a strong history of defining conserved genetic mechanisms critical for neural development and function, the mechanisms used by vertebrates to effect sexual differentiation of the nervous system seem, at first glance, fundamentally different from those in *C. elegans*. As described above, gonadal hormones have long been considered to be the sole arbiter of these processes in vertebrates; their effects, particularly in mammals, are both powerful and well demonstrated. In contrast, steroid hormones have no obvious role in the sex-specific development of the *C. elegans* nervous system, though whether they regulate sex-specific behavior in the worm is an open question. However, the *C. elegans* genome contains no obvious orthologues of the androgen or estrogen receptors so that vertebrate gonadal steroids are unlikely to be involved.

A revision of this orthodox view has come from a series of recent studies in vertebrates that have demonstrated that cell-autonomous sex determination pathways, acting in parallel to gonadal hormones, function in the nervous system to link sexual karyotype to the development of sex-specific characteristics. Arnold has proposed that regulatory genes on the X (that escape dosage compensation) and Y chromosomes may directly organize sex-specific CNS characteristics (Arnold, 2004). Indeed, there is some evidence that the Y-chromosome sex-determining gene *Sry* has such a role in the nervous system (Dewing *et al.*, 2006). However, such a model may capture only one aspect of this process. It is also quite possible that sexual karyotype controls much more complex regulatory networks, such as those characteristic of *C. elegans* and *Drosophila* sex determination, that read the sex-determining signal and set into motion a cascade of interactions that only very indirectly lead to sex-specific gene expression. The potential existence of such a pathway in the mammalian nervous system has intriguing implications for the mechanisms that bring about sex differences in neuroanatomy and neural function; moreover, genes in such a pathway could have central importance in the development of a wide variety of neurological and mental heath conditions, such as autism, mental retardation, and anxiety disorders, that occur with strong sex bias in humans. As conserved regulators of sexual fate with critical functions in *C. elegans* sexual dimorphism, DM-domain factors could be important components of such a mechanism. Further genetic analysis of sex differences in *C. elegans* neural development and behavior is likely to lead to additional candidates regulating similar processes in much more complicated, less experimentally tractable organisms.

Acknowledgments

I thank the members of my research group, particularly K. H. Lee, D. A. Mason, R. M. Miller, and W. R. Mowrey, for insightful discussions and critical reading of the chapter. For generously sharing unpublished results and for comments on the chapter, I am grateful to A. Barrios, G. Kleemann, and S. W. Emmons; H. Schwartz and H. R. Horvitz; J. White and E. M. Jorgensen; A. Whittaker and P. W. Sternberg; K. L. Chow; and T. Liu and M. M. Barr. I also thank WormAtlas (www.wormatlas.org), a valuable resource in researching this chapter. Work in the author's laboratory on neural and behavioral sex differences in *C. elegans* is supported by research grants from Autism Speaks and the National Institutes of Health (NS050268 and DK071645).

References

Aboobaker, A., and Blaxter, M. (2003). Hox gene evolution in nematodes: Novelty conserved. *Curr. Opin. Genet. Dev.* **13,** 593–598.

Ailion, M., and Thomas, J. H. (2000). Dauer formation induced by high temperatures in *Caenorhabditis elegans*. *Genetics* **156,** 1047–1067.

Ambros, V., and Horvitz, H. R. (1984). Heterochronic mutants of the nematode *Caenorhabditis elegans*. *Science* **226,** 409–416.

Ankeny, R. A. (2001). The natural history of *Caenorhabditis elegans* research. *Nat. Rev. Genet.* **2,** 474–479.

Arnold, A. P. (2004). Sex chromosomes and brain gender. *Nat. Rev. Neurosci.* **5,** 701–708.

Bany, I. A., Dong, M. Q., and Koelle, M. R. (2003). Genetic and cellular basis for acetylcholine inhibition of *Caenorhabditis elegans* egg-laying behavior. *J. Neurosci.* **23**, 8060–8069.

Bargmann, C. I. (2006). Chemosensation. *In* "*C. elegans*, WormBook" (The *C. elegans* Research Community, ed.). doi/10.1895/wormbook.1.123.1, http://www.wormbook.org.

Bargmann, C. I., and Horvitz, H. R. (1991a). Control of larval development by chemosensory neurons in *Caenorhabditis elegans*. *Science* **251**, 1243–1246.

Bargmann, C. I., and Horvitz, H. R. (1991b). Chemosensory neurons with overlapping functions direct chemotaxis to multiple chemicals in *C. elegans*. *Neuron* **7**, 729–742.

Barr, M. M., and Garcia, L. R. (2006). Male behavior. *In* "Wormbook" (The *C. elegans* Research Community, ed.). doi/10.1895/wormbook.1.78.1, http://www.wormbook.org.

Barr, M. M., and Sternberg, P. W. (1999). A polycystic kidney-disease gene homologue required for male mating behaviour in *C. elegans*. *Nature* **401**, 386–389.

Barr, M. M., DeModena, J., Braun, D., Nguyen, C. Q., Hall, D. H., and Sternberg, P. W. (2001). The *Caenorhabditis elegans* autosomal dominant polycystic kidney disease gene homologs *lov-1* and *pkd-2* act in the same pathway. *Curr. Biol.* **11**, 1341–1346.

Barriere, A., and Felix, M. A. (2005). High local genetic diversity and low outcrossing rate in *Caenorhabditis elegans* natural populations. *Curr. Biol* **15**, 1176–1184.

Baumeister, R., Liu, Y., and Ruvkun, G. (1996). Lineage-specific regulators couple cell lineage asymmetry to the transcription of the *Caenorhabditis elegans* POU gene *unc-86* during neurogenesis. *Genes. Dev* **10**, 1395–1410.

Bratus, A., and Slota, E. (2006). DMRT1/Dmrt1, the sex determining or sex differentiating gene in Vertebrata. *Folia. Biol. (Krakow)* **54**, 81–86.

Brenner, S. (1974). The genetics of *Caenorhabditis elegans*. *Genetics* **77**, 71–94.

Brown, A. (2003). "In the Beginning Was the Worm: Finding the Secrets of Life in a Tiny Hermaphrodite." Columbia University Press, New York.

Carmi, I., Kopczynski, J. B., and Meyer, B. J. (1998). The nuclear hormone receptor SEX-1 is an X-chromosome signal that determines nematode sex. *Nature* **396**, 168–173.

Chalfie, M., and Au, M. (1989). Genetic control of differentiation of the *Caenorhabditis elegans* touch receptor neurons. *Science* **243**, 1027–1033.

Chan, S. S., Zheng, H., Su, M. W., Wilk, R., Killeen, M. T., Hedgecock, E. M., and Culotti, J. G. (1996). UNC-40, a *C. elegans* homolog of DCC (Deleted in Colorectal Cancer), is required in motile cells responding to UNC-6 netrin cues. *Cell* **87**, 187–195.

Chasnov, J. R., and Chow, K. L. (2002). Why are there males in the hermaphroditic species *Caenorhabditis elegans? Genetics* **160**, 983–994.

Chasnov, J. R., So, W. K., Chan, C. M., and Chow, K. L. (2007). The species, sex, and stage specificity of a *Caenorhabditis* sex pheromone. *Proc. Natl. Acad. Sci. USA* **104**, 6730–6735.

Chisholm, A. (1991). Control of cell fate in the tail region of *C. elegans* by the gene *egl-5*. *Development* **111**, 921–932.

Chow, K. L., and Emmons, S. W. (1994). HOM-C/Hox genes and four interacting loci determine the morphogenetic properties of single cells in the nematode male tail. *Development* **120**, 2579–2592.

Clark, S. G., Chisholm, A. D., and Horvitz, H. R. (1993). Control of cell fates in the central body region of C. elegans by the homeobox gene *lin-39*. *Cell* **74**, 43–55.

Conradt, B., and Horvitz, H. R. (1998). The *C. elegans* protein EGL-1 is required for programmed cell death and interacts with the Bcl-2-like protein CED-9. *Cell* **93**, 519–529.

Conradt, B., and Horvitz, H. R. (1999). The TRA-1A sex determination protein of C. elegans regulates sexually dimorphic cell deaths by repressing the *egl-1* cell death activator gene. *Cell* **98**, 317–327.

Conradt, B., and Xue, D. (2005). Programmed cell death. *In* "Wormbook" (The *C. elegans* Research Community, ed.). doi/10.1895/wormbook.1.32.1, http://www.wormbook.org.

Costa, M., Weir, M., Coulson, A., Sulston, J., and Kenyon, C. (1988). Posterior pattern formation in C. *elegans* involves position-specific expression of a gene containing a homeobox. *Cell* **55**, 747–756.

De Bono, M., Zarkower, D., and Hodgkin, J. (1995). Dominant feminizing mutations implicate protein-protein interactions as the main mode of regulation of the nematode sex-determining gene tra-1. *Genes Dev.* **9,** 155–167.

Desai, C., and Horvitz, H. R. (1989). *Caenorhabditis elegans* mutants defective in the functioning of the motor neurons responsible for egg laying. *Genetics* **121,** 703–721.

Desai, C., Garriga, G., McIntire, S. L., and Horvitz, H. R. (1988). A genetic pathway for the development of the *Caenorhabditis elegans* HSN motor neurons. *Nature* **336,** 638–646.

Dewing, P., Chiang, C. W., Sinchak, K., Sim, H., Fernagut, P. O., Kelly, S., Chesselet, M. F., Micevych, P. E., Albrecht, K. H., Harley, V. R., and Vilain, E. (2006). Direct regulation of adult brain function by the male-specific factor SRY. *Curr. Biol.* **16,** 415–420.

Doniach, T., and Hodgkin, J. (1984). A sex-determining gene, fem-1, required for both male and hermaphrodite development in *Caenorhabditis elegans. Dev. Biol.* **106,** 223–235.

Ellis, H. M. (1985). Genetic control of programmed cell death in the nematode *C. elegans.* Ph.D. Thesis, Massachusetts Institute of Technology.

Ellis, R. E., and Schedl, T. (2006). Sex determination in the germ line. *In* "WormBook" (The *C. elegans* Research Community, ed.). doi/10.1895/wormbook.1.82.1, http://www.wormbook.org.

Emmons, S. W. (2005). Male development. *In* "WormBook" (The *C. elegans* Research Community, ed). doi/10.1895/wormbook.1.33.1http://www.wormbook.org.

Emmons, S. W. (2006). Sexual behavior of the *Caenorhabditis elegans* male. *Int. Rev. Neurobiol.* **69,** 99–123.

Finney, M., and Ruvkun, G. (1990). The *unc-86* gene product couples cell lineage and cell identity in *C. elegans. Cell* **63,** 895–905.

Finney, M., Ruvkun, G., and Horvitz, H. R. (1988). The *C. elegans* cell lineage and differentiation gene *unc-86* encodes a protein with a homeodomain and extended similarity to transcription factors. *Cell* **55,** 757–769.

Garcia, L. R., Mehta, P., and Sternberg, P. W. (2001). Regulation of distinct muscle behaviors controls the *C. elegans* male's copulatory spicules during mating. *Cell* **107,** 777–788.

Giles, A. C., Rose, J. K., and Rankin, C. H. (2006). Investigations of learning and memory in *Caenorhabditis elegans. Int. Rev. Neurobiol.* **69,** 37–71.

Goodman, M. B. (2006). Mechanosensation. *In* "WormBook" (The *C. elegans* Research Community, ed.). doi/10.1895/wormbook.1.62.1, http://www.wormbook.org.

Gray, J. M., Hill, J. J., and Bargmann, C. I. (2005). A circuit for navigation in *Caenorhabditis elegans. Proc. Natl. Acad. Sci. USA* **102,** 3184–3191.

Grote, P., and Conradt, B. (2006). The PLZF-like protein TRA-4 cooperates with the Gli-like transcription factor TRA-1 to promote female development in *C. elegans. Dev. Cell.* **11,** 561–573.

Gruninger, T. R., Gualberto, D. G., LeBoeuf, B., and Garcia, L. R. (2006). Integration of male mating and feeding behaviors in *Caenorhabditis elegans. J. Neurosci.* **26,** 169–179.

Hall, D. H., Lints, R., and Altun, Z. (2006). Nematode neurons: Anatomy and anatomical methods in *Caenorhabditis elegans. Int. Rev. Neurobiol* **69,** 1–35.

Hamelin, M., Zhou, Y., Su, M. W., Scott, I. M., and Culotti, J. G. (1993). Expression of the UNC-5 guidance receptor in the touch neurons of *C. elegans* steers their axons dorsally. *Nature* **364,** 327–330.

Hedgecock, E. M., Culotti, J. G., and Hall, D. H. (1990). The *unc-5*, *unc-6*, and *unc-40* genes guide circumferential migrations of pioneer axons and mesodermal cells on the epidermis in *C. elegans. Neuron* **4,** 61–85.

Hodgkin, J. (1986). Sex determination in the nematode *C. elegans*: Analysis of *tra-3* suppressors and characterization of fem genes. *Genetics* **114,** 15–52.

Hodgkin, J. (1987). A genetic analysis of the sex-determining gene, *tra-1*, in the nematode *Caenorhabditis elegans. Genes Dev.* **1,** 731–745.

Hodgkin, J. A., and Brenner, S. (1977). Mutations causing transformation of sexual phenotype in the nematode *Caenorhabditis elegans*. *Genetics* **86,** 275–287.

Hunter, C. P., and Wood, W. B. (1990). The *tra-1* gene determines sexual phenotype cell-autonomously in *C. elegans*. *Cell* **63,** 1193–1204.

Hunter, C. P., and Wood, W. B. (1992). Evidence from mosaic analysis of the masculinizing gene her-1 for cell interactions in C. *elegans* sex determination. *Nature* **355,** 551–555.

Hunter, C. P., Harris, J. M., Maloof, J. N., and Kenyon, C. (1999). Hox gene expression in a single *Caenorhabditis elegans* cell is regulated by a caudal homolog and intercellular signals that inhibit wnt signaling. *Development* **126,** 805–814.

Ishii, N., Wadsworth, W. G., Stern, B. D., Culotti, J. G., and Hedgecock, E. M. (1992). UNC-6, a laminin-related protein, guides cell and pioneer axon migrations in C. *elegans*. *Neuron* **9,** 873–881.

Jiang, M., Ryu, J., Kiraly, M., Duke, K., Reinke, V., and Kim, S. K. (2001). Genome-wide analysis of developmental and sex-regulated gene expression profiles in *Caenorhabditis elegans*. *Proc. Natl. Acad. Sci. USA* **98,** 218–223.

Kenyon, C. (1986). A gene involved in the development of the posterior body region of C. *elegans*. *Cell* **46,** 477–487.

Kimble, J. (1981). Alterations in cell lineage following laser ablation of cells in the somatic gonad of *Caenorhabditis elegans*. *Dev. Biol.* **87,** 286–300.

Kimble, J., Edgar, L., and Hirsh, D. (1984). Specification of male development in *Caenorhabditis elegans*: The fem genes. *Dev. Biol.* **105,** 234–239.

Kiontke, K., Gavin, N. P., Raynes, Y., Roehrig, C., Piano, F., and Fitch, D. H. (2004). *Caenorhabditis* phylogeny predicts convergence of hermaphroditism and extensive intron loss. *Proc. Natl. Acad. Sci. USA* **101,** 9003–9008.

Klass, M., Wolf, N., and Hirsh, D. (1976). Development of the male reproductive system and sexual transformation in the nematode *Caenorhabditis elegans*. *Dev. Biol.* **52,** 1–18.

Kleemann, G. (2005). Mating behavior and intersexual mating interactions in the nematode *Caenorhabditis elegans*. M.S. Thesis, University of Nebraska.

Kuhara, A., and Mori, I. (2006). Molecular physiology of the neural circuit for calcineurin-dependent associative learning in *Caenorhabditis elegans*. *J. Neurosci.* **26,** 9355–9364.

Lackner, M. R., Nurrish, S. J., and Kaplan, J. M. (1999). Facilitation of synaptic transmission by EGL-30 Gqalpha and EGL-8 PLCbeta: DAG binding to UNC-13 is required to stimulate acetylcholine release. *Neuron* **24,** 335–346.

Lints, R., and Emmons, S. W. (1999). Patterning of dopaminergic neurotransmitter identity among *Caenorhabditis elegans* ray sensory neurons by a TGF-beta family signaling pathway and a Hox gene. *Development* **126,** 5819–5831.

Lints, R., and Emmons, S. W. (2002). Regulation of sex-specific differentiation and mating behavior in C. *elegans* by a new member of the DM domain transcription factor family. *Genes. Dev.* **16,** 2390–2402.

Lipton, J., Kleemann, G., Ghosh, R., Lints, R., and Emmons, S. W. (2004). Mate searching in *Caenorhabditis elegans*: A genetic model for sex drive in a simple invertebrate. *J. Neurosci.* **24,** 7427–7434.

Liu, K. S., and Sternberg, P. W. (1995). Sensory regulation of male mating behavior in *Caenorhabditis elegans*. *Neuron* **14,** 79–89.

Loer, C. M., and Kenyon, C. J. (1993). Serotonin-deficient mutants and male mating behavior in the nematode *Caenorhabditis elegans*. *J. Neurosci.* **13,** 5407–5417.

Madl, J. E., and Herman, R. K. (1979). Polyploids and sex determination in *Caenorhabditis elegans*. *Genetics* **93,** 393–402.

Mah, K. B., and Rankin, C. H. (1992). An analysis of behavioral plasticity in male *Caenorhabditis elegans*. *Behav. Neural. Biol.* **58,** 211–221.

Maruyama, I. N., and Brenner, S. (1991). A phorbol ester/diacylglycerol-binding protein encoded by the *unc-13* gene of *Caenorhabditis elegans*. *Proc. Natl. Acad. Sci. USA* **88,** 5729–5733.
Mathies, L. D., Schvarzstein, M., Morphy, K. M., Blelloch, R., Spence, A. M., and Kimble, J. (2004). TRA-1/GLI controls development of somatic gonadal precursors in *C. elegans*. *Development*. S. **131,** 4333–4343.
Matta, S. G., Balfour, D. J., Benowitz, N. L., Boyd, R. T., Buccafusco, J. J., Caggiula, A. R., Craig, C. R., Collins, M. I., Damaj, M. I., Donny, E. C., Gardiner, P. S., Grady, S. R. *et al.* (2007). Guidelines on nicotine dose selection for in vivo research. *Psychopharmacology (Berl.)* **190,** 269–319.
Meyer, B. J. (2005). X-chromosome dosage compensation. *In* "WormBook" (The *C. elegans* Research Community, ed.). doi/10.1895/wormbook.1.8.1, http://www.wormbook.org.
Morris, J. A., Jordan, C. L., and Breedlove, S. M. (2004). Sexual differentiation of the vertebrate nervous system. *Nat. Neurosci.* **7,** 1034–1039.
Nelson, G. A., Lew, K. K., and Ward, S. (1978). Intersex, a temperature-sensitive mutant of the nematode *Caenorhabditis elegans*. *Dev. Biol.* **66,** 386–409.
Nigon, V., and Dougherty, E. C. (1949). Reproductive patterns and attempts at reciprocal crossing of Rhabditis elegans Maupas, 1900, and Rhabditis briggsae Dougherty and Nigon, 1949 (Nematoda: Rhabditidae). *J. Exp. Zool.* **112,** 485–503.
Nottebohm, F., and Arnold, A. P. (1976). Sexual dimorphism in vocal control areas of the songbird brain. *Science* **194,** 211–213.
O'Hagan, R., and Chalfie, M. (2006). Mechanosensation in *Caenorhabditis elegans*. *Int. Rev. Neurobiol.* **69,** 169–203.
Peden, E. M., and Barr, M. M. (2005). The KLP-6 kinesin is required for male mating behaviors and polycystin localization in *Caenorhabditis elegans*. *Curr. Biol.* **15,** 394–404.
Pierce-Shimomura, J. T., Morse, T. M., and Lockery, S. R. (1999). The fundamental role of pirouettes in *Caenorhabditis elegans* chemotaxis. *J. Neurosci.* **19,** 9557–9569.
Portman, D. S., and Emmons, S. W. (2000). The basic helix-loop-helix transcription factors LIN-32 and HLH-2 function together in multiple steps of a *C. elegans* neuronal sublineage. *Development* **127,** 5415–5426.
Powell, J. R., Jow, M. M., and Meyer, B. J. (2005). The T-box transcription factor SEA-1 is an autosomal element of the X:A signal that determines *C. elegans* sex. *Dev. Cell* **9,** 339–349.
Raisman, G., and Field, P. M. (1971). Sexual dimorphism in the preoptic area of the rat. *Science* **173,** 731–733.
Rankin, C. H. (2002). From gene to identified neuron to behaviour in *Caenorhabditis elegans*. *Nat. Rev. Genet.* **3,** 622–630.
Raymond, C. S., Shamu, C. E., Shen, M. M., Seifert, K. J., Hirsch, B., Hodgkin, J., and Zarkower, D. (1998). Evidence for evolutionary conservation of sex-determining genes. *Nature* **391,** 691–695.
Reinke, V., Gil, I. S., Ward, S., and Kazmer, K. (2004). Genome-wide germline-enriched and sex-biased expression profiles in *Caenorhabditis elegans*. *Development* **131,** 311–323.
Rhind, N. R., Miller, L. M., Kopczynski, J. B., and Meyer, B. J. (1995). xol-1 acts as an early switch in the *C. elegans* male/hermaphrodite decision. *Cell* **80,** 71–82.
Richmond, J. E., Davis, W. S., and Jorgensen, E. M. (1999). UNC-13 is required for synaptic vesicle fusion in *C. elegans*. *Nat. Neurosci.* **2,** 959–964.
Ross, J. M., Kalis, A. K., Murphy, M. W., and Zarkower, D. (2005). The DM domain protein MAB-3 promotes sex-specific neurogenesis in *C. elegans* by regulating bHLH proteins. *Dev. Cell* **8,** 881–892.
Salser, S. J., and Kenyon, C. (1996). A *C. elegans* Hox gene switches on, off, on and off again to regulate proliferation, differentiation and morphogenesis. *Development* **122,** 1651–1661.
Schafer, W. F. (2006). Genetics of egg-laying in worms. *Annu. Rev. Genet.* **40,** 487–509.
Schindelman, G., Whittaker, A. J., Thum, J. Y., Gharib, S., and Sternberg, P. W. (2006). Initiation of male sperm-transfer behavior in *Caenorhabditis elegans* requires input from the ventral nerve cord. *BMC Biol.* **4,** 26.

Schvarzstein, M., and Spence, A. M. (2006). The *C. elegans* sex-determining GLI protein TRA-1A is regulated by sex-specific proteolysis. *Dev. Cell* **11,** 733–740.

Sengupta, P. (2007). Generation and modulation of chemosensory behaviors in *C. elegans*. *Pflugers Arch.* in press, doi:10.1007/s00424–006–0196–9.

Shaham, S., and Bargmann, C. (2002). Control of neuronal subtype identity by the *C. elegans* ARID protein CFI-1. *Genes Dev.* **16,** 972–983.

Shen, M. M., and Hodgkin, J. (1988). *mab-3*, a gene required for sex-specific yolk protein expression and a male-specific lineage in *C. elegans*. *Cell* **54,** 1019–1031.

Simon, J. M., and Sternberg, P. W. (2002). Evidence of a mate-finding cue in the hermaphrodite nematode *Caenorhabditis elegans*. *Proc. Natl. Acad. Sci. USA* **99,** 1598–1603.

Sternberg, P. (2005). Vulval development. *In* "WormBook" (The *C. elegans* Research Community, ed.). doi/10.1895/wormbook.1.6.1, http://www.wormbook.org.

Stewart, A. D., and Phillips, P. C. (2002). Selection and maintenance of androdioecy in *Caenorhabditis elegans*. *Genetics* **160,** 975–982.

Sulston, J. E., and Horvitz, H. R. (1977). Postembryonic cell lineages of the nematode *Caenorhabditis elegans*. *Dev. Biol.* **56,** 110–156.

Sulston, J. E., Albertson, D. G., and Thomson, J. N. (1980). The *Caenorhabditis elegans* male: Postembryonic development of nongonadal structures. *Dev. Biol.* **78,** 542–576.

Sulston, J. E., Schierenberg, E., White, J. G., and Thomson, J. N. (1983). The embryonic cell lineage of the nematode *Caenorhabditis elegans*. *Dev. Biol.* **78,** 542–576.

Thoemke, K., Yi, W., Ross, J. M., Kim, S., Reinke, V., and Zarkower, D. (2005). Genome-wide analysis of sex-enriched gene expression during *C. elegans* larval development. *Dev. Biol.* **284,** 500–508.

Tomioka, M., Adachi, T., Suzuki, H., Kunitomo, H., Schafer, W. R., and Iino, Y. (2006). The insulin/PI 3-kinase pathway regulates salt chemotaxis learning in *Caenorhabditis elegans*. *Neuron* **51,** 613–625.

Trent, C., Tsuing, N., and Horvitz, H. R. (1983). Egg-laying defective mutants of the nematode *Caenorhabditis elegans*. *Genetics* **104,** 619–647.

Troemel, E. R., Chou, J. H., Dwyer, N. D., Colbert, H. A., and Bargmann, C. I. (1995). Divergent seven transmembrane receptors are candidate chemosensory receptors in *C. elegans*. *Cell* **83,** 207–218.

Tsalik, E. L., and Hobert, O. (2003). Functional mapping of neurons that control locomotory behavior in *Caenorhabditis elegans*. *J. Neurobiol.* **56,** 178–197.

van den Berg, M. C., Woerlee, J. Z., Ma, H., and May, R. C. (2006). Sex-dependent resistance to the pathogenic fungus Cryptococcus neoformans. *Genetics* **173,** 677–683.

Vellai, T., McCulloch, D., Gems, D., and Kovacs, A. L. (2006). Effects of sex and insulin/insulin-like growth factor-1 signaling on performance in an associative learning paradigm in *Caenorhabditis elegans*. *Genetics* **174,** 309–316.

Von Stetina, S. E., Treinin, M., and Miller, D. M. (2006). The motor circuit. *Int. Rev. Neurobiol.* **69,** 125–167.

Wakabayashi, T., Kitagawa, I., and Shingai, R. (2004). Neurons regulating the duration of forward locomotion in *Caenorhabditis elegans*. *Neurosci. Res.* **50,** 103–111.

White, J. G. (1988). The anatomy. *In* "The Nematode *Caenorhabditis Elegans*" (W. B. Wood, ed.). Cold Spring Harbor Laboratory Press, Plainview, NY.

White, J. G., Southgate, E., Thomson, J. N., and Brenner, S. (1986). The structure of the nervous system of the nematode *Caenorhabditis elegans*. *Phil. Trans. R. Soc. Lond. B.* **314,** 1–340.

Whittaker, A. J., and Sternberg, P. W. (2004). Sensory processing by neural circuits in *Caenorhabditis elegans*. *Curr. Opin. Neurobiol.* **14,** 450–456.

Wilkins, A. S. (1995). Moving up the hierarchy: A hypothesis on the evolution of a genetic sex determination pathway. *Bioessays* **17,** 71–77.

Wrischnik, L. A., and Kenyon, C. J. (1997). The role of *lin-22*, a *hairy/enhancer of split* homolog, in patterning the peripheral nervous system of C. *elegans*. *Development* **124,** 2875–2888.
Yi, W., Ross, J. M., and Zarkower, D. (2000). *mab-3* is a direct *tra-1* target gene regulating diverse aspects of C. *elegans* male sexual development and behavior. *Development* **127,** 4469–4480.
Zarkower, D. (2001). Establishing sexual dimorphism: Conservation amidst diversity? *Nat. Rev. Genet.* **2,** 175–185.
Zarkower, D. (2006). Somatic sex determination. *In* "WormBook" (The C. *elegans* Research Community, ed.). doi/10.1895/wormbook.1.84.1, http://www.wormbook.org.
Zarkower, D., and Hodgkin, J. (1992). Molecular analysis of the C. *elegans* sex-determining gene tra-1: A gene encoding two zinc finger proteins. *Cell* **70,** 237–249.
Zarkower, D., and Hodgkin, J. (1993). Zinc fingers in sex determination: Only one of the two C. *elegans* Tra-1 proteins binds DNA *in vitro*. *Nucleic Acids Res.* **21,** 3691–3698.
Zhang, Y., Lu, H., and Bargmann, C. I. (2005). Pathogenic bacteria induce aversive olfactory learning in *Caenorhabditis elegans*. *Nature* **438,** 179–184.
Zhao, C. (1995). Developmental control of peripheral sense organs in C. *elegans* by a transcription factor of the bHLH family. Ph.D. Thesis, Albert Einstein College of Medicine.
Zhao, C., and Emmons, S. W. (1995). A transcription factor controlling development of peripheral sense organs in C. *elegans*. *Nature* **373,** 74–78.

2

The Neural and Genetic Substrates of Sexual Behavior in *Drosophila*

Daisuke Yamamoto
Division of Neurogenetics, Graduate School of Life Sciences
Tohoku University, 6-3 Aoba, Aramaki, Aoba-ku, Sendai
Miyagi 980-8578, Japan

ABSTRACT

fruitless (*fru*), originally identified with its mutant conferring male homosexuality, is a neural sex determination gene in *Drosophila* that produces sexually dimorphic sets of transcripts. In the nervous system, Fru is translated only in males. Fru proteins likely regulate the transcription of a set of downstream genes. The expression of Fru proteins is sufficient to induce male sexual behavior in females. A group of *fru*-expressing neurons called "mAL" neurons in the brain shows conspicuous sexual dimorphism. mAL is composed of 5 neurons in females and 30 neurons in males. It includes neurons with bilateral projections in males and contralateral projections in females. Terminal arborization patterns are also sexually dimorphic. These three characteristics are feminized in *fru* mutant males. The inactivation of cell death genes results in the production of additional mAL neurons that are of the male type in the female brain. This suggests that

Advances in Genetics, Vol. 59
0065-2660/07 $35.00
DOI: 10.1016/S0065-2660(07)59002-4
Copyright 2007, Elsevier Inc. All rights reserved.

male-specific Fru inhibits mAL neuron death, leading to the formation of a male-specific neural circuit that underlies male sexual behavior. Fru orchestrates a spectrum of downstream genes as a master control gene to establish the maleness of the brain. © 2007, Elsevier Inc.

I. HYPOTHESIS ON THE MASTER CONTROL GENE FOR BEHAVIOR

Tinbergen (1951) has postulated a hierarchical organization of neural centers to interpret ordered, stereotypic motor output elicited spontaneously or by specific sensory stimuli. This neural organization is referred to as the innate releasing mechanism (IRM) and specific stimuli that trigger a series of motor acts referred to as releasers (Tinbergen, 1951). Tinbergen (1951) proposed such a neural model exclusively on the basis of behavioral observations in various organisms, including his favorite sticklebacks.

The IRM implicates the existence of a few executive neurons whose excitation is sufficient for initiating an entire series of motor acts composing a particular behavior. Such executive neurons are indeed present in the brain, the first example of which was documented in the crayfish by Wiersma and Ikeda (1964), who demonstrated that simple unpatterned electrical stimuli applied to a descending interneuron induce bursting activities in defined sets of motor neurons, resulting in segmentally coordinated swimmeret movements in the crayfish. Such higher-order interneurons that initiate an entire motor program by their activities without overt temporal patterns have been proposed to refer to as the command fiber or command neuron (Wiersma and Ikeda, 1964).

It took an additional 10 years before the cellular substrates for the IRM were identified by extensive intracellular recording in identified neurons (Burrows, 1996; Hoyle, 1970). The wave of neuroethology in the 1970s deterministically revealed the neural circuits underlying complex innate behaviors such as cricket stridulation (Gerhardt and Huber, 2002) and *Tortonia* swimming (Getting, 1989). Neuronal connectivity and its functional properties are principally invariable from individual to individual, and thus are reproducible from experiment to experiment in invertebrate preparations that use identified neurons for analysis. That the IRM is hardwired and built to be invariant across individuals has been demonstrated in these works. Hardwired networks are, by definition, genetically determined; however, they may be modified by environmental factors and experiences of individual animals.

Thus, questions arise as to which genes are really involved in the formation of the neural circuit underlying respective behavior and how they are ordered to establish the hierarchical organization of neural centers in the IRM. These questions encourage the use of genetic and developmental approaches to studying innate behavior. *Drosophila melanogaster* is an obvious choice as the experimental animal in genetics and developmental biology, because a large number of mutants are available, mutagenic and transgenic techniques are readily applicable (Rubin, 1988), and the genetic manipulation of defined cell populations is possible (Ashburner, 1989). All these conditions need to be fulfilled for the genetic dissection of innate behavior.

The genetic principle of the developmental body plan was disclosed first in *Drosophila*, and then extended to other animal species, including humans (Lewis, 1992). The genetic principle of the body plan thus revealed represents a hierarchical organization of transcription factors, which orchestrate morphogenesis along the body axes through the activation or repression of downstream genes responsible for the proliferation, growth, fate specification, and differentiation of cells (Gehring, 1999). The best example of the morphogenetic power of this type of gene hierarchy is the inductive ability of the gene *eyeless* (Halder *et al.*, 1995). A forced expression of *eyeless*$^{+}$ alone in imaginal disks of wings and legs leads to the formation of the "complete" structure of adult compound eyes in these ectopic sites (Halder *et al.*, 1995). Genes that can activate an entire program for morphogenesis, such as *eyeless*, are regarded as master control genes (Gilbert, 1997).

At a glance, the development of neural circuit in the brain looks different from that of the embryonic body or the compound eye. However, many genes involved in embryonic body and eye morphogenesis have been found to mediate neurogenesis (Campos-Ortega and Hartenstein, 1997; Dickson and Hafen, 1993; Yamamoto, 1996). The identity of neurons in *Drosophila* is largely dependent on their lineage (Doe and Technau, 1993), that is from which neuroblast they arise and in which cycle of mitosis they are produced; in contrast, the identity of the photoreceptor in the compound eye is lineage-independent (Wolff and Ready, 1993). A single neuroblast divides asymmetrically to produce a neuroblast and a ganglion mother cell, which then divides symmetrically producing two terminally differentiated neurons or glia (Goodman and Doe, 1993). In the neuroblast lineage NB7-1, for instance, a neuroblast expresses Hunchback (Hb) during the first two cycles of mitosis. Each of the two ganglion mother cells that emerge in these cycles produces a U1 or U2 motor neuron, respectively, together with two sibling cells. Subsequently, Hb expression is abrogated in the neuroblast, which then expresses Krüppel (Kr). The third cycle of mitosis of the neuroblast takes place while the cell is expressing Kr, and the resulting ganglion mother cell produces a U3 motor neuron. The next transcription factor to be expressed in the neuroblast is Pdm, and the fourth cycle of mitosis results in the production of a U4 motor neuron. Thereafter, Cas is expressed in the neuroblast, whose division leads to the formation of a U5 motor neuron (Pearson and Doe, 2003).

The above-mentioned transcription factors play roles in early embryogenesis. *Hb* mRNA is one of the maternal factors deposited in oocytes; its translation is initiated immediately after fertilization, forming an anterior-to-posterior concentration gradient (Gilbert, 1997; Wolpert, 2002). Zygotic *Hb* functions as a gap gene for specifying the anterior part of the embryonic body. *Kr* is another gap gene that specifies the thoracic part of the embryo (Gilbert, 1997; Wolpert, 2002).

These considerations lead to the hypothesis that the hierarchical organization of transcription regulators underlies the formation of a hierarchical architecture of neural centers in the IRM, which ultimately generates a complete series of motor acts constituting a particular behavior. This hypothesis implies the presence of master control genes for the formation of neural circuits that are dedicated to a particular behavior. The activation of such a master switch gene is assumed to ensure the formation of a complete circuit necessary for performing a particular behavior. This is therefore referred to as the master control gene hypothesis for behavior.

In fact, recent studies of *Drosophila fruitless (fru)* provided experimental support to the master control gene hypothesis for behavior (Demir and Dickson, 2005; Kimura *et al.*, 2005). In this chapter, I will provide an overview of the phenotypic characterization of *fru* mutants at the organismal (behavioral) level, the structure and function of *fru* at the molecular level, and the instruction role of *fru* in neural circuit formation at the cellular level. I will also discuss the potential of *fru* as a master control gene for behavior, in light of the conditions proposed above.

II. DISCOVERY OF *fru* MUTANTS AND THEIR PHENOTYPIC CHARACTERISTICS

The *fru* locus was first defined through the isolation of the male sterile mutant *fruity* (later called fru^1) by Gill (1963), who described in an abstract that males of this variant never copulate with females, suggesting behavioral sterility. Gill (1963, 1965) clearly stated that mutant males court both females and males, that is they display bisexual courtship. Hall (1978) was the first to quantify *fru* mutant phenotypes. In his study, he found that *fru* mutant males courted wild-type males seven times more often than wild-type males did. When the total time of courtship was measured and compared, *fru* mutant males engaged in courting other males 100 times longer than wild-type males did. Hall (1978) also recorded the time spent by males for different components of the courtship ritual.

The mating behavior of *D. melanogaster* is made up of discrete components, that is, orientation, tapping, following and unilataral wing vibration, licking, attempted copulation, copulation, and disengagement (Greenspan, 1995; Hall, 1985; Yamamoto and Nakano, 1999). When a male finds a female, he turns his body axis directly toward the female (orientation). Then the male approaches the female and taps her abdomen with his foreleg, on which gustatory receptors are located (tapping). The female usually tends to move away from the male, who, on the other hand, commences to chase her while vibrating his wings (following and unilateral wing vibration). Wild-type males use their either left or right wing at a time, and use their wings alternately every several seconds. The wing vibration generates species-specific sounds known as courtship or love songs, which exert aphrodisiac effects on conspecific females (Ewing and Bennet-Clark, 1967; Schilcher von, 1976b). Courtship songs of *D. melanogaster* have two components: sine song and pulse song. On oscilloscopic recording, the sine song appears as a 160 Hz sine wave sound (Schilcher von, 1976a), whereas the pulse song represents a series of pulse tones with an interpulse interval of ~35 msec (Kyriacou and Hall, 1986; Schilcher von, 1976a). A *D. melanogaster* male typically generates the pulse song and sine song in series; a *D. melanogaster* female tends to reduce locomotion when exposed to courtship songs generated by a conspecific male (Crossley *et al.*, 1995). The male takes this opportunity to lick the female's genitalia (Cobb and Ferveur, 1996; Hall *et al.*, 1980). Subsequently, the male attempts to copulate with the female. When the female is receptive to copulation, she raises her wings, allowing the male to grasp them and can mount her (Hall *et al.*, 1980). The receptive female also opens her vaginal plate for copulation, which persists for about 15–20 min during which the male stays on the back of the female (Baba *et al.*, 1999; Hall *et al.*, 1980). Just prior to disengagement, the female often kicks the male's abdomen on her back with her hind leg (Kuniyoshi *et al.*, 2002). Subsequently, the male releases the genital hold and then dismounts the female.

Hall (1978) found that fru^1 mutant males show all the elementary behaviors mentioned above, except for copulation, toward both females and males. He also noted that they often display bilateral wing vibration, which is never observed in wild-type males (Hall, 1978).

When multiple fru^1 males are placed in a chamber, they begin chasing other males, forming a long line of courting males, the so-called "courtship chain" (Fig. 2.2A). The courtship chain is not unique to *fru* mutants but is also seen among males with enhanced homosexual interactions (Yamamoto *et al.*, 1997), which include the transgenic fly with ubiquitous $white^+$ (w^+) expression (Armstrong *et al.*, 2000; Nilsson *et al.*, 2000; Zhang and Odenwald, 1995), the fly strain with a chromosomal translocation involving the X-2E and 3L-97A regions as induced by light (Sharma, 1977), transgenic flies in which synaptic transmission is temporally blocked in mushroom bodies by the expression of the temperature-sensitive *shibire* minigene (Kitamoto, 2002), and male flies in which a certain brain area has been feminized by the ectopic expression of $transformer^+$ (Villella *et al.*, 2005; see also Ferveur *et al.*, 1995; O'Dell *et al.*, 1995). Enhanced male-to-male courtship has also been reported in the *Voila* mutant in the *prospero* locus (Balakireva *et al.*, 1998; Grosjean *et al.*, 2001) and *quick-to-court* mutant males (Gaines *et al.*, 2000).

Hall (1978) reported that *fru*^{1} mutant males elicit enhanced courtship from wild-type males. However, this courtship stimulation effect of *fru*^{1} does not sufficiently explain the enhanced courtship toward males by *fru*^{1} mutant males because *fru*^{1} mutant males vigorously court wild-type males that hardly induce courtship from wild-type males. The *fru*^{1} chromosome carries an inversion of the 90C–91B1,2 segment, and genetic analysis revealed that the phenotype of *fru*^{1} to court males is associated with the break point at 91B1,2 whereas the phenotype to elicit courtship from males is associated with the break point at 90C. The locus at 91B1,2 is identified as *fru* (Gailey and Hall, 1989). The 90C locus is yet to be identified.

A decade after the publication of the first full paper (Hall, 1978) on *fru*, the P-element insertion alleles of *fru* and their derivatives have been isolated (Castrillon *et al.*, 1993; Gailey and Hall, 1989; Yamamoto *et al.*, 1991, 1996). An extensive phenotypic analysis of these new alleles and deficiencies of the *fru* region revealed variations in courtship anomalies associated with different genetic compositions of the *fru* locus (Villella *et al.*, 1997). The enhancement of male-directed courtship was observed in all genetic combinations that are mutants for *fru*, yet at varying degrees. Mutant combinations involving *fru*^{1} and *fru*^{2} alleles induce high courting activity in males, who court females and males equally or prefer to court females slightly more than males (Anand *et al.*, 2001; Villella *et al.*, 1997). In contrast, the courtship activities of *fru*^{3}, *fru*^{4}, and *fru*^{sat} mutant males and their heteroallelic mutant males are very low; they prefer males than females as courtship targets (Ito *et al.*, 1996; Villella *et al.*, 1997).

Recordings of courtship songs from *fru*^{1} homozygotes or heteroallelic mutants involving the *fru*^{1} allele revealed longer interpulse intervals of pulse songs than those of courtship songs from wild-type males (Villella *et al.*, 1997; Wheeler *et al.*, 1989). Song recording from males of *fru*^{3}, *fru*^{4}, and *fru*^{sat} alleles and their combinatorial mutants has been mostly unsuccessful (Villella *et al.*, 1997; D. Y., unpublished data). The only exception has been reported by Villella *et al.* (1997) who recorded one or two brief bouts of low-amplitude hums similar to the sine song from 8 of 305 *fru*^{3}/*fru*^{4} males. For these reasons, the biophysical properties of courtship songs in these mutant alleles are unknown. The tendency to form a courtship chain among multiple males is high in mutants involving *fru*^{1}, but low in *fru*^{3}, *fru*^{4}, and *fru*^{sat} alleles (Villella *et al.*, 1997). Therefore, the frequency of courtship chain formation seems to correlate with courtship activity in general. Chain formation activity changes with age in the adult stage (Goodwin *et al.*, 2000; Villella *et al.*, 1997). It is practically zero on the day of eclosion and sharply increases thereafter, reaching a plateau on day 4 in any of the *fru*^{1}, *fru*^{3}, *fru*^{4}, and *fru*^{sat} alleles (Lee and Hall, 2000).

Besides chain formation, head-to-head interaction in *fru* mutant males has been reported to develop with aging as in the case of chain formation

(Lee and Hall, 2000). Nilsen *et al.* (2004) pointed out that head-to-head interaction is rarely observed in wild-type males; such an interaction is similar to a "head butt" commonly seen in fighting females. Vrontou *et al.* (2006) demonstrated that the *fru* gene is involved in this gender-selective pattern of aggression.

Males of different *fru* alleles show a wide fertility spectrum: the fru^1, fru^3, fru^4, and fru^{sat} mutants are absolutely sterile, heteroallelic mutants involving fru^1 typically show reduced fertility, and the fru^2 mutant has perfect fertility thus enabling the maintenance of this mutant as a homozygous stock (Anand *et al.*, 2001; Gailey *et al.*, 1991; Villella *et al.*, 1997). It is interesting that some allelic combinations show intragenic complementation in fertility. For example, the fru^1 and fru^{sat} mutants are completely sterile when homozygous for each allele, yet they are fertile when combined as heteroallelic mutants (Villella *et al.*, 1997; D.Y., unpublished data). It is tempting to speculate that transvection underlies this allelic interaction (Judd, 1988).

Even though some heteroallelic *fru* mutants are fertile, their reproductive physiology and behavior are not necessarily normal. In males of such *fru* mutants, copulation duration is highly variable; when assessed using the mean, it prolongs to fourfold than in the wild-type (Lee *et al.*, 2001). These *fru* mutant males often engage in infertile copulation (Lee *et al.*, 2001). These abnormalities in copulation observed in some heteroallelic *fru* mutants may be ascribable to a failure in the transfer of sperm and seminal fluid from the male to the female during copulation (Lee *et al.*, 2001; Villella *et al.*, 2006). The male's internal reproductive organs are innervated by a few serotonergic motor neurons that express Fru. In *fru* mutants, these motor neurons are not stained by the anti-5HT antibody (Lee and Hall, 2001). A malfunction in these neurons has been suggested to cause copulatory defects and infertility in *fru* mutant males (Billeter *et al.*, 2006; Lee *et al.*, 2001). For the development of these serotonergic neurons in full complement, male-type Dsx should be expressed together with Fru (Billeter *et al.*, 2006).

A breakthrough in *fru* research was made from the finding (Gailey *et al.*, 1991; Hall, 2002) that *fru* functions are indispensable for the formation of a male-specific muscle, the muscle of Lawrence (MOL; Fig. 2.1B). The MOL is a large pair of longitudinal muscles present in the tergite of the fifth abdominal (A5) segment of the male adult (Lawrence and Johnston, 1984). The sites of the skeletal attachment of the MOL are closer to the segment borders in both anterior and posterior ends than those of other longitudinal muscles (thus, the MOL is longer than other muscles). The MOL is also different from other abdominal muscles in that it expresses an actin isoform, 79B actin, at a high level equivalent to that seen in thoracic muscles, in contrast to the low 79B actin expression level in conventional abdominal muscles (Courchesne-Smith and Tobin, 1989). Using these criteria, the MOL is unequivocally distinguishable from other abdominal muscles.

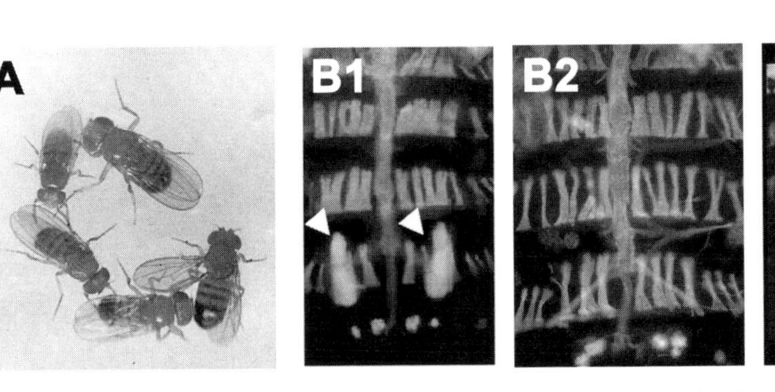

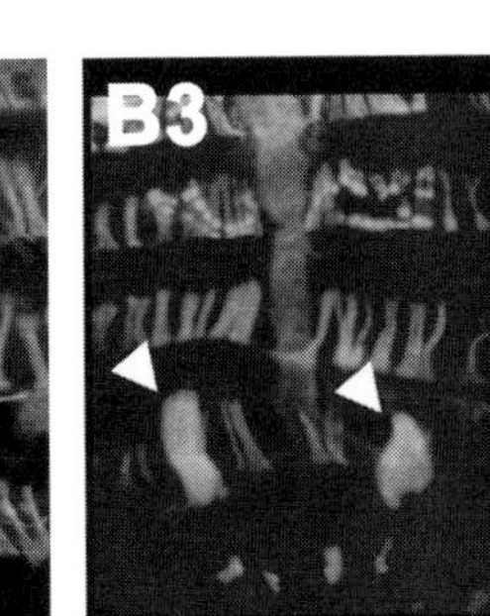

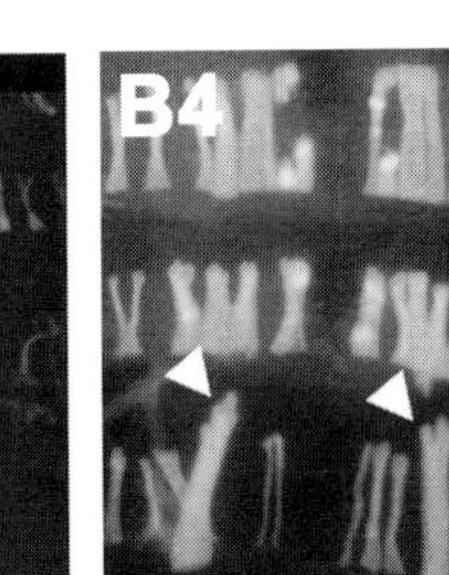

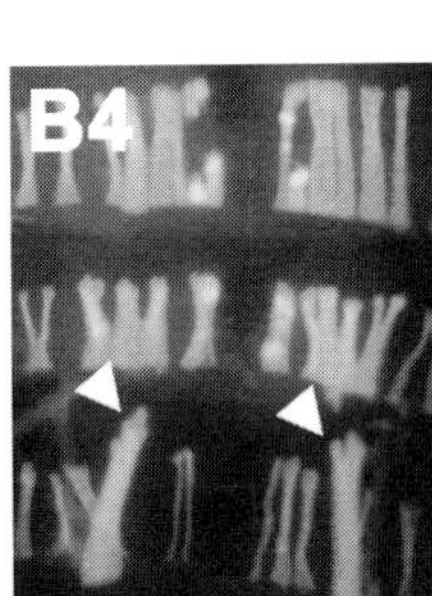

Figure 2.1. Phenotypes of *fru* mutants. (A) Chain formation of courting fru^{sat} males (cited from Ito *et al.*, 1996). Male-specific muscle of Lawrence (MOL; white arrowheads). (B1) The wild-type male has a pair of MOLs in the fifth abdominal segment. (B2) The fru^{sat} male lacks the MOL. (B3) The fru^{sat} male carrying an *hs-fru*$^{+}$ transgene develops the MOL when *hs-fru*$^{+}$ expression is induced during the pupal stage. (B4) The fru^{sat}female acquires the MOL when Fru is expressed in motor neurons by *D42-Gal4–UAS-fru*$^{+}$ (cited from Usui-Aoki *et al.*, 2000). (See Color Insert.)

The selective formation of the MOL in males does not rely on the chromosomal sex of the muscle itself, but of the innervating nerve. Substantial bodies of evidence support this notion. First, the nuclei of myocytes transplanted from a male donor to a female host do not lead to MOL formation in the female, but the same nuclei transplanted from a female donor to a male host lead to MOL formation (Lawrence and Johnston, 1986). Second, myoblasts isolated from female wing disks can be recruited to form the MOL when transplanted into the male abdomen (Kimura *et al.*, 1994). Third, the denervation of the MOL-innervating nerve at the prepupal stage prevents myocytes from developing into the MOL in operated males while not affecting the formation of other abdominal musculatures (Currie and Bate, 1995). These collectively indicate that innervating motor neurons must be chromosomally male for MOL formation. They also imply that the malformation of the MOL observed in *fru* mutant males reflects neural defects such as an inadequate sex determination of neurons.

The process for MOL formation, however, remains poorly understood. Taylor and Knittel (1995) found that nuclei in the MOL are fewer in *fru* mutants than in the wild type, although the total number of myoblasts available for formation of abdominal muscles does not differ between them. Thus, Taylor and Knittel (1995) suggested that an impairment in the male-specific recruitment of myoblasts into MOL myotubes is partially responsible for MOL malformation in *fru* mutants.

The degree of MOL defects varies depending on the *fru* allele, from the complete absence for fru^{sat} to marginal hypotrophy for fru^{2}. There is a good parallel between the degree of reduction in courtship activity and the intensity of the MOL phenotype. Courtship anomalies and MOL malformation are tightly coupled genetically, as documented from the fact that the excision of the P-element from the fru^{sat} genome results in a simultaneous reversion of these two phenotypes (Ito *et al.*, 1996). It is plausible that both phenotypes are associated with a failure in sex determination of the nervous system: courtship anomaly and MOL malformation may have resulted from the "demasculinization" of brain interneurons and motor neurons in the abdominal ganglia, respectively (Hall, 1994; Yamamoto *et al.*, 1996). The neuronal origin of MOL malformation in *fru* mutants was confirmed by the observation that a forced expression of fru^{+} in neurons but not in muscles restores MOL formation in fru^{sat} males (Usui-Aoki *et al.*, 2000) (Fig. 2.1B).

Indeed, MOL formation in males depends on the integrity of two sex determination genes: *Sex-lethal* (*Sxl*) and *transformer* (*tra*). In *Drosophila*, the ratio of the number of X chromosomes to the number of paired autosomes (X/A) determines sex (Penalva and Sánchez, 2003). In the wild type, the A is two (second and third chromosomes) because the small fourth chromosome can be ignored. When the animal has two X chromosomes, the X/A becomes 1.0. When the animal has only one X chromosome (XY or XO), then X/A becomes 0.5.

The animal with an X/A of 0.5 develops as a male whereas that with an X/A greater than 1.0 develops as a female. When X/A is between 0.5 and 1.0, the animal appears intersex (Marín and Baker, 1998; Yamamoto *et al.*, 1998).

The primary female determinant gene is *Sxl*, whose transcription is activated only when X/A exceeds 1.0. Without *Sxl* activation, the animal takes on a default sexual fate, the male. Transcription factors that act on one of the *Sxl* promoters are dimeric proteins, with each composite polypeptide, encoded by a gene on the X chromosome or an autosome, and their correct stoichiometry needs to be established for their proper functioning. Sxl represses the splicing of its binding target, the *tra* primary mRNA (Inoue *et al.*, 1990; Sosnowski *et al.*, 1989). The descendant *tra* mRNA produces the feminizing factor Tra, allowing an XX individual to develop as a female. Without Sxl, the *tra* primary mRNA is spliced so as to produce mRNA that encodes nonfunctional truncated Tra, leading to the development of a male (XY or XO). Tra as a splicing facilitating factor induces the female-specific splicing of its target, the *doublesex* (*dsx*) primary mRNA, with the aid of its cofactor, Tra2 (Goralski *et al.*, 1989; Ota *et al.*, 1981). Tra2 is expressed in both female and male. Without Tra, the *dsx* primary mRNA is spliced at an alternative site, producing the male-type mRNA. The female-type and male-type mRNAs for *dsx* encode the Dsx female protein (DsxF) and Dsx male protein (DsxM), respectively (Burtis and Baker, 1989; Hoshijima *et al.*, 1991; Inoue *et al.*, 1992; Nagoshi and Baker, 1990). These Dsx proteins are transcription factors that activate or repress target genes directly responsible for the production of sexual characteristics in the two sexes.

In females with loss-of-function mutations in *Sxl, tra*, or *tra2*, a male-specific MOL forms (Taylor, 1992). Interestingly, mutations in *dsx* have no effect on MOL formation (Taylor, 1992). *dsx* mutant males develop a normal MOL, whereas *dsx* mutant females do not develop the MOL. These suggest that there must be a second target of Tra that mediates MOL formation (Taylor, 1992; Taylor *et al.*, 1994). *fru* is the obvious candidate for such a target.

Although MOL formation has been the subject of extensive developmental and genetic studies, the physiological functions of the MOL are not yet known. Note that many *Drosophila* species lack the MOL (Gailey *et al.*, 1997). However, in *D. subobscura*, an additional pair of MOLs is present in the A4 segment (Gailey *et al.*, 1997).

The absence or presence of the MOL being unrelated to phylogeny prompted Gailey *et al.* (1997) to assume that the MOL represents a vestigial muscle that is being lost in different phylogenetic branches.

Even more surprising was the finding that MOL-like male-specific muscles are present in the A5 segment of the mosquito (*Anopheles gambiae*) (Gailey *et al.*, 2006). *fru* from *Anopheles* is able to rescue MOL formation in *D. melanogaster fru* mutant males, which are otherwise MOL-free (Gailey *et al.*, 2006). *fru* homologues are present in the bee and beetle, although no obvious counterparts have been found outside *Insecta* (Gailey *et al.*, 2006). Thus, the

MOL provides an exciting opportunity for determining the genetic basis of the evolution of sex-limited characteristics.

When the entire *fru* locus is deleted by combining two overlapping deficiencies, non-sex-specific phenotypes are unraveled. In early neurogenesis, some FasII-positive and BP102-positive connectives and commisures in the central nervous system (CNS) are distorted in fru^{w12}/fru^{sat15} embryos, which is rescued by the overexpression of certain Fru isoforms as driven by the pan-neural expressor *sca-Gal4* (Song *et al.*, 2002). Song *et al.* (2002) also found that pioneer neurons, such as pCC motor neurons, often misroute at the initial stage of axon extension.

The neural expression of some segmentation genes is markedly affected in embryos with severe loss-of-function *fru* mutations: first, there is a delay in the onset of Hb expression in certain neuroblasts (Song *et al.*, 2002) and second, Eve expression in some ganglion mother cells is precociously terminated, followed by an insufficient expression of a glia-specific protein, Reversed polarity (Repo) (Song and Taylor, 2003).

The above findings implicate *fru*'s function in neurogenesis, in general, although its exact role in this process remains elusive. *fru* appears to be also required in the postembryonic development of both sexes. The motor innervation of lateral muscles in pupae with severe *fru* mutations is less extensive than that in pupae of the wild type (Anand *et al.*, 2001). Thus, neurite extension seems to be affected by *fru* mutations in both embryonic and adult stages.

Aside from neural phenotypes, gross morphological deficits are observed in individuals with severe *fru* mutations. Some pupae with lethal *fru* mutations are unable to evert anterior spiracles or retract anterior segments, reminiscent of mutants exhibiting ecdysone signaling (Anand *et al.*, 2001). A few escaper adults often fail to extend their wings accompanied by malformed leg joints (Anand *et al.*, 2001).

These observations indicate that *fru* may play multiple roles in development. This is not surprising in light of the fact that Sxl, a key player in sex determination, also has a general role in neurogenesis that is non-sex-specific (Jan and Jan, 1993). Note, however, that the non-sex-specific roles of *fru* have been investigated using deficiency combinations that could delete genes other than *fru* despite the fact that it is not yet confirmed whether the *fru* function is completely abrogated. Thus, the non-sex-specific roles of *fru* might not really reflect the functions of *fru per se*.

III. MOLECULAR BIOLOGY OF *fru* LOCUS

The molecular cloning of the *fru* locus was accomplished independently by two groups. One group cloned *fru* by chromosome walk starting from the fru^{sat} insertion site by a Northern blot analysis of transcripts derived from the cloned region in wild-type and mutant flies (Ito *et al.*, 1996). Another group isolated a *fru* fragment in search for a Tra-binding consensus sequence in the genome (Ryner *et al.*, 1996). The *fru* thus identified spans about 150 kB of the genome, producing multiple transcript forms (Goodwin *et al.*, 2000; Ito *et al.*, 1996; Ryner *et al.*, 1996; Usui-Aoki *et al.*, 2000). *fru* is transcribed from at least four different promoters, one of which (P1) is responsible for sex-specific functions. All transcripts but P1-derived transcripts share a translation start site, which is used to produce a group of proteins responsible for the non-sex-specific functions of *fru* described in Section II.

Located most distally is the P1 promoter, generating a primary transcript that is subjected to sexually dimorphic splicing in the second exon. The second exon contains the Tra-binding consensus sequence that functions as the 5′ splice site when bound by Tra in females (Heinrichs *et al.*, 1998; Lam *et al.*, 2003; see also Du *et al.*, 1998). In males, an alternative upstream site is used for splicing that excludes a large part of the second exon from the male transcript. This portion of the second exon, which is retained in the female-type mRNA and eliminated in the male-type mRNA, contains stop codons, allowing the male mRNA to have a long open reading frame (ORF) and the female RNA to have an ORF with several stop codons (Ito *et al.*, 1996; Ryner *et al.*, 1996; Usui-Aoki *et al.*, 2000). Thus, the male mRNA can encode proteins that are headed by 101-amino-acid-long male-specific N-termini, beyond the N-termini of non-sex-specific Fru proteins (Ryner *et al.*, 1996; Usui-Aoki *et al.*, 2000). The P1-derived female type of mRNA appears unable to produce any proteins (Fig. 2.2), partly

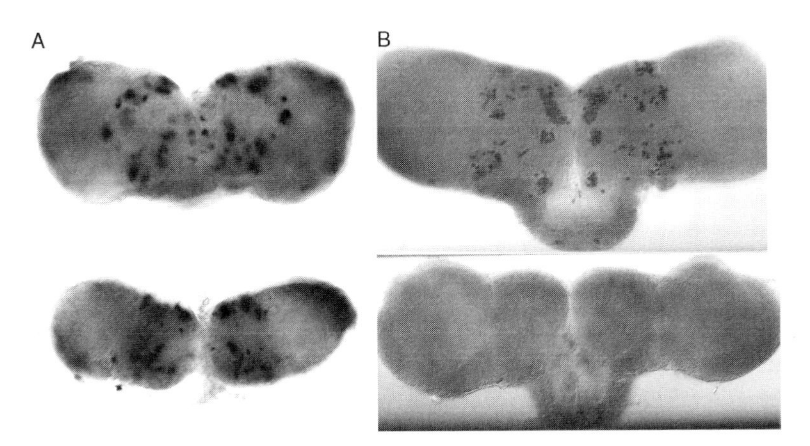

Figure 2.2. Expression of *fru* gene products in pupal brain (cited from Usui-Aoki *et al.*, 2000). (A) *fru* mRNA expression as detected by *in situ* hybridization in wild-type male (upper) and female (lower). (B) Fru expression as detected by immunostaining with anti-Fru antibody in wild-type male (upper) and female (lower). (See Color Insert.)

because of its dicistronic organization. However, no short truncated peptide that could arise from the female type of mRNA has been detected (Lee *et al.*, 2000; Usui-Aoki *et al.*, 2000). The expression level of a reporter gene preceded by the Tra-binding sequence adopted from *fru* has been shown to decrease when Tra and Tra2 are coexpressed in transfected S2 cells (Usui-Aoki *et al.*, 2000). This suggests that Tra binding to female *fru* mRNA represses its translation (Usui-Aoki *et al.*, 2000).

The Fru expression in the CNS appears to be male-specific in most *Drosophila* species (Yamamoto *et al.*, 2004) except in *D. suzukii*, in which Fru is expressed in the CNS of both sexes but with different patterns (Usui-Aoki *et al.*, 2005).

fru mRNAs also differ in their 3′ end. There are five different 8th exons alternatively used. Three of these five 8th exons contain the sequences coding for two tandem Zn-finger motifs that slightly differ among the exons (Ito *et al.*, 1996; Ryner *et al.*, 1996; Usui-Aoki *et al.*, 2000). All Fru proteins share an N-terminal BTB domain (Ito *et al.*, 1996; Ryner *et al.*, 1996; Usui-Aoki *et al.*, 2000). Fru proteins therefore belong to the BTB-Zn-finger protein family, which represents a large group of transcription factors.

The presence of male-specific Fru proteins prompted us to hypothesize that the presence of Fru in the cell leads to the masculinization of the expressing cell, whereas its absence leads to the feminization of the cell (Usui-Aoki *et al.*, 2000). In fact, a forced Fru expression in females using a ubiquitous promoter or a motor neuron-specific Gal4 driver leads to the ectopic induction of a male-specific MOL in females (Usui-Aoki *et al.*, 2000). This unequivocally demonstrates that the presence of Fru is indeed sufficient for neural masculinization. Interestingly, Fru proteins without male-specific N-termini are able to induce the MOL, provided that they have Zn-finger motifs in their C-termini (Usui-Aoki *et al.*, 2000). The idea that a male-specific N-terminus is dispensable for male-specific *fru* function is further supported by the finding that MOL formation can be induced in *Drosophila fru* mutants by a forced expression of the mosquito *fru* homologue, although no sequence of a male-specific N-terminus of Fru is conserved between the fly and mosquito (Gailey *et al.*, 2006). Note that, only the BTB and Zn-finger motifs, besides the Tra-binding sequence, are conserved between the two species. The amino acid sequences of the Fru proteins are highly conserved in their entire length among *Drosophila* species, except in *D. mimica*, where the amino acid identity is less than 80% when compared with the *D. melanogaster* counterpart (Davis *et al.*, 2000a,b; Gailey *et al.*, 2000).

Male-specific Fru proteins are expressed in some 2000 neurons in the brain and ventral ganglia. The expression starts in the late third-instar larval stage, peaks in the mid-pupal stage, and is sustained throughout the adult stage at low levels. It remains to be determined whether each isoform of *fru* has a different expression pattern.

Although Fru proteins are assumed to regulate the transcription of target genes, no experimental support to this has been obtained.

The targets of Fru regulation remain largely unknown. One candidate Fru target is the *takeout* (*to*) gene, which was originally identified for starvation sensitivity and circadian rhythmicity in its transcription (Sarov-Blat *et al.*, 2000; So *et al.*, 2000). Take out is a member of a large family of proteins, some of which function as carriers of small lipophilic compounds such as the juvenile hormone (Du *et al.*, 2003; Robertson *et al.*, 1999). Take out is expressed in the antenna and maxillary palps of both sexes and fat bodies surrounding the brain only in males (Dauwalder *et al.*, 2002). *to* dominantly enhances the *fru* mutant phenotype to reduce the courtship activity of males, which can be restored by the overexpression of to^+ using the *to* 5′ regulatory sequence (Dauwalder *et al.*, 2002). The male-specific expression of *to* in brain fat bodies is under the control of both *dsx* and *fru*. The sex-specific *to* expression appears to be cell-autonomous, because a forced expression of Tra^+ in males using the *to* 5′ regulatory sequence abrogates *to* expression (Dauwalder *et al.*, 2002). Three potential Dsx-binding sites (Erdman *et al.*, 1996) are present within 1-kb upstream of the *to* translation initiation codon, although no Dsx binding in these sites has been demonstrated experimentally (Dauwalder *et al.*, 2002). Because the Fru-binding motif remains unknown, it is not possible to evaluate the possibility that Fru directly controls *to* transcription. It is interesting to note that some of the clock-controlling neurons in the brain express Fru (Kadener *et al.*, 2006; Lee *et al.*, 2006).

Yellow (Y) is reported to be a candidate Fru target (Drapeau *et al.*, 2003). *y* is a recessive visible marker that makes the body color pale, but its precise role in pigmentation remains obscure (Claycomb *et al.*, 2004; Han *et al.*, 2002). In addition to its presence in epidermis, *y* is expressed in and secreted by certain neuroblasts, which are Fru positive (Drapeau, 2003; Drapeau *et al.*, 2003; Radovic *et al.*, 2002). Fru overexpression increases the expression level of Y in these neuroblasts. *y* mutant males show reduced courtship activity (Bastock and Manning, 1955; Burnet *et al.*, 1973; Sturtevant, 1915); this is rescued by overexpressing y^+ in neural cells (Drapeau *et al.*, 2003). However, that *fru* and *y* interact genetically in male courtship has not yet been demonstrated. Drapeau *et al.* (2006) reported that *y* mutant alleles with deficits in mating behavior share a region in the 5′regulatory region of *y*, which they named the mating-success regulatory sequence (MRS) (Drapeau *et al.*, 2006; see also Prud'homme *et al.*, 2006). The MRS is conserved among *Drosophila* species and contains the consensus binding sequences for Dsx and Dorsal (Drapeau *et al.*, 2006). Although the significance of To and Y in regulating mating behavior remains to be elucidated further, it is interesting to see that these proteins function downstream of Dsx or Fru because they potentially mediate nonautonomous sex determination that has not been considered in insects (De Loof and Huybrechts, 1998).

The mechanism controlling *fru* transcription is poorly understood. The 16-kb genomic fragment upstream of the P1 promoter reproduces only 17% of endogenous Fru expression in the CNS, yet it induces MOL formation in females or *fru* mutant males (Billeter and Goodwin, 2004; D. Y., unpublished data). However, the 16-kb 5′ sequence does not rescue courtship defects of *fru* mutant males when used to regulate the expression of a minigene encoding any Fru isoforms (D. Y., unpublished data). No *trans* acting factors of *fru* transcription are known. *eyeless*, however, was found to upregulate *fru* transcription when expressed in the eye disks as detected by microarray analysis (Michaut *et al.*, 2003).

IV. CELLULAR BASIS OF *fru* FUNCTIONS IN MALE COURTSHIP BEHAVIOR

In neurobiology, it is critical to determine the neural circuit underlying respective behavior. It is a demanding issue to describe neural connections formed by *fru*-expressing neurons to illustrate the entire network that determines male courtship behavior. To this end, efforts have been made to identify individual neurons that express Fru (Billeter and Goodwin, 2004; Manoli *et al.*, 2005; Stockinger *et al.*, 2005).

To identify *fru*-expressing neurons, it is crucial to obtain reporter strains that label the entire structure of cells showing endogenous *fru* expression because a strict nuclear localization of Fru prevents the visualization of neurites by tissue staining with anti-Fru antibodies.

Manoli *et al.* (2005) and Stockinger *et al.* (2005) independently generated homologous recombination knockin transgenic strains, in which the *Gal4*-coding sequence is inserted downstream of the P1 promoter, whereas the *fru*-coding sequences are inactivated. They recapitulated the pattern of the cell body localization of *fru*-expressing neurons in the CNS, which has been observed by anti-Fru antibody staining. An unexpected finding in these experiments is the strong Gal4 expression in the peripheral nervous system, where no endogenous Fru expression has been reported, with exceptions of motor nerves projecting to the internal reproductive organ and MOL (Billeter and Goodwin, 2004; Lee and Hall, 2001). Many sensory neurons that play olfactory and gustatory roles in the antenna, labellum, labrum, maxillary palp, and foreleg express *fru-Gal4* (Manoli *et al.*, 2005; Stockinger *et al.*, 2005). Mechanosensory neurons in the wing joint and male genitalia are also positive for Gal4 (Manoli *et al.*, 2005). These peripheral sensory neurons express Fru, as determined by anti-Fru antibody staining (Manoli *et al.*, 2005). A similar peripheral Gal4 expression pattern has been observed in the fly line in which the *fru*^{4} P-element was replaced with the *P-Gal4* element (Dornan *et al.*, 2005), or in the strains that express Gal4 driven by the 5′ sequence upstream of the P1 promoter (D. Y., unpublished data).

The projection of olfactory afferents in the brain has been studied (Stockinger *et al.*, 2005). Sensory neurons labeled by the *fru-Gal4* reporter terminate in a few specific glomeruli of the antennal lobe (Stockinger *et al.*, 2005). Those glomeruli, which are innervated by *fru-Gal4*-positive sensory neurons, are DA1, VA1v, and VL2a (Stockinger *et al.*, 2005). We previously showed that DA1 and VA1v are significantly larger in males than in females, and that no other glomeruli are sexually dimorphic (Kondoh *et al.*, 2003). We also demonstrated that the ectopic expression of tra^{+} in male fly epidermis reduces the size of DA1 and VA1v compared with that in females. These indicate that sexual dimorphism in DA1 and VA1v sizes is determined by innervating sensory afferents (Kondoh *et al.*, 2003). Stockinger *et al.* (2005) confirmed the sexual dimorphism of DA1 and VAv1 and further showed that VL2A is slightly larger in males than in females. Only these three sexually dimorphic glomeruli receive innervation by sensory neurons labeled by *fru-Gal4* (Stockinger *et al.*, 2005).

Demir and Dickson (2005) produced knockin lines that express either male-specific *fru* mRNAs or female-specific *fru* mRNAs regardless of the sex of the fly. Male flies that express only female-specific *fru* mRNAs had female-sized DA1, VA1v, and VL2A, whereas female flies that express only male-specific *fru* mRNAs had male-sized DA1, VA1v, and VL2 (Stockinger *et al.*, 2005). These indicate that the *fru*-dependent sex determination of sensory neurons is responsible for the sexual difference in glomerular size (Stockinger *et al.*, 2005).

Kurtovic *et al.* (2007) showed that the *fru*-positive olfactory neurons projecting to the DA1 glomerulus express the *Or67d* olfactory receptor that is responsive to a male-derived courtship-inhibiting pheromone, ll-cis-vaccenyl acetate (Ejima *et al.*, 2007; Jallon *et al.*, 1981).

The most striking observation in "splice-fixed" knockin flies is that females expressing only male-specific *fru* mRNAs display a male-type courtship behavior, including tapping, wing vibration, and licking, although they do not show copulation attempts or copulation (Demir and Dickson, 2005). The last two behavioral elements require the male genitalia and abdominal musculature, which are formed by a Fru-independent, Dsx-dependent mechanism; and therefore, female flies with male-type *fru* mRNAs could not perform them. These observations prompted them to claim that male-specific *fru* mRNAs can instruct the entire male-type courtship behavior.

Thus far, no *fru* isoforms have been shown to rescue courtship defects in *fru* mutant males, or to induce male-type behavior such as wing vibration in females when expressed (Demir and Dickson, 2005). This implies that a delicate balance in the expression profile of distinct *fru* isoforms is to be established first to instruct male courtship behavior. The overexpression of a single *fru* isoform cannot attain such a condition. However, the "splice-fixed" knockin flies fulfill this requirement. This view is reinforced by the fact that the selective loss of one type of isoform is sufficient to confer *fru* mutant phenotypes. However, the supplementation of the lost isoform by transgenic overexpression cannot fully reverse abnormalities in mutants (Billeter *et al.*, 2006).

When the pheromonal compositions of males are feminized by a forced *tra*$^{+}$ expression in enocytes, males were courted by females with male-type *fru* mRNA expression (Demir and Dickson, 2005). Thus, the sexual roles were reversed in this case.

Females that express male-type *fru* mRNAs were less receptive when courted by males and less frequently chosen by males as targets of courtship (Demir and Dickson, 2005). This suggests that females with male-specific *fru* mRNAs are masculinized and recognized by males under such conditions.

Another approach to alter *fru* expression was taken by Manoli and Baker (2004) in an attempt to manipulate male courtship behavior. They generated transgenic flies that express dsRNA against the male-specific *fru* mRNA (*fruIR*). A forced expression of *fruIR* in median bundle neurons abrogated Fru expression as detected by anti-Fru antibody staining, accompanied by marked changes in male courtship behavior (Manoli and Baker, 2004). All the steps of mating behavior proceeded so quickly that some of them appeared skipped in males expressing *fruIR* in median bundle neurons. From these results, Manoli and Baker (2004) concluded that a subset of median bundle neurons trigger the fixed action pattern of courtship behavior. However, this conclusion needs to be taken with caution for the following reasons: First, no specificity of the *fruIR* action is critically evaluated. Even though the endogenous *fru* expression is downregulated in these flies, it does not mean that other mRNAs (i.e., those unrelated to *fru*) are unaffected. Second, the effects of *fruIR* outside median bundle neurons cannot be completely eliminated. Third, the behavioral outcome of *fru* knockdown induced by *fruIR* is different from known *fru* mutant phenotypes. No *fru* mutants have shown hyperactive, disordered courtship behavior. Conversely, *fruIR*-expressing males do not exhibit homosexual courtship, which is a hallmark of *fru* mutant males. These issues should be addressed to define the roles of median bundle neurons in male courtship behavior.

Because no obvious sexual difference was found in neurons-expressing *fru* reporters (Manoli *et al.*, 2005; Stockinger *et al.*, 2005) or in median bundle neurons (Manoli and Baker, 2004), it was suggested that the maleness of the neural circuitry instructed by *fru* resides in its function rather than structure. Subsequently, the conspicuous sexual difference in *fru*-expressing neural architecture was demonstrated with a Gal4 enhancer trap line using single-cell labeling (Kimura *et al.*, 2005) by the MARCM method (Lee and Luo, 1999). The Gal4 enhancer trap line *NP21* has a *P-Gal4* insertion in the second intron of *fru*, in which Gal4 expression simulates the endogenous *fru* expression (Kimura *et al.*, 2005). *NP21* by itself is *fru* allele (Kimura *et al.*, 2005). Two sets of Gal4-expressing neurons exhibit sexual dimorphism in *NP21*. A group of *fru*-expressing interneurons in the optic lobe exists only in males. Another group of *fru*-expressing neurons located just above the antennal lobe contains different numbers of cells in females and males (Kimura *et al.*, 2005). This cluster was named mAL by Lee *et al.* (2000), who classified *fru*-expressing neurons into 20 groups. mAL is unique in that *fru* expression in these neurons is undetectable in fru^{1} mutant males, which retained significant *fru* expression in the cells of the other groups (Lee *et al.*, 2000). *fru* alleles that show markedly reduced courtship activity (fru^{sat}, fru^{3}, and fru^{4}) lack *fru* expression from almost all neurons that express *fru* in wild-type flies (Lee *et al.*, 2000).

mAL contains 30 cells in males and 5 cells in females (Kimura *et al.*, 2005). When the three major genes that induce apoptosis were inactivated in mAL, the number of cells contained in this cluster increased to nearly 30 in females, no changes were observed in males (Kimura *et al.*, 2005). This result indicates that 25 neurons are actively eliminated in males.

By staining the entire structures of mAL neurons, a marked sexual difference in their projection patterns was observed (Kimura *et al.*, 2005) (Fig. 2.3). In females, each of the five mAL neurons extends from its soma a single neurite, which crosses the midline and bifurcates. One of the branches projects to the lateral

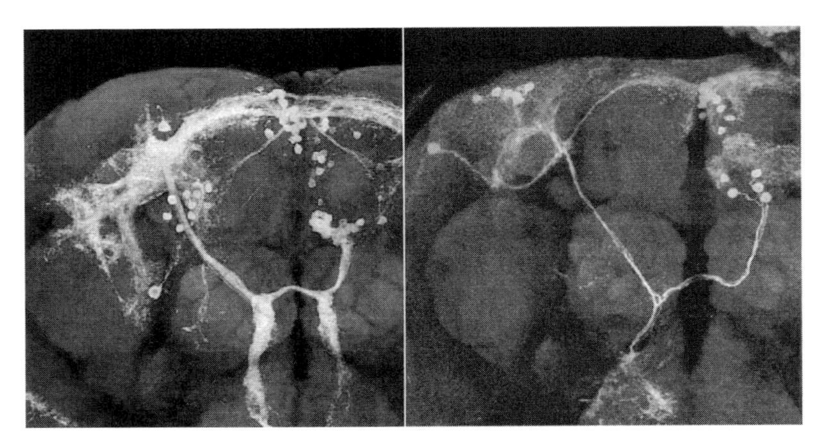

Figure 2.3. Sexual dimorphism in mAL neurons that express *fru* (cited from Kimura *et al.*, 2005). The mAL cluster is composed of 30 neurons per hemisphere in males (left) and 5 neurons in females (right). mAL neurons in males have both contralateral and ipsilateral neurites (left), whereas those in females extend only contralateral neurites (right). Dendritic branches in the subesophageal ganglion (lowest part of the neurons) are horsetail-like in males (left) or forked in females (right) (See Color Insert.).

protocerebrum. Another branch projects to the subesophageal ganglion. On the other hand, there are two classes of neurons in males. One class of mAL neurons in males has a similar overall structure to those in females, with a contralateral projection. Another class of mAL neurons in males distinctly differs from those in females in that it has an ipsilateral projection in addition to a contralateral projection.

Both classes of mAL neurons in males are clearly different from the homologous neurons in females in arborization pattern in the subesophageal ganglion. All mAL neurons in males develop "horsetail-like" terminals, whereas mAL neurons of females develop fork-shaped terminals. In *fru* mutant males, there are only five mAL neurons as in the case of wild-type females. Furthermore, these five neurons in *fru* mutant males have only contralateral projections, whose terminals in the subesophageal ganglion are Y-shaped. Thus, in *fru* mutant males, mAL is feminized in all three aspects of sexual dimorphism (Kimura *et al.*, 2005).

When cell death is prevented, additional mAL neurons are produced in females, as mentioned above. These neurons that escaped death develop ipsilateral projections that have never been observed in wild-type females. However, the terminal structure in the subesophageal ganglion is Y-shaped, indicating that the neurons retain female characteristics in this part (Kimura *et al.*, 2005).

Thus, in females, programmed cell death eliminates some mAL neurons that, if they survive, will develop male-like projections. In males, Fru prevents cell death, thereby generating mAL neurons with male-typical projections (Kimura *et al.*, 2005). In addition, Fru determines terminal branching patterns in the subesophageal ganglion to be of the male type.

In this manner, *fru* can make developmentally homologous neurons different between females and males, thereby establishing a sexually dimorphic neural circuit that likely produces sex-specific behavior.

It should be emphasized, however, that no mAL neurons have been clearly demonstrated to be involved in sexual behavior. A presynaptic marker, synaptotagmin, is localized in the terminals in the lateral protocerebrum and not in those of the subesophageal ganglion, indicating that the terminals in the subesophageal ganglion represent input sites and those in the lateral protocerebrum represent output sites. Since the subesophageal ganglion receives gustatory afferents from the periphery, it is tempting to speculate that pheromonal information conveyed by sensory afferents converges on mAL neurons in the subesophageal ganglion and is integrated with other inputs. Decisions to court or not to court could be made by these mAL neurons, which either send a command to trigger male-type courtship behavior via output synapses in the lateral protocerebrum or not. This hypothesis can now be tested by the genetic manipulation of identified neurons that are amenable to activity monitoring by imaging techniques (Ng *et al.*, 2002; Suh *et al.*, 2004).

V. CONCLUSIONS

Recent studies of the *fru* gene clearly show that a splicing difference in its products determines the sexual type of mating behavior. The expression of male-type *fru* mRNAs in females is sufficient to switch female behavior into male behavior. This manipulation results in the complete masculinization of mating behavior rather than in a sexually intermediate state of mating behavior. The role of *fru* in determining mating behavior appears to be comparable to that of *eyeless* in the development of the compound eye because both genes orchestrate the activities of the entire set of downstream genes required for establishing the system for behavior or development. In this sense, *fru* may be regarded to be a master control gene for behavior.

It does not necessarily mean that the neural circuit underlying courtship behavior is constructed exclusively by *fru*-expressing neurons. Although sensory afferent neurons originating from antennal sensilla express *fru*, second-order projection neurons that receive synaptic inputs from them do not express *fru*. It is, however, possible that these projection neurons with dendrites in particular glomeruli of the antennal lobe are inductively affected by *fru*-expressing input fibers. Because the MOL is induced by *fru*-expressing motor neurons, similar inductive actions of *fru*-expressing neurons on synaptic partners may be postulated. In addition to such direct interactions between *fru*-expressing and non-*fru*-expressing cells, humoral mediators of sexual differentiation may function downstream of *fru*. This possibility is suggested by the finding that Take out is regulated by Fru. Take out is assumed to function as a carrier of lipophilic compounds, based on the fact that a Take out family protein is a carrier for juvenile hormones (Du *et al.*, 2003), and that Take out is detected in circulating heamolymph of *Drosophila* (Lazareva *et al.*, 2006). It is conceivable that Fru exerts its action as a sex determinant by regulating circulating hormones. This neuroendocrine mechanism might coordinate different sex determination pathways running parallel, that is the *fru*, *dsx*, and *dissatisfaction* pathways (Finley *et al.*, 1997, 1998).

It is also important to stress that some neurons involved in sex-specific functions are shared by both sexes. In this context, note that a complete series of male courtship behavior is elicited in mutant females for the *retained* (*retn*) gene, which functions independent of *fru* (Ditch *et al.*, 2005). The male-typical behavior in *retn* mutant females can be seen only after aging, implying that the neural circuit present in the female brain has been suppressed in youth, and that it is released from the suppression by some type of deterioration process in old females. If this is the case, the circuits for male and female behavior sit together in the brain, yet only one of them is active. Sex determination genes have sustained expression in the CNS during the adult stage (Bopp *et al.*, 1991), tempting us to speculate that they play some role in the suppression of the neural circuit for alternative sex. The critical period for Tra and Tra2 actions in the determination of behavioral sex is a controversial issue (Arthur *et al.*, 1998; Belote and Baker, 1987), which needs further elucidation. Although *retn*-expressing neurons as visualized by a reporter are different from *fru*-expressing neurons, they share important features, that is both have dendritic arborizations in the subesophageal ganglion and projections extending to the protocerebrum. Retn could function to secure the proper operation of the *fru*-mediated as well as dsx-mediated switching mechanism of behavioral sex types through neural interaction, rather than genomic interaction (Shirangi *et al.*, 2006). Fru, on the other hand, plays an instructive role in the construction of a switching mechanism of behavioral sex during development. This hypothesis invites further experimental trials that will ultimately lead to our understanding of the molecular and cellular mechanisms determining the femaleness and maleness of the brain.

Acknowledgments

This work was supported by Special Promoted Research grant no. 1802012 from the Ministry of Education, Culture, Sports, Science, and Technology in Japan. I thank Y. Fujita for secretarial assistance.

References

Anand, A., Villella, A., Ryner, L. C., Carlo, T., Goowin, S. F., Song, H. L., Gailey, D. A., Morales, J. C., Hall, J. C., Baker, B. S., and Taylor, B. J. (2001). Molecular genetic dissection of the sex-specific and vital functions of the *Drosophila melanogaster* sex determination gene *fruitless*. *Genetics* **158,** 1569–1595.

Armstrong, X. A. J., Kaiser, K., and O'Dell, K. M. C. (2000). The effects of ectopic *white* and *transformer* expression on *Drosophila* courtship behavior. *J. Neurogenet.* **14,** 227–243.

Arthur, B. I., Jr., Jallon, J.-M., Caflish, B., Choffat, Y., and Nöthiger, R. (1998). Sexual behaviour in *Drosophila* is irreversibly programmed during a critical period. *Curr.Biol.* **8,** 1187–1190.

Ashburner, M. (1989). "*Drosophila:* A Laboratory Handbook," p. 1331. Cold Spring Harbor Laboratory Press, New York.

Baba, K., Takeshita, A., Majima, K., Ueda, R., Kondo, S., Juni, N., and Yamamoto, D. (1999). The *Drosophila* Bruton's tyrosine kinase (Btk) homolog is required for adult survival and male genital formation. *Mol. Cell. Biol.* **19,** 4405–4413.

Balakireva, M., Stocker, R. F., Gendre, N., and Ferveur, J.-F. (1998). *Voila*, a new courtship variant that affects the nervous system: Behavioral, neural, and genetic characterization. *J. Neurosci.* **18,** 4335–4343.

Bastock, M., and Manning, A. (1955). The courtship of *Drosophila melanogaster*. *Behaviour* **8,** 85–111.

Belote, J. M., and Baker, B. S. (1987). Sexual behavior: Its genetic control during development and adulthood in *Drosophila melanogaster*. *Proc. Natl. Acad. Sci. USA* **84,** 8026–8030.

Billeter, J. C., and Goodwin, S. F. (2004). Characterization of *Drosophila fruitless-Gal4* transgene expression in male-specific *fruitless* neurons and innervation of male reproductive structures. *J. Comp. Neurol.* **475,** 270–287.

Billeter, J. C., Villella, A., Allendorfer, J. B., Dorman, A. J., Richardson, M., Gailey, D. A., and Goodwin, S. F. (2006). Isoform-specific control of male neuronal differentiation and behavior in *Drosophila* by the *fruitless* gene. *Curr. Biol.* **16,** 1063–1076.

Bopp, D., Bell, L. R., Cline, T. W., and Schedl, P. (1991). Developmental distribution of female-specific *Sex-lethal* proteins in *Drosophila melanogaster*. *Genes Dev.* **5,** 403–415.

Burnet, B., Connolly, K. J., and Harrison, B. (1973). Phenocopies of pigmentary and behavioral effects of the *yellow* mutant in *Drosophila* induced by α-dimethyltyrosine. *Science* **181,** 1059–1060.

Burrows, M. (1996). "The Neurobiology of an Insect Brain." Oxford University Press, Oxford.

Burtis, K. C., and Baker, B. S. (1989). *Drosophila doublesex* gene controls somatic sexual differentiation by producing alternatively spliced mRNAs encoding related sex-specific polypeptides. *Cell* **56,** 997–1010.

Campos-Ortega, J., and Hartenstein, V. (1997). "The Embryonic Development of *Drosophila melanogaster*," 2nd Edn., p. 405. Springer, Berlin.

Castrillon, D. H., Gönzy, P., Alexander, S., Rawson, R., Eberhart, C. G., Viswanathan, S., DiNardo, S., and Wasserman, S. A. (1993). Toward a molecular genetic analysis of spermatogenesis in *Drosophila melanogaster*: Characterization of male-sterile mutants generated by single *P* element mutagenesis. *Genetics* **135,** 489–505.

Claycomb, J. M., Benasutti, M., Bosco, G., Fenger, D. D., and Orr-Weaver, T. (2004). Gene amplification as a developmental strategy: Isolation of two developmental amplicons in *Drosophila*. *Dev. Cell* **6,** 145–155.

Cobb, M., and Ferveur, J.-F. (1996). Evolution and genetic control of mate recognition and stimulation in *Drosophila*. *Behav. Process.* **35,** 35–54.

Courchesne-Smith, C. L., and Tobin, S. L. (1989). Tissue-specific expression of the 79B actin gene during *Drosophila* development. *Dev. Biol.* **133,** 313–321.

Crossley, S.A, Bennet-Clark, H. C., and Evert, H. T. (1995). Courtship song components affect male and female *Drosophila* differently. *Anim. Behav.* **50,** 827–839.

Currie, D. A., and Bate, M. (1995). Innervation is essential for the development and differentiation of a sex specific adult muscle in *Drosophila melanogaster*. *Development* **121,** 2549–2557.

Dauwalder, B., Tsujimoto, S., Moss, J., and Mattox, W. (2002). The *Drosophila takeout* gene is regulated by the somatic sex-determination pathway and affects male courtship behavior. *Genes Dev.* **16,** 2879–2892.

Davis, T., Kunihira, J., Yoshino, E., and Yamamoto, D. (2000a). Genomic Organisation of the neural sex determination gene fruitless (fru) in the Hawaiian species *Drosophilia silvestris* and the conservation of the fru BTB protein-building domain throughout evolution. *Hereditas* **132,** 67–78.

Davis, T., Kurihara, J., and Yamamoto, D. (2000b). Genomic organisation and characterisation of the neural sex-determination gene *fruitless (fru)* in the Hawaiian species *Drosophila heteroneura*. *Gene* **246,** 143–149.

De Loof, A., and Huybrechts, R. (1998). Insects do not have sex hormones: A myth? *Gen. Comp. Endocrinol.* **111,** 245–260.

Demir, E., and Dickson, B. J. (2005). Fruitless splicing specifies male courtship behavior in *Drosophila*. *Cell* **121,** 1–10.

Dornan, A. J., Gailey, D. A., and Goodwin, S. F. (2005). Gal4 enhancer trap targeting of the *Drosophila* sex determination gene *fruitless*. *Genesis* **42,** 236–246.

Drapeau, M. D. (2003). A novel hypothesis on the biochemical role of the *Drosophila* Yellow protein. *Biochem. Biophys. Res. Commun.* **311,** 1–3.

Drapeau, M. D., Radovic, A., Wittkopp, P. J., and Long, A. D. (2003). A gene necessary for normal male courtship, *yellow*, acts downstream of *fruitless* in the *Drosophila melanogaster* larval brain. *J. Neurobiol.* **55,** 53–72.

Drapeau, M. D., Cyran, S. A., Viering, M. M., Geyer, P. K., and Long, A. D. (2006). A *cis*-regulatory sequence within the *yellow* locus of *Drosophila melanogaster* required for normal male mating success. *Genetics* **172,** 1009–1030.

Dickson, B. J., and Hafen, E. (1993). Genetic dissection of eye development in Drosophila. *In* "The Development of Drosophila melanogaster" (M. Bate and A. Martinez Arias, eds.), pp. 1327–1352. Cold Spring Harbor Laboratory Press, New York.

Ditch, L. M., Shirangi, T., Pitman, J. L., Latham, K. L., Dinley, K. D., Edeen, P. T., Taylor, B. J., and McKeown, M. (2005). *Drosophila retained/dead ringer* is necessary for normal pathfinding, female receptivity and repression of *fruitless* independent male courtship behaviors. *Development* **132,** 155–164.

Doe, C. Q., and Technau, G. M. (1993). Identification and cell lineage of individual neural precursors in the *Drosophila* CNS. *Trends Neurosci.* **16,** 510–514.

Du, C., McGuffin, E., Dauwalder, B., Rabinow, L., and Mattox, W. (1998). Protein phosphorylation plays an essential role in the regulation of alternative splicing and sex determination in *Drosophila*. *Mol. Cell* **2,** 741–750.

Du, J., Hiruma, K., and Riddiford, L. M. (2003). A novel gene in the *takeout* gene family is regulated by hormones and nutrients in *Manduca* larval epidermis. *Insect Biochem. Mol. Biol.* **33,** 803–814.

Ejima, A., Smith, B. P., Lucas, C., van der Goes van Naters, W., Miller, C. J., Carlson, J. R., Levine, J. D., and Griffith, L. C. (2007). Generalization of courtship learning in *Drosophila* is mediated by *cis*-vaccenyl acetate. *Curr. Biol.* **17,** 599–605.

Erdman, S. E., Chen, H. J., and Burtis, K. C. (1996). Functional and genetic characterization of the oligomerization and DNA binding properties of the *Drosophila Doublesex* proteins. *Genetics* **144,** 1639–1952.

Ewing, A. W., and Bennet-Clark, H. C. (1967). Stimuli provoked by courtship of male. *Drosophila melanogaster*. *Nature* **215,** 669–671.

Ferveur, J.-F., Störtkuhl, K. F., Stocker, R. F., and Greenspan, R. J. (1995). Genetic feminization of brain structures and changed orientation in male *Drosophila*. *Science* **267,** 902–905.

Finley, K. D., Taylor, B. J., Milstein, M., and McKeown, M. (1997). *dissatisfaction*, a gene involved in sex-specific behavior and neural development of *Drosophila melanogaster*. *Proc. Natl. Acad. Sci. USA* **94,** 913–918.

Finley, K. D., Edeen, P. T., Foss, M., Gross, E., Ghbeish, N., Palmer, R. H., Taylor, B. J., and McKeown, M. (1998). *dissatisfaction* encodes a Tailless-like nuclear receptor expressed in a subset of CNS neurons controlling *Drosophila* sexual behavior. *Neuron* **21,** 1363–1374.

Gailey, D. A., and Hall, J. C. (1989). Behavior and cytogenetics of *fruitless* in *Drosophila melanogaster*: Different courtship defects caused by separate, closely linked lesions. *Genetics* **121,** 773–785.

Gailey, D. A., Taylor, B. J., and Hall, J. C. (1991). Elements of the *fruitless* locus regulate development of the muscle of Lawrence, a male-specific structure in the abdomen of *Drosophila melanogaster* adults. *Development* **113,** 879–890.

Gailey, D. A., Ohshima, S., Santiago, S. J., Montez, J. M., Arellano, A. R., Robillo, J., Villarimo, C. A., Roberts, L., Fine, E., Villella, A., and Hall, J. C. (1997). The muscle of Lawrence in *Drosophila*: A case of repeated evolutionary loss. *Proc. Natl. Acad. Sci. USA* **96,** 4543–4547.

Gailey, D. A., Ho, S. K., Ohshima, S., Liu, J. H., Eyassu, M., Washington, M. A., Yamamoto, D., and Davis, T. (2000). A Phylogeny of the Drosophilidae using the sex-behaviour gene *fruitless*. *Hereditas* **133,** 81–83.

Gailey, D. A., Billeter, J.-C., Liu, J. H., Bauzon, F., Allendorfer, J. B., and Goodwin, S. F. (2006). Functional conservation of the *fruitless* male sex-determination gene across 250 myr of insect evolution. *Mol. Biol. Evol.* **23,** 633–643.

Gaines, P., Tompkins, L., Woodard, C. T., and Carlson, J. R. (2000). *quick-to-court*, a *Drosophila* mutant with elevated levels of sexual behavior, is defective in a predicted coiled-coil protein. *Genetics* **154,** 1627–1637.

Gehring, W. J. (1999). "Master Control Genes in Development and Evolution: The Homeobox Story." Yale University Press, London.

Gerhardt, H. C., and Huber, F. (2002). "Acoustic Communication in Insects and Amurans: Common Problems and Diverse Solutions." University of Chicago Press, Chicago.

Getting, P. A. (1989). Emerging principles governing the operation of neural networks. *Annu. Rev. Neurosci.* **12,** 185–204.

Gilbert, S. F. (1997). "Developmental Biology," 7th edn. Sinauer Associates, Inc., Sunderland, Massachusetts.

Gill, K. S. (1963). Amutation causing abnormal courtship and mating behavior in males of. *Drosophila melanogaster*. *Am. Zool.* **3,** 507.

Gill, K. S. (1965). A mutation causing abnormal mating behavior. *Drosophila Information Service* **38,** 33.

Goodman, C., and Doe, C. Q. (1993). Embryonic development of the *Drosophila* central nervous system. *In* "The Development of *Drosophila melanogaster*" (M. Bate and A. Martinez Arias, eds.), pp. 1131–1206. Cold Spring Harbor Laboratory Press, New York.

Goodwin, S., Taylor, B., Villella, A., Foss, M., Ryner, L. C., Baker, B. S., and Hall, J. C. (2000). Aberrant splicing and altered spatial expression patterns in *fruitless* mutants of *Drosophila melanogaster*. *Genetics* **154,** 725–745.

Goralski, T. J., Edström, J.-E., and Baker, B. S. (1989). The sex determination locus *transformer-2* of *Drosophila* encodes a polypeptide with similarity to RNA binding proteins. *Cell* **56,** 1011–1018.

Greenspan, R. J. (1995). Understanding the genetic construction of behavior. *Sci. Am.* **272,** 72–76.

Grosjean, Y., Balakireva, M., Dartevelle, L., and Ferveur, J.-F. (2001). PGal4 excision reveals the pleiotropic effects of *Voila*, a *Drosophila* locus that affects development and courtship behavior. *Genet. Res. Camb.* **77,** 239–250.

Halder, G., Callaert, P., and Gehring, W. J. (1995). Induction of ectopic eyes by targeted expression of the eyeless gene in *Drosophila*. *Science* **267,** 1788–1792.

Hall, J. C. (1978). Courtship among males due to a male-sterile mutation in. *Drosophila melanogaster*. *Behav. Genet.* **8,** 125–141.

Hall, J. C. (1985). Genetic analysis of behavior in insects. *In* "Comprehensive Insect Physiology Biochemistry and Pharmacology" (G. A. Kerkut and L. I. Gilbert, eds.), Vol. 9, pp. 287–373. Pergamon Press, Oxford.

Hall, J. C. (1994). The mating of a fly. *Science* **264,** 1702–1714.

Hall, J. C. (2002). Courtship lite: A personal history of reproductive behavioral neurogtenetics in *Drosophila*. *J. Neurogenet.* **16,** 135–163.

Hall, J. C., Siegel, R. W., Tompkins, L., and Kyriacou, C. P. (1980). Neurogenetics of courtship in *Drosophila*. *Stadler Symp.* **12,** 43–82.

Han, Q., Fang, J., Ding, H., Johnson, J. K., Christeinsen, B. M., and Li, J. (2002). Identification of *Drosophila melanogaster* yellow-f abd yellow-f2 proteins as dopachrome-conversion enzymes. *Biochem. J.* **368,** 333–340.

Heinrichs, V., Ryner, L. C., and Baker, B. S. (1998). Regulation of sex-specific selection of *fruitless* 5′ splice sites by *transformer* and *transformer-2*. *Mol. Cell. Biol.* **18,** 450–458.

Hoshijima, K., Inoue, K., Higuchi, I., Sakamoto, H., and Shimura, Y. (1991). Control of *doublesex* alternative splicing by Transformer and Transformer-2 in *Drosophila*. *Science* **252,** 833–836.

Hoyle, G. (1970). Cellular mechanisms underlying behavior—neuroethology. *Adv. Insect Physiol.* **7,** 349–444.

Inoue, K., Hoshijima, K., Sakamoto, H., and Shimura, Y. (1990). Binding of the *Drosophila Sex-lethal* gene product to the alternative splice site of *transformer* primary transcript. *Nature* **344,** 461–463.

Inoue, K., Hoshijima, K., Higuchi, I., Sakamoto, H., and Shimura, Y. (1992). Binding of the *Drosophila* transformer and transformer-2 proteins to the regulatory elements of doublesex primary transcript for sex-specific RNA processing. *Proc. Natl. Acad. Sci. USA* **89,** 8092–8096.

Ito, H., Fujitani, K., Usui, K., Shimizu-Nishikawa, K., Tanaka, S., and Yamamoto, D. (1996). Sexual orientation in *Drosophila* is altered by the satori mutation in the sex-determination gene fruitless that encodes a zinc finger protein with a BTB domain. *Proc. Natl. Acad. Sci. USA* **93,** 9687–9692.

Jallon, J.-M., Antony, C., and Benamar, O. (1981). Un anti-aphrodisiafue produit par les mâles *Drosophila melanogoster* et transferé aux femalles lors de la copulation. *C.R. Acad. Sci. Paris* **292,** 1147–1149.

Jan, Y. N., and Jan, L. Y. (1993). HLH proteins, fly neurogenesis, and vertebrate myogenesis. *Cell* **75,** 827–830.

Judd, B. H. (1988). Transvection: Allelic cross talk. *Cell* **53,** 841–843.

Kadener, S., Villella, A., Kula, E., Palm, K., Pyza, E., Botas, J., Hall, J. C., and Robash, M. (2006). Neurotoxic protein expression reveals connections between the circadian clock and mating behavior in *Drosophila. Proc. Natl. Acad. Sci. USA* **103,** 13537–13542.

Kimura, K.-I., Usui, K., and Tanimura, T. (1994). Female myoblasts can participate in the formation of a male-specific muscle in *Drosophila. Zool. Sci.* **11,** 247–251.

Kimura, K.-I., Ote, M., Tazawa, T., and Yamamoto, D. (2005). Fruitless specifies sexually dimorphic neural circuitry in the *Drosophila* brain. *Nature* **438,** 229–233.

Kitamoto, T. (2002). Conditional disruption of synaptic transmission induces male-male courtship behavior in *Drosophila. Proc. Natl. Acad. Sci. USA* **99,** 13232–13237.

Kondoh, Y., Kaneshiro, K. Y., Kimura, K.-I., and Yamamoto, D. (2003). Evolution of sexual dimorphism in the olfactory brain of Hawaiian Drosophila. *Proc. R. Soc. Lond. B.* **270,** 1005–1013.

Kuniyoshi, H., Baba, K., Ueda, R., Kondo, S., Awano, W., Juni, N., and Yamamoto, D. (2002). lingerer, a *Drosophila* gene involved in initiation and termination of copulation, encodes a set of novel cytoplasmic proteins. *Genetics* **162,** 1775–1789.

Kurtovic, A., Widmer, A., and Dickson, B. J. (2007). A single class of olfactory neurons mediates behavioural responses to a *Drosophila* sex pheromone. *Nature* **446,** 542–546.

Kyriacou, C. P., and Hall, J. C. (1986). Interspecific genetic control of courtship song production and reception in. *Drosophila. Science* **232,** 494–497.

Lam, B. J., Bakshi, A., Ekinci, F. Y., Webb, J., Graveley, B. R., and Hertel, K. J. (2003). Enhancer-dependent 5′-splice site control of *fruitless* pre-mRNA splicing. *J. Biol. Chem.* **278,** 22740–22747.

Lawrence, P. A., and Johnston, P. (1984). The genetic specification of a segment of *Drosophila* muscle. *Cell* **36,** 775–782.

Lawrence, P. A., and Johnston, P. (1986). The muscle pattern of a segment of *Drosophila* may be determined by neurons and not by contributing myoblasts. *Cell* **45,** 505–513.

Lazareva, A., Mattox, W., Hardin, P., and Dauwalder, B. (2007). A role for the adult fat body in *Drosophila* male courtship behavior. *PloS Genet.* **3**(1), e16. doi: 10.1371/ journal. p gen. 0030016, 115–122.

Lee, G., and Hall, J. C. (2000). A newly uncovered phenotype associated with the *fruitless* gene of *Drosophila melanogaster*: Aggression-like head interactions between mutant males. *Behav. Genet.* **30,** 263–275.

Lee, G., and Hall, J. C. (2001). Abnormalities of male-specific FRU protein and serotonin expression in the CNS of *fruitless* mutants in *Drosophila. J. Neurosci.* **21,** 513–526.

Lee, G., Foss, M., Goodwin, S. F., Carlo, T., Taylor, B. J., and Hall, J. C. (2000). Spatial, temporal, and sexually dimorphic expression patterns of the *fruitless* gene in the *Drosophila* central nervous system. *J. Neurobiol.* **43,** 404–426.

Lee, G., Villella, A., Taylor, B. J., and Hall, J. C. (2001). New reproductive anomalies in fruitless-mutant *Drosophila* males: Extreme lengthening of mating durations and infertility correlated with defective serotonergic innervation of reproductive organs. *J. Neurobiol.* **47,** 121–149.

Lee, G., Bahn, J. H., and Park, J. H. (2006). Sex- and clock-controlled expression of the *neuropeptide F* gene in *Drosophila*. *Proc. Natl. Acad. Sci. USA* **103,** 12580–12585.

Lee, T., and Luo, L. (1999). Mosaic analysis with a repressive cell marker for studies of gene function in neuronal morphogenesis. *Neuron* **22,** 451–461.

Lewis, E. B. (1992). Clusters of master control genes regulate the development of higher organisms. *J. Am. Med. Assoc.* **267,** 1524–1531.

Manoli, D. S., and Baker, B. S. (2004). Median bundle neurons coordinate behaviours during *Drosophila* male courtship. *Nature* **430,** 564–569.

Manoli, D. S., Foss, M., Villella, A., Taylor, B. J., Hall, J. C., and Baker, B. S. (2005). Male-specific *fruitless* specifies the neural substrates of *Drosophila* courtship behaviour. *Nature* **436,** 395–400.

Marín, I., and Baker, B. S. (1998). The evolutionary dynamics of sex determination. *Science* **281,** 1990–1994.

Michaut, L., Flister, S., Neeb, M., White, K. P., Certa, U., and Gehring, W. J. (2003). Analysis of the eye developmental pathway in *Drosophila* using DNA microarrays. *Proc. Natl. Acad. Sci. USA* **100,** 4024–4029.

Nagoshi, R. N., and Baker, B. S. (1990). Regulation of sex-specific RNA splicing at the *Drosophila doublesex* gene: *cis*-acting mutations in exon sequences alter sex-specific RNA splicing patterns. *Genes Dev.* **4,** 89–97.

Ng, M., Roorda, R. D., Lima, S. Q., Zemelman, B. V., Morcillo, P., and Miesenböck, G. (2002). Transmission of olfactory information between three populations of neurons in the antennal lobe of the fly. *Neuron* **36,** 463–474.

Nilsen, S. P., Chan, Y.-B., Huber, R., and Kravitz, E. A. (2004). Gender-selective patterns of aggressive behavior in. *Drosophila melanogaster*. *Proc. Natl. Acad. Sci. USA* **101,** 12342–12347.

Nilsson, E. E., Asztalos, Z., Lukacsovich, T., Awano, W., Usui-Aoki, K., and Yamamoto, D. (2000). *fruitless* is in the regulatory pathway by which ectopic mini-white and transformer induce bisexual courtship in *Drosophila*. *J. Neurogenet.* **13,** 213–232.

O'Dell, K. M. C., Armstrong, J. D., Yang, M. Y., and Kaiser, K. (1995). Functional dissection of the *Drosophila* mushroom bodies by selective feminization of genetically defined subcompartments. *Neuron* **15,** 55–61.

Ota, T., Fukunaga, A., Kawabe, M., and Oishi, K. (1981). Interactions between sex transformation mutants of *Drosophila melanogaster*. I. Hemolymph vitellogenins and gonad morphology. *Genetics* **99,** 429–441.

Pearson, B. J., and Doe, C. Q. (2003). Regulation of neuroblast competence in *Drosophila*. *Nature* **425,** 624–628.

Penalva, L. O. F., and Sánchez, L. (2003). RNA binding protein Sex-lethal (Sxl) and control of *Drosophila* sex determination and dosage compensation. *Microbiol. Mol. Biol. Rev.* **67,** 343–359.

Prud'homme, B., Gompel, N., Rokas, A., Kassner, V. A., Williams, T. M., Yeh, S.-D., True, JU. R., and Carroll, S. B. (2006). Repeated morphological evolution through *cis*-regulatory changes in a pleiotropic gene. *Nature* **440,** 1050–1053.

Radovic, A., Wittkopp, P. J., Long, A. D., and Drapeau, M. D. (2002). Immunohistochemical colocalization of Yellow and male-specific Fruitless in *Drosophila melanogaster* neuroblasts. *Biochem. Biophys. Res. Commun.* **293,** 1262–1264.

Robertson, H. M., Martos, R., Sears, C. R., Todres, E. Z., Walden, K. K. O., and Nardi, J. B. (1999). Diversity of odorant binding proteins revealed by an expressed sequence tag project on male *Manduca sexta* moth antennae. *Insect Biochem. Mol. Biol.* **8,** 501–518.

Rubin, G. M. (1988). *Drosophila melanogaster* as an experimental organism. *Science* **240,** 1453–1459.
Ryner, L. C., Goodwin, S. F., Casterillon, D. H., Anand, A., Villella, A., Baker, B. S., Hall, J. C., Taylor, B. J., and Wasserman, S. A. (1996). Control of male sexual behavior and sexual orientation in *Drosophila* by the *fruitless* gene. *Cell* **87,** 1079–1089.
Sarov-Blat, L., So, W. V., Liu, L., and Rosbash, M. (2000). The *Drosophila takeout* gene is a novel molecular link between circadian rhythms and feeding behavior. *Cell* **101,** 647–656.
Schilcher von, F. (1976a). The role of auditory stimuli in the courtship of. *Drosophila melanogaster*. *Anim. Behav*. **24,** 18–26.
Schilcher von, F. (1976b). The function of pulse song and sine song in the courtship of. *Drosophila melanogaster*. *Anim. Behav*. **24,** 622–625.
Sharma, R. P. (1977). Light-dependent homosexual activity in males of a mutant of. *Drosophila melanogaster*. *Experientia* **33,** 171–173.
Shirangi, T. R., Taylor, B. J., and Mckeown, M. (2006). A double-switch system regulates male courtship behavior in male and female *Drosophila melanogaster*. *Nat. Genet*. **38,** 1435–1439.
So, W. V., Sarov-Blat, L., Kotarski, C. K., McDonald, M. J., Allada, R., and Rosbash, M. (2000). takeout, a novel *Drosophila* gene under circadian clock transcription regulation. *Mol. Cell. Biol*. **20,** 6935–6944.
Song, H. J., and Taylor, B. J. (2003). *fruitless* gene is required to maintain neuronal identity in evenskipped-expressing neurons in the embryonic CNS of *Drosophila*. *J. Neurobiol*. **55,** 115–133.
Song, H. J., Bitller, T., Reynaud, E., Carlo, T., Spana, E. P., Perrimon, N., Goodwin, S. F., Baker, B. S., and Taylor, B. J. (2002). The fruitless gene is required for the proper formation of axonal tracts in the embryonic central nervous system of. *Drosophila*. *Genetics* **162,** 1703–1724.
Sosnowski, B. A., Belote, J. M., and McKeown, M. (1989). Sex-specific alternative splicing of RNA from the transformer gene results from sequence-dependent splice site blockage. *Cell* **58,** 449–459.
Stockinger, P., Kvitsiani, D., Rotkopf, S., Tirian, L., and Dickson, B. J. (2005). Neural circuitry that governs *Drosophila* male courtship behavior. *Cell* **121,** 795–807.
Sturtevant, A. H. (1915). Experiments on sex recognition and the problem of sexual selection in. *Drosophila*. *J. Anim. Behav*. **5,** 351–366.
Suh, G. S. B., Wong, A. M., Hergarden, A. C., Wang, J. W., Simon, A. F., Benzer, S., Axel, R., and Anderson, D. J. (2004). A single population of olfactory sensory neurons mediates an innate avoidance behaviour in *Drosophila*. *Nature* **431,** 854–859.
Taylor, B. J. (1992). Differentiation of a male-specific muscle in *Drosophila melanogaster* does not require the sex-determining gene *doublesex* or *intersex*. *Genetics* **132,** 179–191.
Taylor, B. J., and Knittel, L. M. (1995). Sex-specific differentiation of a male-specific abdominal muscle, the muscle of Lawrence, is abnormal in hydroxyurea-treated and in *fruitless* male flies. *Development* **121,** 3079–3088.
Taylor, B. J., Villella, A., Ryner, L. C., Baker, B. S., and Hall, J. C. (1994). Behavioral and neurobiological implications of sex-determining factors in *Drosophila*. *Dev. Genet*. **15,** 275–296.
Tinbergen, N. (1951). "The Study of Instinct." Oxford University Press, London.
Usui-Aoki, K., Ito, H., Ui-Tei, K., Takahashi, K., Lukacsovich, T., Awano, W., Nakata, H., Piao, E. E., Nilsson, E. E., Tomida, J., and Yamamoto, D. (2000). Formation of the male-specific muscle in female *Drosophila* by ectopic *fruitless* expression. *Nat. Cell Biol*. **2,** 500–506.
Usui-Aoki, K., Mikawa, Y., and Yamamoto, D. (2005). Species-specific patterns of sexual dimorphism in the expression of Fruitless protein, a neural masculinizing factor in. *Drosophila*. *J. Neurogenet*. **19,** 109–121.
Villella, A., Gailey, D. A., Berwald, B., Ohshima, S., Barnes, P. T., and Hall, J. C. (1997). Extended reproductive roles of the *fruitless* gene in *Drosophila melanogaster* revealed by behavioral analysis of *fru* mutants. *Genetics* **147,** 1107–1130.

Villella, A., Ferri, S. L., Krystal, J. D., and Hall, J. C. (2005). Functional analysis of *fruitless* gene expression by transgenic manipulations of *Drosophila* courtship. *Proc. Natl. Acad. Sci. USA* **102**(46), 16550–16557.

Villella, A., Peyre, J.-B., Aigaki, T., and Hall, J. C. (2006). Defective transfer of seminal-fluid materials during matings of semi-fertile *fruitless* mutants in *Drosophila*. *J. Comp. Physiol. A.* **192,** 1253–1269.

Vrontou, E., Nilsen, S. P., Demir, E., Kravitz, E. A., and Dickson, B. J. (2006). *fruitless* regulates aggression and dominance in *Drosophila*. *Nat. Neurosci.* **9,** 1469–1471.

Wheeler, D. A., Kulkarni, S. J., Gailey, D. A., and Hall, J. C. (1989). Spectral analysis of courtship songs in behavioral mutants of *Drosophila melanogaster*. *Behav. Genet.* **19,** 503–528.

Wiersma, C. A. G., and Ikeda, K. (1964). Interneurons commanding swimmeret movements in the crayfish, Procambarus clarki (Girard). *Comp. Biochem. Physiol.* **12,** 509–525.

Wolff, T., and Ready, D. (1993). Pattern formation in the *Drosophila* retina. *In* "The Development of Drosophila melanogaster" (M. Bate and A. Martinez Arias, eds.), pp. 1277–1325. Cold Spring Harbor Laboratory Press, New York.

Wolpert, L. (2002). "Principles of Development," 2nd edn. Oxford University Press, Oxford.

Yamamoto, D. (1996). "Molecular Dynamics of Developing *Drosophila* Eye." Springer Verlag, New York.

Yamamoto, D., and Nakano, Y. (1999). Sexual behavior mutants revisited: Molecular and cellular basis of *Drosophila* mating. *Cell. Mol. Life Sci.* **56,** 634–646.

Yamamoto, D., Sano, Y., Ueda, R., Togashi, S., Ttsurumura, S., and Sato, K. (1991). Newly isolated mutants of *Drosophila melanogaster* defective in mating behavior. *J. Neurogenet.* **7,** 152.

Yamamoto, D., Ito, H., and Fujitani, K. (1996). Genetic dissection of sexual orientation: Behavioral, cellular and molecular approaches in. *Drosophila melanogaster*. *Neurosci. Res.* **26,** 95–107.

Yamamoto, D., Jallon, J.-M., and Komatsu, A. (1997). Genetic dissection of sexual behavior in *Drosophila melanogaster*. *Annu. Rev. Entomol.* **42,** 551–585.

Yamamoto, D., Fujitani, K., Usui, K., Ito, H., and Yamamoto, D. (1998). From behavior to development: Genes for sexual behavior define the neuronal sexual switch in. *Drosophila*. *Mech. Dev.* **73,** 135–146.

Yamamoto, D., Usui-Aoki, K., and Shima, S. (2004). Male-specific expression of the Fruitless protein is not common to all *Drosophila* species. *Genetica* **120,** 267–272.

Zhang, S. D., and Odenwald, W. F. (1995). Misexpression of the *white (w)* gene triggers male-male courtship in *Drosophila*. *Proc. Natl. Acad. Sci. USA* **92,** 5525–5529.

3

Sexual Differentiation of the Vocal Control System of Birds

Manfred Gahr
Max Planck Institute for Ornithology, Seewiesen, Germany

ABSTRACT

Birds evolved neural circuits of various complexities in relation to their capacity to produce learned or unlearned vocalizations. These vocalizations, in particular those that function in the realm of reproduction, are frequently sexually

Advances in Genetics, Vol. 59
Copyright 2007, Elsevier Inc. All rights reserved.
0065-2660/07 $35.00
DOI: 10.1016/S0065-2660(07)59003-6

dimorphic, both in vocal learners (songbirds, parrots, some hummingbirds) and vocal nonlearners (all other birds). In many cases, the development and/or the adult differentiation of vocalizations of sociosexual function is sensitive to sex hormones, androgens and estrogens. The underlying mechanisms have been studied in detail in songbirds, a bird group that comprises about half of all bird species. Next to unlearned calls, songbirds produce learned songs that require forebrain vocal control areas that express receptors for androgens and estrogens. These forebrain vocal areas are sexually dimorphic in many species, but a clear relation between the degree of "brain sex" and sex differences in vocal pattern is lacking, except that a minimum number of vocal neurons is necessary to sing learned songs. Genetic brain-intrinsic mechanisms are likely to determine the neuron pools that develop into forebrain song control areas. Subsequently, gonadal steroid hormones, androgens and estrogens, modulate the fate of these neurons and thus the functionality of the vocal control systems. Further action of gonadal hormones, and may be other factors signaling the sociosexual and physical environment, affect the phenotype of vocal control areas in adulthood. Despite the clear evidence of hormone dependency of both adult vocalizations and phenotypes of vocal neuron pools, their causal relation is little understood.
© 2007, Elsevier Inc.

I. INTRODUCTION

In all bird species, male and female communicate with sounds that function in the realms of reproduction, territorial defense, maintenance of social pairs and groups, species, sex and kin recognition, begging, and alarming. The more complex and longer vocalizations that are frequently composed of at least two different motor units (syllables) are called songs, although there is no unifying definition to distinguish between songs and the nonsongs (named calls) of all bird species. The function of song of male birds is closely linked to reproductive success and the result of sexual selection (Searcy and Yasukawa, 1996). The main functions of song of males are mate attraction and defense of breeding territory against other males (Catchpole and Slater, 1995). Individual variation in song characteristics does affect reproductive success through mate choice and male–male competition (Andersson, 1994). There is evidence of sexual selection for song traits such as repertoire size, song rate, structure of song motor units (syllables), and speed of syllable repetitions (for review see Podos *et al.*, 2004). Current theory predicts that when senders and receivers have different evolutionary interests, as in sexual selection, signals must be subject to some costly constraint to constitute stable, honest indicators of quality (Grafen, 1990). Individual variation in vocalizations will, therefore, depend on the condition of the adult male and its developmental history (Rowe and Houle, 1996), but the costs of vocalizations are not well understood (Gil and Gahr, 2002). The function of female songs is not studied in detail but appears to be (feeding) territorial or social or, in few species such as the dunnocks (*Prunella modularis* and *Prunella collaris*), even sexual (Langmore and Davies, 1997; Langmore *et al.*, 1996; Ritchison, 1983). Last, there are a few bird species like the buttonquail (*Turnix suscitator*) (Madge and McGowan, 2002) and the black coucal (*Centropus grillii*) (Goymann *et al.*, 2004), with reversal of the common sexual roles of birds. Females defend breeding territories and court for males and vocal efforts in such species appear to be sex reversed.

In three avian taxa—that are only distantly related—the songbirds (suborder Oscines of order Passeriformes), the parrots (order Psittaciformes), and the hummingbirds (family Trochilidae of the order Apodiformes) songs and in some cases contact calls of social and sexual partners are learned. The amount of vocal learning from external models might considerably vary between species and the features to be learned are likely species specific (Güttinger, 1979, 1981; Leitner *et al.*, 2002; Marler, 1991, 1997). Among songbirds, males of all species produce learned vocalizations while females of only a limited number of songbird species do so. Among hummingbirds, there are species of which neither male nor female sing and thus are unlikely to produce learned sounds. Thus, all bird species produce innate vocalization of varying complexity, while vocal learning has evolved at least three times among birds. Further, the capacity to develop learning mechanisms had been lost several times among hummingbird species and in a sex-specific manner among many songbird species.

Vocal learning requires first the formation of an auditory memory called template that guides auditory-motor learning (Konishi, 1965). Experimental and correlative works strongly suggest that the integrity of the forebrain vocal control systems (see below) is necessary for auditory-motor learning and the production of learned vocalizations while the production of innate sounds can be "done" with mid- and hindbrain areas (Benton *et al.*, 1998; Bottjer *et al.*, 1984; Brainard and Doupe, 2000; Halle *et al.*, 2003; Leonardo and Konishi, 1999; Nottebohm *et al.*, 1976; Scharff and Nottebohm, 1991; Scharff *et al.*, 2000; Simpson and Vicario, 1990). The site of template formation is unclear but might involve auditory areas different from the vocal control system (for review see Bolhuis and Gahr, 2006). Thus, from a proximate perspective, we need to distinguish between bird species in which learned vocalization are under sexual selection and species in which sexual selection works on vocalization that are not learned. In the first case, a successful "sexy" singer [in most cases a male, although females might learn their songs in some species, e.g., the Northern Cardinal (*Cardinalis cardinalis*; Yamaguchi, 2001)] needs at least to develop a forebrain vocal control system, needs to modify the neural vocal system through learning, and needs to adapt the system to its actual physiological condition that in turn depends on its genetic background and the environment. Since the behavior that makes a male successful takes place in adulthood, while song development and learning are either restricted to or at least start during ontogeny (Marler, 1991, 1997), we have to consider sexual differentiation of the vocal system as a result of life history. The same is true for species in which sexual quality is coded in non-learned sounds, with the modification that these species do not develop a forebrain vocal system and require less complex syringeal muscles than vocal learners.

The action of gonadal steroid hormones, androgens and estrogens, specifies the sexual differentiation of brain and behavior on brain-intrinsic genetic mechanisms (Agate *et al.*, 2003; Carruth *et al.*, 2002; Gahr, 2003; Gahr and Metzdorf, 1999). These genetic mechanisms are thought to control the area-specific expression of androgen and estrogen receptors in the developing brain (Gahr and Balaban, 1996; Gahr and Metzdorf, 1999). Then, the gonadal hormones first specify the global development of sexually determined brain areas such as song system, frequently called organizational action, and subsequently modify more detailed phenotypes, frequently called activational action (Arnold and Gorski, 1984). One major cellular mechanism of steroid action is the change of protein synthesis via activating their cognate receptors. Genetic and environment-driven mechanisms that control the expression of androgen and estrogen receptors and those that control the cerebral availability of active androgens and estrogens appear crucial for sexual differentiation of brain and behavior throughout life.

In relation to the life history process of adult sexual behavior, I discuss behavioral evidence of vocal sexual behavior of birds (Section II) and its endocrine correlates (Section III). Then I summarize the hormone-sensitivity of vocal control networks (Section IV) and the potential link between neural and vocal sexual phenotype (Section V). In Section VIA, the genetic brain-intrinsic mechanisms of sexual differentiation of vocal control systems are discussed. Subsequently, I investigate the action of gonadal steroid hormones for sexual ontogeny of vocal brain areas (Section VIB) and the role of gonadal steroid hormones in sculpting the sex-typical song system in adulthood (Section VII).

II. SEX-SPECIFIC AND SEX-TYPICAL VOCALIZATIONS

As indicated above, males and females of all bird species communicate acoustically. In relation, sex-specific or sex-typical vocalizations are known for all avian orders and are likely to occur in all birds. Sex-specific vocalizations are those that are produced by one sex of a species, which suggest sex-specific obligate or facultative constraints of the sound-producing system. Facultative sex-specific vocalizations mean that certain sounds are normally only produced by one sex (mainly the male) but can be induced experimentally in the opposite sex (mainly the female) or occur under rare circumstances in the opposite sex.

The most pertinent examples of obligate sex-specific vocalizations are the songs of songbird species such as the zebra finch (*Taeniopygia guttata*) or the orange bishop (*Euplectus franciscanus*) in which only the male is singing (Arai *et al.*, 1989; Nottebohm and Arnold, 1976). Further examples are the songs of the reproductive period of many songbirds of the temperate zones such as the chaffinch (*Fringilla coelebs*) or the European robin (*Erithacus rubecula*) that are only produced by the males (Hoelzel, 1986; Kling and Stevenson-Hinde, 1977). Females of these species may sing simpler versions of the males' song outside of the breeding season, that is, in a nonreproductive context (Hoelzel, 1986). In cases that testosterone treatment of such females induces male-like singing or calling, vocal sex difference appears to be facultative and thus depends on the physiological conditions and the context, in which males and females are studied. For example, captive female European starlings (*Sturnis vulgaris*) stop singing when nest-boxes are present in the aviary (Henry and Hausberger, 2001). In some songbird species such as the dunnocks (Langmore and Davies, 1997; Langmore *et al.*, 1996), females regularly produce songs in the breeding season that share some syntactical features of the male song or some of its syllable (vocal units) repertoire. In duetting species too, females sing during the breeding season (Farabaugh, 1982). However, there are only few thorough studies that analyzed if natural or testosterone-induced vocalizations of females are in all characteristics male typical, that is, it remains to be seen if obligate sex-specific vocal pattern are the exception or widespread among songbirds.

Sex-specific vocalizations are, however, not restricted to songbirds and the other groups of learning birds, that is, concern also innate vocalizations. In the Least flycatcher (*Empidonax minimus*), a suboscine passerine, song structure appears to differ between males and females (Kasumovic *et al.*, 2003). In non-passerines as divers as the Western Screech-Owl (*Otus kennicottii*) (Herting and Belthoff, 2001), the Leach's Storm Petrel (*Oceanodroma leucorhoa*) (Taoka *et al.*, 1989), the White-tailed Hawk (*Buteo albicaudatus*) (Farquhar, 1993), the whooping crane (*Grus americana*) (Carlson and Trost, 1992), the Collared Dove (*Streptopelia decaocto*) (Ballintijn and Ten Cate, 1997), the King Penguin (*Apendodytes patagonicus*) (Robisson, 1992), and domestic fowl (*Gallus gallus*) (Andrew, 1963), sex differences in call structure have been documented. In some of these examples, sex differences not only concern mating calls but alarm calls and flight calls.

Sex-typical vocalizing means that sounds of similar structure are produced regularly by either sex but are uttered more frequently in one sex. However, since vocal control networks are likely to code for the structure of vocalizations but not for the quantity of singing, I do not elaborate on this issue. For example, bilateral lesions of the HVC of songbirds lead to the deterioration of the entire song structure while the birds still initiate singing postures, that is, are motivated to sing (Nottebohm *et al.*, 1976). Areas relevant for the motivation to sing are expected in hypothalamic-limbic regions (Riters and Ball, 1999) that I consider not part of vocal control circuits, independent of whether sounds are learned or innate.

Last, there are species in which males and females produce similar vocalizations or have similar-sized vocal repertoires in a reproductive context; examples are many duetting species such as the forest weaver (*Ploceus bicolor*) (Seibt *et al.*, 2002; Wickler and Seibt, 1980) or the East African shrike (*Lanarius funebris*) (Gahr *et al.*, 1998). Interestingly in the latter species, females and males sing different syllable types despite a similar syllable number but are able to learn the syllables of the opposite sex (Wickler and Seibt, 1988; Wickler and Sonnenschein, 1989), that is, sex-specific vocalizations are not due to endocrine or (neuro)physiological constraints but learned. In another duetting species, the colonial white-browed sparrow weaver (*Plocepasser mahali*) duet singing, too, is not sex typical or sex specific. Both male and female sparrow weavers share the entire repertoire of duet syllables and readily engage in duet- and chorus singing (Voigt *et al.*, 2006). However, the dominant male of such a weaver colony sings a second type of song, called solo song that cannot be produced by the females or the subordinate males of the colony (Voigt *et al.*, 2006). Thus, sex-specific vocalization depends on sex and social status in this species.

III. THE ENDOCRINOLOGY OF BIRDS' VOCALIZATION

Obligate sex-specific vocalizations are either due to brain-intrinsic genetic mechanisms or result from irreversible ontogenetic action of testosterone and its androgenic and estrogenic metabolites. Facultative sex-specific vocalizations might be due to the action of elevated levels of gonadal steroids that could occur (repeatedly) at various time-windows throughout life. Although there is very good experimental evidence for sex steroids affecting sex-specific vocalizations as in the case of songbirds such as the zebra finch (Gurney, 1982; Gurney and Konishi, 1980; Simpson and Vicario, 1991) and the canary (*Serinus canaria*) (Weichel *et al.*, 1989), there are very little developmental data that document a sex difference in ontogenetic hormone production as a possible cause for sex-specific vocal development. Even for well-studied zebra finch, it is unclear if the developmental production of testosterone or estrogens could cause the sex-specific development of their vocal system (Adkins-Regan *et al.*, 1994; Hutchison *et al.*, 1984; Schlinger and Arnold, 1992). This lack of evidence might be due to technical difficulties measuring low hormone plasma levels in very small blood samples, which are a necessity in order to collect blood samples repeatedly from the same individual during ontogeny.

The endocrine correlates of vocalizations concern mainly the testosterone dependency of adult male vocalizations. These vocalizations change in structure after castration and, in many cases, can be induced in adult females by testosterone treatment. Supporting data of testosterone-dependent vocal pattern comes from species of a wide variety of avian taxa including vocal learners [songbirds (Heid *et al.*, 1985; Kern and King, 1972; Leonard, 1939; Shoemaker, 1939), parrots (Brockway, 1968, 1969; Nespor *et al.*, 1996)], and nonlearners, the suboscine passerines (Kroodsma, 1985), chicken (Andrew, 1963; Marler *et al.*, 1962), Japanese quails (Beani *et al.*, 2000), partridges (Fusani *et al.*, 1994), night herons (Noble and Wurm, 1940), doves (Bennet, 1940), and gulls (Groothuis and Meeuwissen, 1992; Terkel *et al.*, 1976). As mentioned already above, little attention has been paid to verify if testosterone-induced vocalizations of females are indeed "male-typical," that is, testosterone-treated female canaries sing male-like songs (Leonard, 1939; Shoemaker, 1939), which in average are composed of much fewer syllables compared to male canaries (Fusani *et al.*, 2003; Hartley and Suthers, 1990). In these cases, it needs to be seen if females have neurophysiological limits that do not allow producing an entirely masculine song or if these deficits are due to procedural shortcomings concerning the hormone therapies or the behavioral settings.

In relation with these testosterone-dependent vocalizations, seasonal changes in song structure correlate with periods of dramatically increased testis size and testosterone production (Leitner *et al.*, 2001). Although the overall relationship between testosterone and singing is well established, there are only a few studies of how variation in steroid levels relate to individual variation in song. Experimental manipulations show that testosterone levels in the dark-eyed junco (*Junco hyemalis*) (Ketterson *et al.*, 1992) influence song rates. However, two correlative studies provide negative evidence for such a relationship in other species: testosterone levels are not related to song output in the barn swallow (*Hirunda rustica*) (Saino and Moller, 1995) or to repertoire size in the red-winged blackbird (*Agelaius phoenicus*) (Weatherhead *et al.*, 1993). This lack of correlation is to be expected if there are individual differences in the responsiveness of the brain vocal systems to testosterone, which is mediated by, among other processes, converting testosterone into estrogens by brain aromatase (Fusani and Gahr, 2006; Fusani *et al.*, 2003; Schlinger and Arnold, 1991). Further, if song characteristics are dependent on testosterone levels during song development, identifying these patterns in adult birds might not be relevant.

That estrogens are active metabolites of testosterone controlling some aspects of sexual vocalizations has been demonstrated in songbirds (zebra finch, starling, canary) and nonsongbirds (ring dove, *Streptopelia risoria*). In the songbirds, estrogen treatment during development mimics testosterone treatment and masculinizes in part song development (Casto and Ball, 1996; Gurney, 1982; Gurney and Konishi, 1980; Lohmann and Gahr, 2000; Simpson and Vicario, 1991; Weichel *et al.*, 1989). In adult canaries, estrogens derived from testosterone are required to sing songs with high syllable repetition rates (Fusani and Gahr, 2006; Fusani *et al.*, 2003), a feature that is important for sexual quality of canaries' songs (Kreutzer and Vallet, 1991). In adult male ring doves, estrogens derived from testosterone are important to stimulate sexual vocal displays, but this might be a motivational rather than a motor action of estrogens (Hutchison and Steimer, 1984).

In many tropical and some temperate zone birds, singing and vocalization can be heard year-round, which suggest that such vocalizations do not depend on acute availability of gonadal hormones. In respect, we need to differentiate between vocalizations that are the same in structure year-round and those that change seasonally in structure. The first category is obviously independent of the acute action of gonadal hormones but might require organizational action of sex hormones during some developmental stage. There are, however, no good arguments to this, except the assumption that sex hormones are always necessary to develop the vocal control system (of songbirds) to a functional state (see Section VII). Examples of the second category are males of species that produce so-called "autumnal" songs outside the breeding season when circulating levels of gonadal hormones are undetectable. For example, individual wild canaries sing songs of different temporal structure and use different syllables compared to the breeding season when testosterone levels are low (Leitner *et al.*, 2001). Further, autumnal songs can be induced in captivity by castration (Heid *et al.*, 1985). Clearly, some forms of singing and hence neural mechanisms of vocal production are possible without activational action of gonadal hormones.

Endocrine correlates of female-typical calling and singing are widely unknown but might be hormone-independent in many species. However, an alternative suggestion is that females release low levels of circulating testosterone, which is possible since testosterone is a precursor during ovarian estrogen production. In relation, females that sing periodically or circumstantially are thought to secrete higher levels of testosterone in these periods. Such examples might be the female song sparrow (*Melospizia melodia*) (Arcese *et al.*, 1988) and female dunnocks, in which competition for male reproductive investment elevates testosterone (Langmore *et al.*, 2002). However, in most species with singing females, these songs are either produced outside the breeding season (e.g., the European robin) when the ovaries are degenerated or year-round (e.g., many duetting species). In both cases, testosterone dependency of female vocalization is unlikely. Similarly, gonadal estrogens are unlikely to control female vocal behavior. Although estrogens have been shown to mediate some of the activational action of testosterone on vocalizations (Fusani and Gahr, 2006; Fusani *et al.*, 2003), these actions are cerebral and estrogens are derived in the brain. Hence, systemic estrogen treatments of females (or males) either induce female courtship display or are not superior to testosterone treatment for vocal performance (Brockway, 1968, 1969; Chiba and Hosokawa, 2006; Kern and King, 1972; Terkel *et al.*, 1976).

IV. STEROID SENSITIVITY OF THE VOCAL CONTROL SYSTEM OF AFFERENT AUDITORY REGIONS AND MODULATORY SYSTEMS

A. The anatomy of avian vocal control networks

In songbirds, neural vocal control is achieved by a chain of interconnected brain areas in the fore-, mid-, and hindbrain (Nottebohm *et al.*, 1976; Vates *et al.*, 1997; Vu *et al.*, 1994; Wild, 1997; Yu *et al.*, 1996). Vocal learning of songbirds correlates with the differentiation of forebrain vocal control areas, the robust nucleus of the arcopallium (RA), lateral magnocellular nucleus of the anterior nidopallium (lMAN), HVC (used as proper name), medial magnocellular nucleus of the anterior nidopallium (mMAN), Area X, and nucleus interface of the nidopallium (NIF) (Fig. 3.1). Albeit under intense study, the specific role of any forebrain nucleus for learning and production of vocal pattern is unclear. Hummingbirds and parrots, too, evolved forebrain networks, which are composed of multiple areas (for review: Gahr, 2003). In parakeets, some of these areas do not seem homologous to the forebrain vocal control nuclei of oscines, while the hodologic and molecular properties of others suggest similarities in the neural organization of forebrain vocal control pathways of budgerigars and songbirds (Durand *et al.*, 1997; Jarvis and Mello, 2000; Striedter, 1994). Similarly, the singing hummingbird species such as Anna's hummingbird (*Calypte anna*) and Amazilia hummingbird (*Amazilia amazilia*) differentiate an lMAN-like, an HVC-like, and an RA-like area (Gahr, 2000). Areas similar to these reported by Gahr (2000) were found in the Sombre hummingbirds (*Aphantochroa cirrhochloris*) and Rufous-breasted hermit (*Glaucis hirsuta*) to be active during vocalizing as indicated by the expression of immediate early genes, a molecular activity marker (Jarvis *et al.*, 2000). The HVC-, RA-, and lMAN-like areas are rudimentary in adult male Ruby-throated hummingbirds (*Archilochus colubris*) and Allen's hummingbirds (*Selasphorus sasin*) that do not sing (Gahr, 2000). Areas comparable to vocal areas of songbirds, parrots, and hummingbirds are not found in vocal nonlearning swifts and suboscines, the taxonomic sister groups of hummingbirds and songbirds, respectively (Gahr, 2000). These areas are further missing in owls, doves, gulls, and gallinaceous species, avian taxa that do not learn vocalizations (Gahr, 2000). Common to all three vocal-learning avian taxa is a projection of archistriatal neurons on brain stem nuclei, in particular to the medullary syringeal motonucleus (nucleus hypoglossus pars tracheosyringealis, nXIIts) and the respiratory pre-motonucleus RAm (nucleus retroambigualis) (Gahr, 2000; Striedter, 1994; Wild, 1997)(Fig.3.1).

The nucleus hypoglossus appears to be the common motonucleus of the syrinx in all bird taxa. Similarly, the vocal-respiratory areas of the brain stem and parts of the midbrain nucleus intercollicularis innervating the brain stem areas are common to all birds (Wild, 1997; Wild *et al.*, 1997).

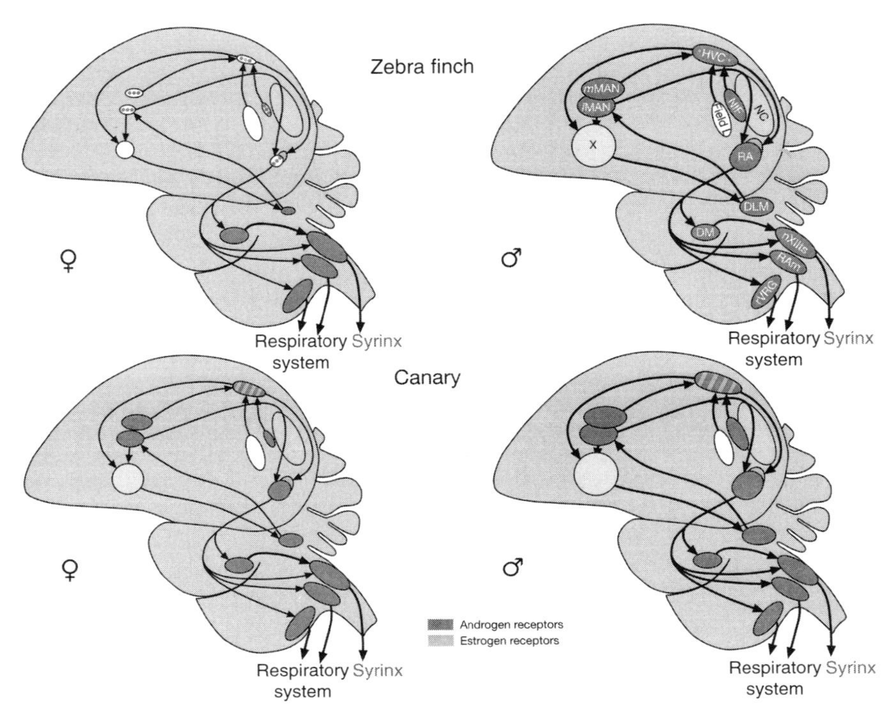

Figure 3.1. The vocal control system of songbirds and degrees of sexual dimorphisms. This schematic diagram of a composite view of parasagittal sections of a songbird brain gives approximate positions of vocal nuclei and brain regions and their content of androgen receptors (red, rose) and estrogen receptors (green). All structures are bilateral—for reasons of clarity, only those in one-half of the brain are depicted. Further, some thalamic brain areas that appear important for coordination of the left and right vocal control network are omitted. The volumes of vocal control areas of adult zebra finches are highly sexually dimorphic, while those of canaries are to a lesser extent, indicated in the relative size of the areas. Despite these sex differences, all areas and connections as well as the sex hormone receptors are present in the female vocal control system and syrinx. The HVC of canaries contains higher amounts of estrogen receptors compared to the zebra finch. Area X and NC (rose) contains androgen receptors in only some animals. Androgen receptors are found in the caudal nidopallium including the caudomedial nidopallium. Abbreviations: DLM, nucleus dorsolateralis anterior, pars medialis; DM, dorsomedial nucleus of the midbrain nucleus intercollicularis; HVC, acronym used as a proper name; formerly known as high vocal center; Field L; lMAN, lateral magnocellular nucleus of the anterior nidopallium; mMAN, medial magnocellular nucleus of the anterior nidopallium; NC, caudal nidopallium; NIF, nucleus interface of the nidopallium; nXIIts, tracheosyringeal portion of the nucleus hypoglossus; RA, robust nucleus of the arcopallium; RAm, nucleus retroambigualis; rVRG, rostroventral respiratory group; Area X. (See Color Insert.)

Auditory input to the forebrain vocal areas is likely provided by the forebrain field L, which represents the auditory cortex of birds and adjacent areas of the caudal nidopallium including the caudomedial nidopallium (NCM) representing secondary auditory cortices (for review see Bolhuis and Gahr, 2006). There is, however, no strong direct synaptic input of the primary or secondary auditory cortex to the vocal areas (Fortune and Margoliash, 1995).

Next to this indirect connection with the auditory system of songbirds, forebrain vocal areas obtain input of catecholaminergic neurons located mainly in the midbrain central gray (HVC) and area ventralis of Tsai (HVC, RA, Area X) (Appeltants *et al.*, 2001, 2002) and cholinergic input of the ventral paleostriatum (Sakaguchi *et al.*, 2000) or via uvaeform nucleus of the thalamus (Akutagawa and Konishi, 2005).

In songbirds, all areas of the vocal control circuit are found in males and females, although there are considerable sex differences in various species (Fig. 3.1; see below, Table 3.1). In particular, forebrain nuclei are degenerated or cannot be delineated anatomically in adult females of those songbird species in which females do not utter learned sounds such as the zebra finch and the Orange bishop (Arai *et al.*, 1989; Nottebohm and Arnold, 1976). Nevertheless, although some areas in the zebra finch such as Area X are not visible in histological stainings, neurons in the potential region of Area X are connected with HVC, as in the case of males (Fig. 3.1 and Fig. 3.2). This suggests that even in case of species with nonsinging sexes, the general vocal control network of songbirds develops. Similarly, in nonsinging female hummingbirds, forebrain vocal areas are rudimentary (Gahr, 2000). In the budgerigar, the only parrot studied in detail, vocal areas are present in females with some sexually dimorphic features (Brauth *et al.*, 2005). In avian taxa that do not learn their vocalizations, sex differences in vocal areas are not studied in much detail. It appears that sex differences in vocalizations of these species are mainly due to steroid-dependent sexually dimorphic differentiation of the syrinx (Ballintijn and Ten Cate, 1997; Beani *et al.*, 1995; Takahashi and Noumura, 1987) but might involve as well sex differences in subareas of the midbrain nucleus intercollicularis (for review see Beani *et al.*, 1995).

B. Steroid sensitivity of avian vocal control networks

Comparative studies of the distribution of estrogen and androgen receptors expressing cells in vertebrate brains have shown that the brain regions, which typically contain such cells, are evolutionarily conserved (e.g., hypothalamic-preoptic areas) or are linked to taxa-specific sexual behaviors (Kim *et al.*, 1978; Pfaff, 1980). Vocal control areas are an example of the latter (Bernard *et al.*, 1999; Gahr, 2000; Gahr *et al.*, 1993; Metzdorf *et al.*, 1999). The classical androgen and estrogen receptors are ligand-dependent transcription factors. Although gonadal steroids act on neurons *in vitro* via a wide range of nongenomic

Table 3.1. Sex Differences in the Size of Vocal Control Areas Do not Correlate with Sex Differences in Song Repertoires

Species	RA	HVC	Repertoire		References
			Male	Female	
I. Only the male sings					
Zebra finch	11.9–5.5	13.6–5.0	1.2	0	Gahr, M., unpublished data; Gurney, 1981, 1982; Immelmann, 1969; Nottebohm and Arnold, 1976
Bengalese finch	8.3	>30	11	0	Soma *et al.*, 2006; Tobari *et al.*, 2005
Marsh wren	5	10	53.6	0	Brenowitz *et al.*, 1994; Canady *et al.*, 1984
Orange bishopbird	29	>29	1	0	Arai *et al.*, 1989
Dark-eyed juncos	3.6	4.5	2.7–5	0	Deviche and Gulledge, 2000; Williams and MacRoberts, 1978
Carolina Wren	>30	>30	32	0	Morton, 1987; Nealen and Perkel, 2000
II. Female song repertoire size differs from the male					
Canary	3.0–2.7	4.3–2.7	30–20	35–5	Nottebohm, 1980; Pesch and Güttinger, 1985; Gahr, M., unpublished data
Red-winged blackbird	4.7	3.2	5.9	2	Beletsky, 1983; Kirn *et al.*, 1989
White-browed robin chat	2.4	2.9	40	3.5	Brenowitz *et al.*, 1985; Todt *et al.*, 1981
Rufous and white wren	2.2	1.7	10.8	8.5	Brenowitz and Arnold, 1986; Mennill and Vehrencamp, 2005

European Starling	1.7	1.6	46.4	24	Bernard *et al.*, 1993; Hausberger *et al.*, 1995
III. Similar repertoire size in male and female					
Red-checked cordon blue	1.4	1.5	1.3	1.2	Gahr and Güttinger, 1985; Gahr and Güttinger, 1986, and unpublished data
White-crowned sparrow	3.7	2.4	1	1	Baptista *et al.*, 1993; Baker *et al.*, 1984
White-throated sparrow	1.5	1.9	1	1	DeVoogd *et al.*, 1995; Falls and Kopachena, 1994
Buff-breasted wren	1.5	1.3	13	11.7	Brenowitz *et al.*, 1985; Farabaugh, 1982
Bay wren	1.1	1.5	16.5	15.2	Brenowitz and Arnold, 1986; Farabaugh, 1982
Bush shrike	2.0	1.8	4.6	4.5	Gahr *et al.*, 1998; Wickler and Seibt, 1980
IV. Hormone-induced male-like singing of females					
Masculinized canary	2.1–1.9	2.3–2.0	30–20	35–5	Nottebohm, 1980; Pesch and Güttinger, 1985; Gahr, M., unpublished data
Masculinized zebra finch	1.9–1.6	1.8–1.6	1.2	1.2	Gahr, M., unpublished data; Gurney, 1981, 1982

Listed are the male-to-female ratios of the volume of the HVC and the RA. Repertoire size (means) comprises number of song types or syllable types. For the canary and the zebra finch, the range of values is given because these species have been studied in several investigations with quite different results. Only species of which the female singing is analyzed in some detail are included. Statistical comparisons (Gahr *et al.*, 1998) show that sex differences in species in which only the males sing are greater than in species in which females also sing, but sex differences in HVC and RA are independent of the vocal behavior in species with singing females.

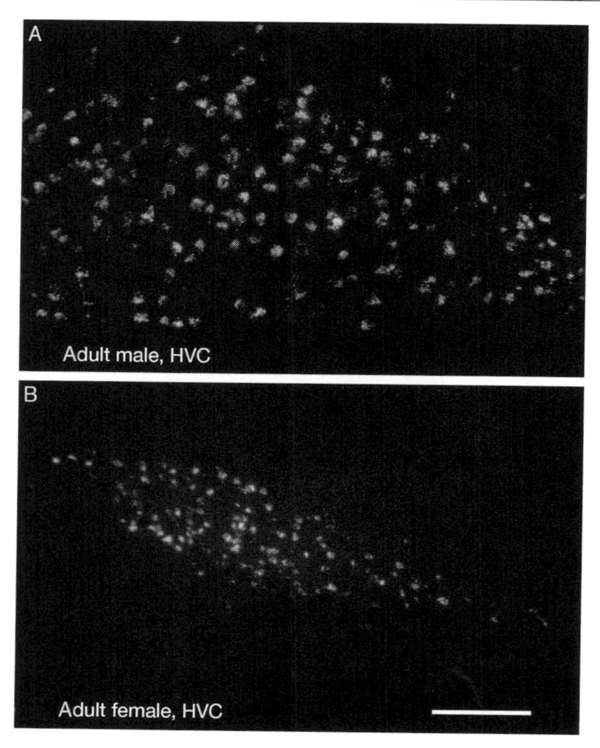

Figure 3.2. The HVC of adult male and female zebra finches delineated by neurons projecting to Area X. Area X was injected in adulthood with rhodamin-labeled latex microspheres that are retrogradely transported. Note that HVC is much smaller in female zebra finches but nevertheless has the same general connectivity as the male HVC (Gahr, unpublished data). Scale bar = 80 μm.

mechanisms not involving their cognate receptors, there are only few examples involving such mechanisms in sexual differentiation (Beyer *et al.*, 2002). Since the latter has not been studied in any detail in the bird vocal model, I consider androgen and estrogen action on vocal control areas and their afferences based on the presence of classical androgen and estrogen receptors α and β. Next to these receptors, the action of sex hormones on the brain depends on its neural availability. The androgen testosterone can be converted by the enzyme 5α-reductase into the androgen 5-dihydrotestosterone (both androgens bind to the androgen receptor) and into the estrogen 17β-estradiol (a potent ligand of both estrogen receptor α and β) via the enzyme aromatase. Since these enzymes are expressed in certain brain regions and cell types, active androgens and estrogens can be formed both in the gonads and directly in the brain (Metzdorf *et al.*, 1999;

Saldanha *et al.*, 2000a; Schlinger and Arnold, 1991). Further, there is increasing evidence that the brain of birds, including songbirds, might express the entire enzymatic cascade to form androgens and estrogens from cholesterol, similar to the gonads (London *et al.*, 2006).

After the first detection of androgen-binding sites in vocal areas (Arnold *et al.*, 1976), androgen receptors were found in all forebrain song control areas of the zebra finch and the canary, which we studied in detail (Fig. 3.1) (Gahr, 2000; Metzdorf *et al.*, 1999). In Area X of these species, androgen receptors occur in only some individuals for unknown reasons (Gahr, 2004; Kim *et al.*, 2004). Estrogen receptor α is only expressed in HVC and around the dorsal aspect of RA of canaries and zebra finches (Gahr *et al.*, 1987, 1993; Metzdorf *et al.*, 1999), while estrogen receptor β is not expressed in any of the vocal areas (Bernard *et al.*, 1999). The expression of the receptors in other songbird species has been studied in less detail. Nevertheless, the comparative works of several laboratories suggest that the expression of androgen receptors is a general feature of all forebrain vocal areas of songbirds while estrogen receptors occur mainly in HVC and, there, in a species-typical pattern (for review see Gahr, 2000). For example, among HVC neurons, about 0.5% contain estrogen receptor α in the zebra finch male while in the male canary about 15% contain these receptors (Gahr *et al.*, 1987, 1993). In the brain stem, androgen receptors occur in all respiratory-vocal areas. While the presence of such receptors in the inspiratory hindbrain premotor region (rVRG) is common in all bird taxa studied, androgen receptors in syringeal motor neurons and expiratory premotor neurons are a feature of songbirds (Gahr, 2000; Gahr and Wild, 1997). The only exception known so far is the hummingbirds. Similar to the songbirds, androgen receptors are found in forebrain vocal areas and in all vocal-respiratory hindbrain areas of two singing hummingbird species that differentiate forebrain vocal areas (see above). This suggests that in hummingbirds, like in songbirds, vocal learning is linked to the differentiation of androgen-sensitive forebrain vocal areas (Gahr, 2000). The expression of estrogen receptors in forebrain vocal areas is, however, an oscine-specific avian feature (Gahr, 2000). Aromatase and 5α-reductase are not abundant or entirely lacking in forebrain vocal areas except the NIF and mMAN that express the aromatase gene and might be the source of presynaptic aromatase in HVC (Fusani *et al.*, 2000; Metzdorf *et al.*, 1999; Peterson *et al.*, 2005).

Androgen and estrogen receptors are readily expressed in nonlimbic forebrain regions, including the caudal nidopallium of the zebra finch and canary (Gahr *et al.*, 1993; Metzdorf *et al.*, 1999). Further, in the caudal nidopallium, aromatase is highly expressed in songbirds and in one species of suboscines, the golden-colored manakin (*Manacus vitellinus*) but is missing in tyrannid suboscines (Metzdorf *et al.*, 1999; Saldanha *et al.*, 2000b; Schlinger and Arnold, 1991). These androgen receptor and aromatase expressions in the forebrain might overlap with auditory regions of the caudal nidopallium except field L, although

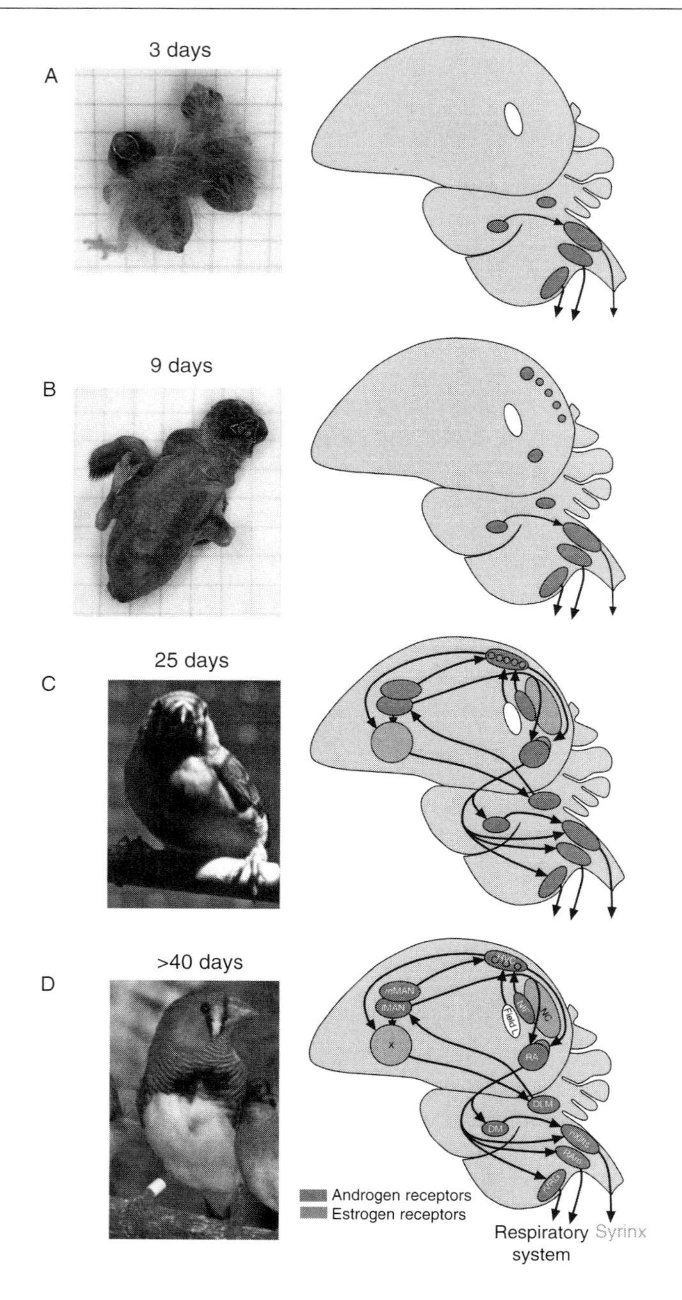
3 days
A
9 days
B
25 days
C
>40 days
D
mMAN
lMAN
X
HVC
NIf
Field L
NC
RA
DLM
DM
nXIIts
RAm
Androgen receptors
Estrogen receptors
Respiratory system
Syrinx

HVC size or the size of other vocal control regions, besides the observation that learned songs require a minimum size of neuron pools that compose the forebrain vocal areas.

Among possibilities that might account for the often-poor correspondence between behavioral and neural sexual dimorphism is that vocal areas might be involved in tasks other than singing such as sound-based mate selection. This is suggested for the HVC of reproductively active female canaries and the RA of male zebra finches (Brenowitz *et al.*, 1991; Del Negro *et al.*, 2000; Vicario *et al.*, 2001). However, HVC size of female canaries correlates with their preference for sexy songs (Leitner and Catchpole, 2002), but seasonal changes in vocal control nuclei of male song sparrows do not correlate with song discrimination (Reeves *et al.*, 2003).

VI. SEXUAL DIFFERENTIATION OF THE VOCAL CONTROL SYSTEM OF SONGBIRDS

A. Brain-intrinsic mechanisms

The sexual differentiation of the vocal control system appears to depend on genetically determined brain-intrinsic mechanisms and thus is not just a slave of the sexual determination and development of the avian gonads (Fig. 3.5) (Agate *et al.*, 2003; Gahr and Metzdorf, 1999). In the following text, we study the expression pattern of androgen receptors in the developing HVC of the zebra finch to argue for sex-hormone independent setup of this pattern. The forebrain areas of the zebra finch are first detectable between posthatching days 5 and 10 (Fig. 3.3). In the zebra finch, the size, neuron number, and androgen receptor expression of the HVC are already sexually dimorphic on posthatching day 9 (Gahr and Metzdorf, 1999) (Fig. 3.6). No androgen or estrogen receptors are expressed in the HVC region before posthatching day 9 (Gahr, 1996; Gahr and Metzdorf, 1999; Kim *et al.*, 2004; Perlman *et al.*, 2003), although androgen and estrogen receptors are found in various neural tissues outside HVC before (Gahr, 1996; Godsave *et al.*, 2002; Kim *et al.*, 2004; Perlman *et al.*, 2003). Slice cultures of the caudale neostriatum, where HVC develops, of posthatching day 5 males and females in hormone-free conditions show that the sexually dimorphic expression of androgen receptor mRNA in HVC is independent of the direct action of steroids on this nucleus or any of its immediate presynaptic or postsynaptic partners (Gahr and Metzdorf, 1999). Therefore, gonadal steroids do not appear to be directly involved in the initial sex difference in the expression pattern of androgen receptors, neuron number, and size of the HVC. Similarly, works with quail-chicken brain chimeras suggest brain-intrinsic determination of sex hormone expressing neuron pools (Gahr and Balaban, 1996). The brain-autonomous mechanisms that determine

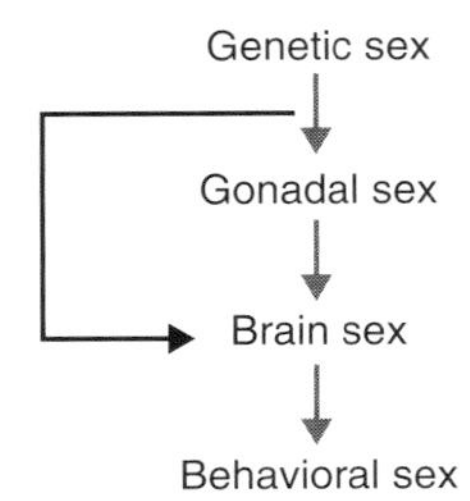

Figure 3.5. A model for the sexual differentiation of the brain and behavior in higher vertebrates such as songs of songbirds. Most current models of sexual differentiation of brain and behavior (*gray arrows*) suggest that the epigenetic action of gonadal steroids controls brain sex entirely. In difference, this model proposes brain sex-determining factors that control brain sex by brain-autonomous mechanisms (*black arrow*), independent of or in concert with gonadal steroid-dependent mechanisms. Environmental and behavioral determination of gonadal sex (for review, see Crews, 1993) are neglected for simplicity (from Gahr and Metzdorf, 1999).

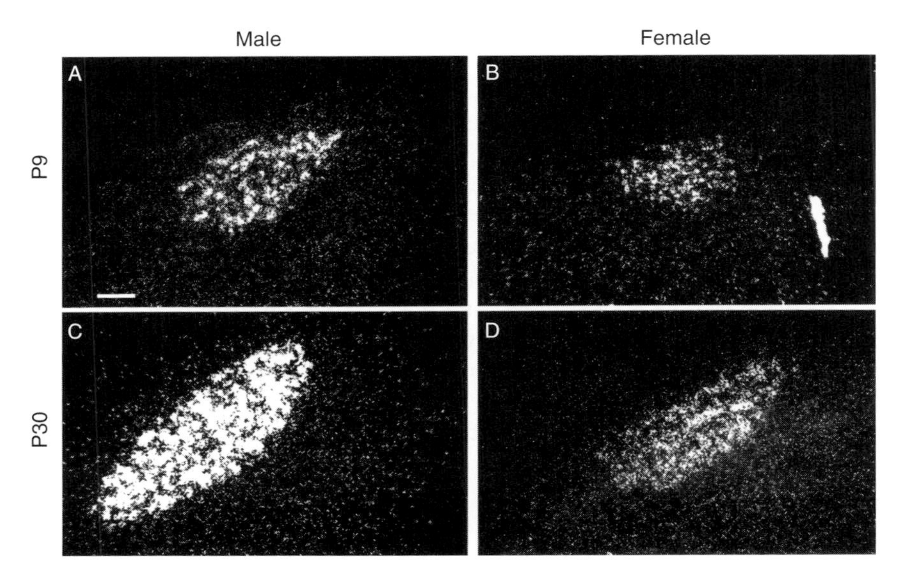

Figure 3.6. The HVC of the zebra finch is highly sexually dimorphic at posthatching day 9 (P9). The AR mRNA distribution was used to map the HVC. In the caudal nidopallium of male (A) and female (B) zebra finches, AR mRNA is first found at P9. At this time, the male AR mRNA-defined HVC is 2.2 times larger ($p < 0.001$; $t = 3.922$) and contains 2.2 times more ($p < 0.001$; $t = 3.922$) AR mRNA-expressing cells than the female HVC. At P30, the male HVC (C) is 3.3 times larger ($p < 0.001$; $t = 3.922$) and contains three times more ($p < 0.001$; $t = 3.922$) AR mRNA-expressing cells than the female HVC (D). Scale bar, 50 μm (from Gahr and Metzdorf, 1999).

forebrain vocal neurons are unknown. The finding of a gynandromorphic zebra finch (Agate *et al.*, 2003) supports the above argument of brain-autonomous determination of vocal neuron pools such as HVC independent of the general hormonal milieu (Fig. 3.5). Further, microarray analysis indicate sexually dimorphic gene expression in the embryonic brain of chicken and in the zebra finch brain at posthatching age (Agate *et al.*, 2004; Scholz *et al.*, 2006; Wade *et al.*, 2005). It needs to be seen if these sex differences are brain-intrinsic or the consequence of steroids' action on developing vocal neurons.

This notion of genetically determined brain-intrinsic mechanisms is further supported by experimental evidence in Japanese quails. Although this experiment does not concern vocal control circuits per se, it is thus far the only experiment that suggests brain-intrinsic mechanisms of sexual development (Gahr, 2003). In gallinaceous birds such as quails and chicken, female brain sex is thought to develop due to estrogen-dependent demasculinization of a default male brain phenotype (Adkins and Adler, 1972; Wilson and Glick, 1970). To test this concept of gonad-dependent brain sex, male-to-female (MF), female-to-male (FM), male-to-male (MM), and female-to-female (FF) isotopic isochronic transplantation of the brain primordium rostral to the otic capsules was performed at the second embryonic day in Japanese quails (Gahr, 2003). In these brain chimeras, the forebrain including the hypothalamus originates from the donor. MM, FF, and MF chimeras showed sexual behavior according to the host genetic sex. FM males showed no mounting, only rudimentary crowing behavior, and have a female-like medial preoptic nucleus (Gahr, 2003), a region that is normally larger in males than in females (Panzica *et al.*, 2001).

Such mechanisms of brain-autonomic sexual differentiation might explain puzzling area-specific sexual dimorphisms within the vocal network that are difficult to explain by epigenetic hormone-driven mechanisms (see below). For example, the size of the vocal area lMAN is not sexually dimorphic in the Bengalese finch (*Lonchura striata*) and the cowbird (*Molothrus ater*), while vocal areas HVC and RA are highly dimorphic, in register with the lack of females' song in these species (Hamilton *et al.*, 1997; Tobari *et al.*, 2005).

B. Endocrine mechanisms

Juvenile songbirds do not utter song precursors, so-called sub-songs, until a species-specific ontogenetic time-window. In the zebra finch, this time-window coincides with the formation of the HVC to RA synapses (around posthatching day 30) that is delayed relative to the interconnection of the other neuron pools determined to become vocal areas (Fig. 3.3) (Konishi and Akutagawa, 1985; Mooney, 1992; Mooney and Rao, 1994). It is assumed that this connectional property causes the transition from the nonfunctional to the functional song system concerning its motor function. In animals, which naturally or experimentally lack this connection, many HVC and RA neurons die and the survivors do not develop the neuronal phenotypes that are typical for HVC and RA of adult singing songbirds (Konishi and Akutagawa, 1985, 1987; Sakaguchi, 1996).

That gonadal steroids influence the fate of determined vocal neurons to reach this functional state is best evidenced by the sexually dimorphic development of the HVC and RA in some songbird species. In juvenile female zebra finches and starlings, estrogen treatment shifts the differentiation of these areas (measured in terms of area volume and neuron size) in male direction (Casto and Ball, 1996; Gurney, 1981; Gurney and Konishi, 1980; Pohl-Apel, 1985; Simpson and Vicario, 1991). These anatomical features nevertheless remain different from the male phenotypes reflecting the above-discussed genetic brain-autonomous sex differences (Gahr and Metzdorf, 1999). Gene expression of male (but not female) HVC neurons is sensitive to estrogens as early as posthatching day 15 (Dittrich *et al.*, 1999) and estrogen treatment masculinizes partially the volume (and probably neuron numbers) of female HVC between posthatching day 20 and 30 (Gahr and Metzdorf, 1999). The formation of HVC to RA synapses is inducible if estrogen treatment occurs in the first weeks posthatching (Konishi and Akutagawa, 1985). *In vitro* experiments with brain slice cultures of both male and female forebrains support the estrogen dependency of the RA to HVC connection (Holloway and Clayton, 2001). In case of elevated levels of intracerebral estrogens, the HVC to RA connection makes the vocal system functional, which leads to abrupt changes of its neuronal phenotypes as indicated by the upregulation of brain-derived neurotrophic factor (BDNF) (Dittrich *et al.*, 1999) and the song-system-nuclear-antigen (Akutagawa and Konishi, 2001). Due to the initial sexually dimorphic HVC size (see above Section VI), 60–70% of the HVC size of adult males can be explained by this initial size and general growth of the forebrain, while 30–40% of the adult HVC-volume is hormone dependent (Gahr and Metzdorf, 1999). The latter includes both neuronal spacing (Gahr and Metzdorf, 1999) and neuronal recruitment (Kirn and DeVoogd, 1989). The neurons of the functional "connected" HVC maturate in terms of neurochemical and electrophysiological properties until about 80–100 days posthatching (Adret and Margoliash, 2002), the onset of reproductive activities that involve stereotyped singing. Further, the variance in HVC size increases about threefold between 30 days of age and adulthood (Gahr and Metzdorf, 1999). In case of low levels of intracerebral estrogens, HVC neurons shall enter a cell death fate at a certain time point, in the latest after 35 days of posthatching life (Konishi and Akutagawa, 1988), which coincides with the loss of estrogen receptors in the HVC region (Gahr and Konishi, 1988).

The *in vitro* experiments (Holloway and Clayton, 2001), the results of intracerebral implantation of estrogens near HVC (Grisham *et al.*, 1994), the presence of estrogen receptors in and near HVC of juveniles (Gahr, 1996; Gahr and Konishi, 1988), and HVC lesions (Herrmann and Arnold, 1991) point to the HVC as the origin of estrogen-dependent differentiation of forebrain vocal areas. However, estrogen treatment at posthatching day 11 before the HVC to Area X connection is established nevertheless induces androgen expression in Area X, suggesting additional area-specific mechanisms of estrogen-sensitive sexual development (Kim *et al.*, 2004). Further, androgens might have area-specific effects independent of estrogen-dependent forebrain differentiation since androgens maturate the bursting properties of lMAN neurons already at posthatching day 25 (Livingston and Mooney, 2001). The time-course of hormone-sensitive development of the vocal system is likely species specific (Gahr, 1997; Gahr *et al.*, 1996).

VII. SEXUAL DIFFERENTIATION OF VOCAL CONTROL AREAS IN ADULTHOOD

Hormone-driven behavioral differentiation in adulthood is generally thought to be due to transient hormone-induced alterations of vocal area size that, in turn, is explained by changes of neuron numbers, neuronal size, neuron spacing, or of the size of terminal fields of projection neurons (Nottebohm, 1980, 1981; Nottebohm *et al.*, 1986; but Gahr, 1990, 1997). Differences in the relative importance of these mechanisms for volume changes of different vocal areas are obvious; for example, RA does not recruit newborn neurons in adult songbirds while HVC does (for review see Gahr *et al.*, 2002). Further, as detailed above, in some species there are seasonal vocal changes despite a lack of seasonal changes of the gross morphology of vocal areas (Leitner *et al.*, 2001; but Nottebohm, 1981). Here I discuss if we can dissociate between neuronal properties and size of a vocal area, the canary HVC, in relation to different hormone-dependent vocal entities. In canaries, songs are composed of syllables that are repeated identically several times (so-called tours) before switching to the next syllable. Canaries need to learn parts of the syllable repertoire, while the temporal organization of the song is innate (Güttinger, 1979, 1981).

Fusani *et al.* (2003) induced singing in adult female canaries through testosterone treatment, but inhibited in one group of such females the aromatization of testosterone into estrogens with an aromatase inhibitor (Fig. 3.7). Testosterone-induced development of male-like song in female canaries is accompanied by an increase in the expression and enzymatic activity of aromatase in the telencephalon near HVC (Fusani *et al.*, 2001), the only vocal areas with higher levels of estrogen receptors (Fig. 3.1). After 3–4 weeks of testosterone treatment, females develop a male-like song, with exception that such females sing few different syllables. In correlation with the male-like songs (long tours), the HVC size of singing females is similarly increased and different from untreated nonsinging control females (Fusani *et al.*, 2003). The estrogen-deprived singing

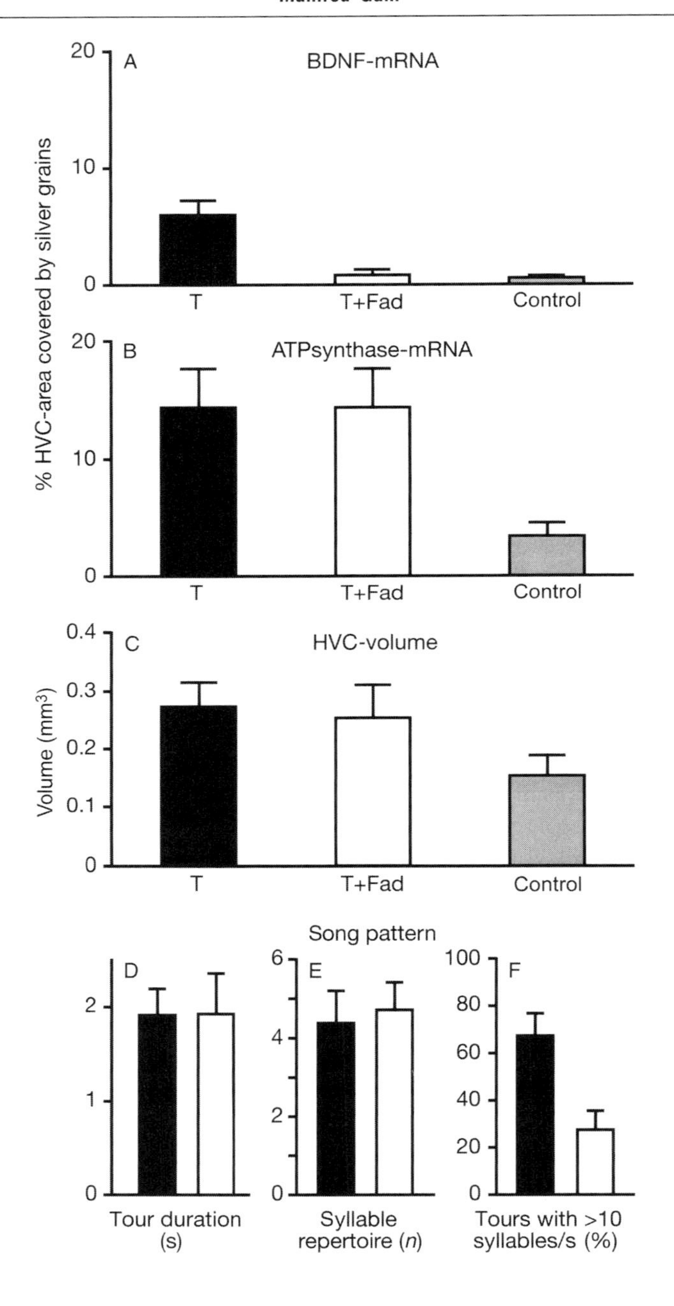

A
BDNF-mRNA
B
ATPsynthase-mRNA
% HVC-area covered by silver grains
20
10
0
T
T+Fad
Control
C
HVC-volume
Volume (mm3)
0.4
0.3
0.2
0.1
0
Song pattern
D
E
F
Tour duration (s)
Syllable repertoire (n)
Tours with >10 syllables/s (%)

females differ, however, in that they produce more tours with lower repetition rates compared to the nondeprived singing females (Fusani *et al.*, 2003). This difference correlates with the decreased expression of BDNF, while another gene, ATPsynthase involved in cellular metabolism, is expressed similarly in both groups. The data suggest that certain neural phenotypes of vocal neurons are under androgenic control (HVC size, ATPsynthase expression) while others (BDNF expression) are under estrogenic control. The song activity of both groups of testosterone-induced singing females was similar in this experiment, suggesting that the difference in BDNF expression level is not due to the singing activity, as proposed previously for male canaries (Rasika *et al.*, 1999). Although BDNF is well known to affect the differentiation and functions of central neurons (Bonhoeffer, 1996), these works do not suggest that the BDNF level is the cause for the observed behavioral differences between the above experimental singers. The work of Fusani *et al.* (2003) suggests, however, that androgen-dependent overall HVC morphology (reflected in HVC-volume) is necessary for the production of certain song pattern of canaries while other (in this case estrogen dependent) song features require neuronal properties that are independent of vocal area size. Similar results were obtained for male canaries (Rybak and Gahr, 2004). Clearly, if and how various sexually dimorphic or sex hormone-dependent neural properties relate to vocalization in various species requires electrophysiological and molecular approaches.

In adulthood, next to androgen and estrogen receptors, vocal neurons express receptors for non-gonadal hormones, in particular the pineal hormone melatonin (Gahr and Kosar, 1996; Jansen *et al.*, 2005), and are sensitive for modulatory action of monoaminergic systems (Appeltants *et al.*, 2001, 2002; Sakaguchi *et al.*, 2000). These neurochemical phenotypes suggest that vocal phenotypes are under the control of multiple environmental signals that directly act on vocal neurons, bypassing in part the gonadal system. Such mechanisms might explain the impact of the sociosexual and physical environment on the differentiation of vocal areas and on song development and performance that

---

Figure 3.7. Androgen- and estrogen-dependent differentiation of the HVC and the song of adult canaries. The expression level of BDNF-mRNA (A) is higher in testosterone-treated (T) female canaries than in testosterone-treated females in which estrogen formation (T + Fad) is inhibited, and than in control females (control). In contrast, the expression level of ATPsynthase-mRNA (B) and the volume of HVC (C) is increased in both T and T + Fad females compared to control females but do not differ between T and T + Fad females. Tour duration (D) and syllable repertoire (E) are similar in T and T + Fad females but the latter sing significantly less tours of high syllable repetition rates (F). Control females do not sing. Thus, androgens appear to control HVC-volume, ATP-synthase expression, tour duration, and repertoire size, while estrogens are required for BDNF expression and high-speed singing (redrawn from Fusani *et al.*, 2003).

appears independent of gonadal hormones (Bentley *et al.*, 1999; Deviche and Gulledge, 2000; Gulledge and Deviche, 1998; Tramontin *et al.*, 1999). The most direct evidence suggesting non-gonadal plasticity of vocalization comes from works on adult male zebra finches: melatonin affects their song pattern and the electrophysiology of vocal neurons transiently (Jansen *et al.*, 2005).

References

Adkins, E. K., and Adler, N. T. (1972). Hormonal control of behavior in Japanese quail. *J. Comp. Physiol. Psychol.* **81,** 27–36.

Adkins-Regan, E., Mansukhani, V., Seiwert, C., and Thompson, R. (1994). Sexual differentiation of brain and behavior in the zebra finch: Critical periods for effects of early estrogen treatment. *J. Neurobiol.* **25**(7), 865–877.

Adret, P., and Margoliash, D. (2002). Metabolic and neural activity in the song system nucleus robustus archistriatalis: Effect of age and gender. *J. Comp. Neurol.* **454**(4), 409–423.

Agate, R. J., Grisham, W., Wade, J., Mann, S., Wingfield, J., Schanen, C., Palotie, A., and Arnold, A. P. (2003). Neural, not gonadal, origin of brain sex differences in a gynandromorphic finch. *Proc. Natl. Acad. Sci. USA* **100,** 4873–4878.

Agate, R. J., Choe, M., and Arnold, A. P. (2004). Sex differences in structure and expression of the sex chromosome genes CHD1Z and CHD1W in zebra finches. *Mol. Biol. Evol.* **21,** 384–396.

Airey, D. C., and DeVoogd, T. J. (2000). Greater song complexity is associated with augmented song system anatomy in zebra finches. *Neuroreport* **11,** 2339–2344.

Akutagawa, E., and Konishi, M. (2001). A monoclonal antibody specific to a song system nuclear antigen in estrildine finches. *Neuron* **31,** 545–556.

Akutagawa, E., and Konishi, M. (2005). Connections of thalamic modulatory centers of the vocal control system of the zebra finch. *Proc. Natl. Acad. Sci. USA* **102,** 14086–14091.

Andersson, M. (1994). "Sexual Selection." Princeton University Press, Princeton, New Jersey.

Andrew, R. J. (1963). Effect of testosterone on the behavior of the domestic chick. *J. Comp. Physiol. Pschol.* **56,** 933–940.

Appeltants, D., Ball, G. F., and Balthazart, J. (2001). The distribution of tyrosine hydroxylase in the canary brain: Demonstration of a specific and sexual dimorphic catecholaminergic innervation of the telencephalic song control nuclei. *Cell Tissue Res.* **304,** 237–259.

Appeltants, D., Ball, G. F., and Balthazart, J. (2002). The origin of catecholaminergic inputs to the song control nucleus RA in canaries. *Neuroreport* **13,** 649–653.

Appeltants, D., Ball, G. F., and Balthazart, J. (2003). Song activation by testosterone is associated with an increased catecholaminergic innervation of the song control system in female canaries. *Neuroscience* **121,** 801–814.

Arai, O., Taniguchi, I., and Saito, N. (1989). Correlations between the size of song control nuclei and plumage color change in orange bishop birds. *Neurosci. Lett.* **98,** 144–148.

Arcese, P., Stoddard, P. K., and Hiebert, M. S. (1988). The form and function of song in female Song Sparrows. *Condor* **90,** 44–50.

Arnold, A. P., and Gorski, R. A. (1984). Gonadal steroid induction of structural sex differences in the central nervous system. *Annu. Rev. Neurosci.* **7,** 413–442.

Arnold, A. P., Nottebohm, F., and Pfaff, D. W. (1976). Hormone concentrating cells in vocal control and other areas of the brain of the zebra finch (*Poephila guttata*). *J. Comp. Neurol.* **165,** 487–511.

Baker, M. C., Bottjer, S. W., and Arnold, A. P. (1984). Sexual dimorphism and lack of seasonal changes in vocal control regions of the white-crowned sparrow brain. *Brain Res.* **295,** 85–89.

Ballintijn, M. R., and Ten Cate, C. (1997). Sex differences in the vocalizations and syrinx of the Collared Dove (*Streptopelia decaocto*). *Auk* **114,** 22–39.

Baptista, L. F., Trail, P. W., Dewolfe, B. B., and Morton, M. L. (1993). Singing and its functions in female white-crowned sparrows. *Anim. Behav.* **46,** 511–524.

Beani, L., Panzica, G., Briganti, F., Persichella, P., and Dessi-Fulheri, F. (1995). Testosterone-induced changes of call structure, midbrain and syrinx anatomy in partridges. *Physiol. Behav.* **58,** 1149–1157.

Beani, L., Briganti, F., Campanella, G., Lupo, C., and Dessi-Fulgheri, F. (2000). Effect of androgens on structure and rate of crowing in the Japanese quail (*Coturnix japonica*). *Behaviour* **137,** 417–435.

Beletsky, L. D. (1983). Aggressive and pair-bond maintenance songs of female red-winged blackbirds (Agelaius phoeniceus). *Z. Tierpsychol.* **62,** 47–54.

Bennet, M. A. (1940). The vocal hierarchy in ring doves. II. The effects of testosterone propionate. *Ecology* **21,** 148–165.

Bentley, G. E., Van't Hof, T. J., and Ball, G. F. (1999). Seasonal neuroplasticity in the songbird telencepahlon: A role for melatonin. *Proc. Natl. Acad. Sci. USA* **96,** 4674–4679.

Benton, S., Nelson, D. A., Marler, P., and DeVoogd, T. J. (1998). Anterior forebrain pathway is needed for stable song expression in adult white-crowned sparrow (*Zonotrichia leucophrys*). *Behav. Brain Res.* **96,** 135–150.

Bernard, D. J., Casto, J. M., and Ball, G. F. (1993). Sexual dimorphism in the volume of song control nuclei in European starlings: Assessment by a Nissl stain and autoradiography for muscarinic cholinergic receptors. *J. Comp. Neurol.* **334,** 559–570.

Bernard, D. J., Bentley, G. E., Balthazart, J., Turek, F. G., and Ball, G. F. (1999). Androgen receptor, estrogen receptor alpha, and estrogen receptor beta show distinct patterns of expression in forebrain song control nuclei of European starlings. *Endocrinology* **140,** 4633–4643.

Beyer, C., Ivanonva, T., Karolczak, M., and Kuppers, E. (2002). Cell-type specificity of non-classical estrogen signalling in the developing midbrain. *J. Steroid Biochem. Mol. Biol.* **81,** 319–325.

Bolhuis, J. J., and Gahr, M. (2006). Neural mechanisms of birdsong memory. *Nat. Rev. Neurosci.* **7,** 347–357.

Bonhoeffer, T. (1996). Neurotrophins and activity-dependent development of the neocortex. *Curr. Opin. Neurobiol.* **6,** 119–126.

Bottjer, S. W., Miesner, E. A., and Arnold, A. P. (1984). Forebrain lesions disrupt development but not maintenance of song in passerine birds. *Science* **224,** 901–902.

Bottjer, S. W., Roselinsky, H., and Tran, N. B. (1997). Sex differences in neuropeptide staining of song-control nuclei in zebra finch brains. *Brain Behav. Evol.* **50**(5), 284–303.

Brainard, M. S., and Doupe, A. J. (2000). Interruption of a basal ganglia-forebrain circuit prevents plasticity of learned vocalizations. *Nature* **404,** 762–766.

Brauth, S. E., Liang, W., Amateau, S. K., and Robert, T. F. (2005). Sexual dimorphism of vocal control nuclei in Budgerigars (*Melopsittacus undulates*) revealed with Nissl and NADPH-d staining. *J. Comp. Neurol.* **484,** 15–27.

Brenowitz, E. A., and Arnold, A. P. (1986). Interspecific comparisons of the size of neural song control regions and song complexity in duetting birds: Evolutionary implications. *J. Neurosci.* **6,** 2875–2879.

Brenowitz, E. A., and Lent, K. (2002). Act locally and think globally: Intracerebral testosterone implants induce seasonal-like growth of adult avian song control circuits. *Proc. Natl. Acad. Sci. USA* **99,** 12421–12426.

Brenowitz, E. A., Arnold, A. P., and Levin, R. N. (1985). Neural correlates of female song in tropical duetting birds. *Brain Res.* **343,** 104–112.

Brenowitz, E. A., Nafis, B., Wingfield, J. C., and Kroodsma, D. E. (1991). Seasonal changes in avian song nuclei without seasonal changes in song repertoire. *J. Neurosci.* **11,** 1367–1374.

Brenowitz, E. A., Nalls, B., Kroodsma, D. E., and Horning, C. (1994). Female marsh wrens do not provide evidence of anatomical specializations of song nuclei for perception of male song. *J. Neurobiol.* **25,** 197–208.

Brockway, B. F. (1968). Influences of sex hormones on the loud and soft warbling of male budgerigars. *Anim. Behav.* **16,** 5–12.

Brockway, B. F. (1969). Roles of budgerigar vocalizations in the integration of breeding behaviour. *In* "Bird Vocalizations" (R. A. Hinde, ed.), pp. 131–158. Cambridge University Press, London.

Burgess, L. H., and Handa, R. J. (1993). Hormonal regulation of androgen receptor mRNA in the brain and anterior pituitary gland of the male rat. *Mol. Brain Res.* **19,** 31–38.

Canady, R. A., Kroodsma, D. E., and Nottebohm, F. (1984). Population differences in complexity of a learned skill are correlated with the brain space involved. *Proc. Natl. Acad. Sci. USA* **81,** 6232–6234.

Carlson, G., and Trost, C. H. (1992). Sex determination of the Whooping Crane by analysis of vocalization. *Condor* **84,** 532–536.

Carruth, L. L., Reisert, I., and Arnold, A. P. (2002). Sex chromosome genes directly affect brain sexual differentiation. *Nat. Neurosci.* **5,** 933–934.

Casto, J. M., and Ball, G. F. (1996). Early administration of 17beta-estradiol partially masculinizes song control regions and alpha2-adrenergic receptor distribution in European starlings (*Sturnus vulgaris*). *Horm. Behav.* **30,** 387–406.

Catchpole, C. K., and Slater, P. J. B. (1995). "Bird Song: Biological Themes and Variations." Cambridge University Press, Cambridge, UK.

Chen, X., Agate, R. J., Itoh, Y., and Arnold, A. P. (2005). Sexually dimorphic expression of trkB, a Z-linked gene, in early posthatch zebra finch brain. *Proc. Natl. Acad. Sci. USA* **102,** 7730–7735.

Chiba, A., and Hosokawa, N. (2006). Effects of androgens and estrogens on crowings on distress callings in male Japanese quail, *Coturnix japonica*. *Horm. Behav.* **49,** 4–14.

Crews, D. (1993). The organizational concept and vertebrates without sex chromosomes. *Brain Behav. Evol.* **42,** 202–214.

Del Negro, C., and Edeline, J. M. (2001). Differences in auditory and physiological properties of HVc neurons between reproductively active male and female canaries (Serinus canaria). *Eur.J. Neurosci.* **14,** 1377–1389.

Del Negro, C., Kreutzer, M., and Gahr, M. (2000). Sexually stimulating signals of canary (*Serinus canaria*) songs: Evidence for a female-specific auditory representation in the HVC nucleus during the breeding season. *Behav. Neurosci.* **114,** 526–542.

DeVoogd, T. J., and Nottebohm, F. (1981a). Sex differences in dendritic morphology of song control nucleus in the canary: A quantitative Golgi-study. *J. Comp. Neurol.* **196,** 309–316.

DeVoogd, T. J., and Nottebohm, F. (1981b). Gonadal hormones induce dendritic growth in the adult brain. *Science* **214,** 202–204.

DeVoogd, T. J., Krebs, J. R., Healy, S. D., and Purvis, A. (1993). Relations between song repertoire size and the volume of brain nuclei related to song: Comparative evolutionary analyses amongst oscine birds. *Proc. R. Soc. Lond. B Biol. Sci.* **254,** 75–82.

DeVoogd, T. J., Houtman, A. M., and Falls, J. B. (1995). White-throated sparrow morphs that differ in song production rate also differ in the anatomy of some song-related brain areas. *J. Neurobiol.* **28,** 202–213.

Deviche, P., and Gulledge, C. C. (2000). Vocal control region sizes of an adult female songbird change seasonally in the absence of detectable circulating testosterone concentrations. *J. Neurobiol.* **42,** 202–211.

Dittrich, F., Feng, Y., Metzdorf, R., and Gahr, M. (1999). Estrogen-inducible, sex- of brain-derived neurotrophic factor mRNA in a forebrain song control nucleus of the juvenile zebra finch. *Proc. Natl. Acad. Sci. USA* **96,** 8221–8246.

Durand, S. E., Heaton, J. T., Amateau, S. K., and Brauth, S. E. (1997). Vocal control pathways through the anterior forebrain of a parrot. (*Melopsittacus undulatus*). *J. Comp. Neurol.* **377,** 179–206.

Falls, J. B., and Kopachena, J. G. (1994). White throated sparrows (*Zonotrichia albicollis*). *Birds North. Am.* **128,** 1–32.

Farabaugh, S. M. (1982). The ecological and social significance of duetting. *In* "Acoustic Communication in Birds" (D. E. Kroodsma and E. H. Miller, eds.), Vol. 2, pp. 85–125. Academic Press, New York.

Farquhar, C. C. (1993). Individual and intersexual variation in alarm calls of the white-tailed hawk. *Condor* **95,** 234–239.

Fortune, E. S., and Margoliash, D. (1995). Parallel pathways and convergence onto HVc and adjacent neostriatum of adult zebra finches (*Taeniopygia guttata*). *J. Comp. Neurol.* **360,** 413–441.

Fusani, L., and Gahr, M. (2006). Hormonal influence on song structure and organization: The role of estrogen. *Neuroscience* **138,** 939–946.

Fusani, L., Beani, L., and Dessi-Fulgheri, F. (1994). Testosterone affects the acoustic structure of the male call in the grey partridge (*Perdix Perdix*). *Behaviour* **128,** 301–310.

Fusani, L., Van't Hof, T., Hutchison, J. B., and Gahr, M. (2000). Seasonal expression of androgen receptors, oestrogen receptors and aromatase in the canary brain in relation to circulating androgens and oestrogens. *J. Neurobiol.* **43,** 254–268.

Fusani, L., Hutchison, J. B., and Gahr, M. (2001). Testosterone regulates the activity and expression of aromatase in the canary neostriatum. *J. Neurobiol.* **49,** 1–8.

Fusani, L., Metzdorf, R., Hutchison, J. B., and Gahr, M. (2003). Aromatase inhibition affects testosterone-induced masculinization of song and the neural song system in female canaries. *J. Neurobiol.* **54,** 370–379.

Gahr, M. (1990). Delineation of a brain nucleus: Comparison of cytochemical, hodological, and cytoarchitectural views of the song control nucleus HVC of the adult canary. *J. Comp. Neurol.* **294,** 30–36.

Gahr, M. (1994). Brains structure: Causes and consequences of brain sex. *In* "The Differences Between the Sexes" (R. V. Short and E. Balaban, eds.), pp. 273–302. Cambridge, University Press.

Gahr, M. (1996). Developmental changes in the distribution of oestrogen receptor mRNA expressing cells in the forebrain of female, male and masculinized female zebra finches. *Neuroreport* **7,** 2469–2473.

Gahr, M. (1997). How should brain nuclei be delineated? Consequences for developmental mechanisms and for correlations of area size, neuron numbers and functions of brain nuclei. *Trends Neurosci.* **20,** 58–62.

Gahr, M. (2000). Neural song control system of hummingbirds: Comparison to swifts, vocal learning (songbirds) and nonlearning (suboscines) passerines, and vocal learning (budgerigars) and nonlearning (dove, owl, gull quail, chicken) nonpasserines. *J. Comp. Neurol.* **426,** 182–196.

Gahr, M. (2003). Male Japanese quails with female brains do not show male sexual behaviors. *Proc. Natl. Acad. Sci. USA* **100,** 7959–7964.

Gahr, M. (2004). Hormone-dependent neural plasticity in the juvenile and adult song system: What makes a successful male? *Ann. N. Y. Acad. Sci.* **1016,** 684–703.

Gahr, M., and Balaban, E. (1996). The development of a species difference in the local distribution of brain estrogen receptive cells. *Dev. Brain Res.* **92,** 182–189.

Gahr, M., and Garcia-Segura, L. M. (1996). Testosterone-induced increase of gap-junctions in HVC neurons of adult female canaries. *Brain Res.* **712,** 69–73.

Gahr, M., and Güttinger, H. R. (1985). Korrelation zwischen der sexual dimorphen Gehirndifferenzierung und der Verhaltensausprägung bei Prachtfinken (Estrildidae). *J. Ornithol.* **126,** 310.

Gahr, M., and Güttinger, H. R. (1986). Functional aspects of singing in male and female Uraeginthus bengalus (Estrildidae). *Ethology* **72,** 123–131.
Gahr, M., and Konishi, M. (1988). Developmental changes in estrogen-sensitive neurons in the forebrain of the zebra finch. *Proc. Natl. Acad. Sci. USA* **85,** 7380–7383.
Gahr, M., and Kosar, E. (1996). Identification, distribution, and developmental changes of a melatonin binding site in the song control system of the zebra finch. *J. Comp. Neurol.* **367,** 308–318.
Gahr, M., and Metzdorf, R. (1997). Distribution and dynamics in the expression of androgen and estrogen receptors in vocal control systems of songbirds. *Brain Res. Bull.* **44,** 509–517.
Gahr, M., and Metzdorf, R. (1999). The sexually dimorphic expression of androgen receptors in the song nucleus hyperstriatalis ventrale pars caudale of the zebra finch develops independently of gonadal steroids. *J. Neurosci.* **19,** 2628–2636.
Gahr, M., and Wild, J. M. (1997). Localization of androgen receptor mRNA-containing cells in avian respiratory-vocal nuclei: An *in situ* hybridization study. *J. Neurobiol.* **33,** 865–876.
Gahr, M., Flugge, G., and Güttinger, H. R. (1987). Immunocytochemical localization of estrogen-binding neurons in the songbird brain. *Brain Res.* **402,** 173–177.
Gahr, M., Güttinger, H. R., and Kroodsma, D. E. (1993). Estrogen receptors in the avian brain: Survey reveals general distribution and forebrain areas unique to songbirds. *J. Comp. Neurol.* **327,** 112–122.
Gahr, M., Metzdorf, R., and Aschenbrenner, S. (1996). The ontogeny of the canary HVC revealed by the expression of androgen and oestrogen receptors. *Neuroreport* **8,** 311–315.
Gahr, M., Sonnenschein, E., and Wickler, W. (1998). Sex differences in the size of the neural song control regions in a dueting songbird with similar song repertoire size of males and females. *J. Neurosci.* **18,** 1124–1131.
Gahr, M., Leitner, S., Fusani, L., and Rybak, F. (2002). What is the adaptive role of neurogenesis in adult birds? *Prog. Brain Res.* **138,** 233–254.
Gil, D., and Gahr, M. (2002). The honesty of bird song: Multiple constraints for multiple traits. *Trends Ecol. Evol.* **17,** 133–140.
Gil, D., Naguib, M., Riebel, K., Rutstein, A., and Gahr, M. (2006). Early condition, song learning and the volume of song brain nuclei in the zebra finch (*Taeniopygia guttata*). *J. Neurobiol.* **66,** 1602–1612.
Godsave, S. F., Lohmann, R., Vloet, R. P., and Gahr, M. (2002). Androgen receptors in the embryonic zebra finch hindbrain suggest a function for maternal androgens in perihatching survival. *J. Comp. Neurol.* **453,** 57–70.
Goymann, W., Wittenzellner, A., and Wingfield, J. C. (2004). Competing females and caring males. Polyandry and sex-role reversal in African black coucals (*Centropus grillii*). *Ethology* **110,** 807–823.
Grafen, A. (1990). Biological signals as handicaps. *J. Theor. Biol.* **144,** 517–546.
Grisham, W., Mathews, G. A., and Arnold, A. P. (1994). Local intracerebral implants of estrogen masculinize some aspects of the zebra finch song system. *J. Neurobiol.* **25,** 185–196.
Groothuis, T. G. G., and Meeuwissen, G. (1992). The influence of testosterone on the development and fixation of the form of displays in two age classes of young black-headed gulls. *Anim. Behav.* **43,** 189–208.
Gulledge, C. C., and Deviche, P. (1998). Photoperiod and testosterone independently affect vocal control region volumes in adolescent male songbirds. *J. Neurobiol.* **36,** 550–558.
Gurney, M. E. (1981). Hormonal control of cell form and number in the zebra finch song system. *J. Neurosci.* **1,** 658–673.
Gurney, M. E. (1982). Behavioral correlates of sexual differentiation in the zebra finch song system. *Brain Res.* **231,** 153–173.
Gurney, M. E., and Konishi, M. (1980). Hormone-induced sexual differentiation of brain and behavior in zebra finches. *Science* **208,** 1380–1383.

Güttinger, H. R. (1979). The integration of learnt and genetically programmed behaviour: A study of hierarchical organization in songs of canaries, greenfinches and their hybrids. *Z. Tierpsychol.* **49,** 285–303.

Güttinger, H. R. (1981). Self-differentiation of song organization rules by deaf canaries. *Z. Tierpsychol.* **56,** 323–340.

Halle, F., Gahr, M., and Kreutzer, M. (2003). Effects of unilateral lesions of HVC on song patterns of males domesticated canaries. *J. Neurobiol.* **56,** 303–314.

Hamilton, K. S., King, A. P., Sengelaub, D. R., and West, M. J. (1997). A brain of her own: A neural correlate of song assessment in a female songbird. *Neurobiol. Learn. Mem.* **68,** 325–332.

Hartley, R. S., and Suthers, R. A. (1990). Lateralization of syringeal function during song production in the canary. *J. Neurobiol.* **21,** 1236–1248.

Hausberger, M., Richard-Yris, M. A., Henry, L., Lepage, L., and Schmidt, I. (1995). Song sharing reflects the social organization in a captive group of European starlings (*Sturnus vulgaris*). *J. Comp. Psychol.* **109,** 222–241.

Heid, P., Güttinger, H. R., and Pröve, E. (1985). The influence of castration and testosterone replacement on the song architecture of canaries (*Serinus canaria*). *Z. Tierpsychol.* **69,** 224–236.

Henry, L., and Hausberger, M. (2001). Differences in the social context of song production in captive male and female European starlings. *C. R. Acad. Sci. III.* **324,** 1167–1174.

Herrmann, K., and Arnold, A. P. (1991). Lesions of HVC block the developmental masculinizing effects of estradiol in the female zebra finch song system. *J. Neurobiol.* **22,** 29–39.

Herting, B. L., and Belthoff, J. R. (2001). Bounce and double trill songs of male and female western Screech-Owls: Characterization and usefulness for classification of sex. *Auk* **118,** 1095–1101.

Hoelzel, A. R. (1986). Song characteristics and response to playback of male and female Robins *Erithacus rubecula*. *Ibis* **128,** 115–127.

Holloway, C. C., and Clayton, D. F. (2001). Estrogen synthesis in the male brain triggers development of the avian song control pathway *in-vitro*. *Nat. Neurosci.* **4,** 170–175.

Hutchison, J. B., and Steimer, T. (1984). Androgen metabolism in the brain: Behavioural correlates. *In* "Sex Differences in the Brain" (G. J. de Vries, J. P. C. de Bruin, H. B. M. Uylings, and M. A. Corner, eds.), Vol. 61, pp. 23–51. Elsevier Science, Amsterdam.

Hutchison, J. B., Wingfield, J. C., and Hutchison, R. E. (1984). Sex differences in plasma concentration of steroids during the sensitive period for brain differentiation in the zebra finch. *J. Endocrinol.* **103,** 363–369.

Immelmann, K. (1969). Song development in the zebra finch and other estrildid finches. *In* "Bird Vocalizations" (R. A. Hind, ed.), pp. 61–74. Cambridge University Press, London.

Jacobs, E. C., Arnold, A. P., and Campagnoni, A. T. (1999). Developmental regulation of the distribution of aromatase- and estrogen-receptor-mRNA-expressing cells in the zebra finch brain. *Dev. Neurosci.* **21,** 453–472.

Jansen, R., Metzdorf, R., Roest, M., Ter Maat, A., and Gahr, M. (2005). Melatonin-dependent song pattern of adult male zebra finches. *FASEB J.* **19,** 848–850.

Jarvis, E. D., and Mello, C. V. (2000). Molecular mapping of brain areas involved in parrot vocal communication. *J. Comp. Neurol.* **419,** 1–31.

Jarvis, E. D., Ribeiro, S., da Silva, M. L., Ventura, D., Vielliard, J., and Mello, C. V. (2000). Behaviourally driven gene expression reveals song nuclei in hummingbird brain. *Nature* **406,** 628–632.

Kasumovic, M. M., Ratcliffe, L. M., and Boag, P. T. (2003). Song structure may differ between male and female least flycatchers. *Wilson Bull.* **115,** 241–245.

Kern, M. D., and King, J. R. (1972). Testosterone-induced singing in female white-crowned sparrows. *Condor* **74,** 204–209.

Ketterson, E. D., Nolan, V., Jr., Wolf, L., Ziegenfus, C., Dufty, A. M., Ball, G. F., and Johnsen, T. S. (1992). Testosterone and avian life histories: Effects of experimentally elevated testosterone on behavior and correlates of fitness in the dark-eyed junco (*Junco hyemalis*). *Am. Nat.* **140,** 980–999.

Kim, Y., Stumpf, W. E., Sar, M., and Martinez-Vargas, M. C. (1978). Estrogen and androgen target cells in the brain of fishes, reptiles and birds: Phylogeny and ontogeny. *Am. Zool.* **18,** 425–433.

Kim, Y. H., Perlman, W. R., and Arnold, A. P. (2004). Expression of androgen receptor mRNA in zebra finch song system: Developmental regulation by estrogen. *J. Comp. Neurol.* **469,** 535–547.

Kirn, J. R., and DeVoogd, T. J. (1989). Genesis and death of vocal control neurons during sexual differentiation in the zebra finch. *J. Neurosci.* **9,** 3176–3187.

Kirn, J. R., Clower, R. P., Kroodsma, D. E., and DeVoogd, T. J. (1989). Song related brain regions in the red-winged blackbird are affected by sex and season but not repertoire size . *J. Neurobiol.* **20,** 139–163.

Kling, J. W., and Stevenson-Hinde, J. (1977). Development of song and reinforcing effects of song in female chaffinches. *Anim. Behav.* **25,** 215–220.

Konishi, M. (1965). The role of auditory feedback in the control of vocalization in the white-crowned sparrow. *Z. Tierpsychol.* **22,** 770–783.

Konishi, M., and Akutagawa, E. (1981). Androgen increases protein-synthesis within the avian brain vocal control-system. *Brain Res.* **222,** 442–446.

Konishi, M., and Akutagawa, E. (1985). Neuronal growth, atrophy and death in a sexually dimorphic song nucleus in the zebra finch brain. *Nature* **315,** 145–147.

Konishi, M., and Akutagawa, E. (1987). Hormonal control of cell death in a sexually dimorphic song nucleus in the zebra finch. *In* "Selective Neuronal Death Ciba Foundation Symposium 126," pp. 173–185. John Wiley and Sons, New York.

Konishi, M., and Akutagawa, E. (1988). A critical period for estrogen action on neurons of the song control system in the zebra finch. *Proc. Natl. Acad. Sci. USA* **85,** 7006–7007.

Kreutzer, M., and Vallet, E. (1991). Differences in the response of captive female canaries to variation in conspecific and heterospecific songs. *Behaviour* **117,** 106–116.

Kroodsma, D. E. (1985). Development and use of two song forms by the Eastern Phoebe. *Wilson Bull.* **97,** 21–29.

Langmore, N. E., and Davies, N. B. (1997). Female dunnocks use vocalizations to compete for males. *Anim. Behav.* **53,** 881–890.

Langmore, N. E., Davies, N. B., Hatchwell, B. J., and Hartley, I. R. (1996). Female song attracts males in the alpine accentor *Prunella collaris*. *Proc. R. Soc. Lond. B* **263,** 141–146.

Langmore, N. E., Cockrem, J. F., and Candy, E. J. (2002). Competition for male reproductive investments elevates testosterone levels in female dunnocks, *Prunella modularis*. *Proc. R. Soc. Lond. B* **269,** 2473–2478.

Leitner, S., and Catchpole, C. (2002). Female canaries that respond and discriminate more between male songs of different quality have a larger song control nucleus (HVC) in the brain. *J. Neurobiol.* **52,** 294–301.

Leitner, S., and Catchpole, C. (2004). Syllable repertoire and the size of the song control system in captive canaries (*Serinus canaria*). *J. Neurobiol.* **60,** 21–27.

Leitner, S., Voigt, C., Garcia-Segura, L. M., Van't Hof, T., and Gahr, M. (2001). Seasonal activation and inactivation of song motor memories in free living canaries is not reflected in neuroanatomical changes of forebrain song areas. *Horm. Behav.* **40,** 160–168.

Leitner, S., Nicholson, J., Leisler, B., DeVoogd, T. J., and Catchpole, C. K. (2002). Song and the song control pathway in the brain can develop independently of exposure to song in the sedge warbler. *Proc. R. Soc. Lond. B* **269,** 2519–2524.

Leonard, S. L. (1939). Induction of singing in female canaries by injections of male hormone. *Proc. Soc. Exp. Biol.* **21,** 229–230.

Leonardo, A., and Konishi, M. (1999). Decrystallization of adult birdsong by perturbation of auditory feedback. *Nature* **399,** 466–470.

Lisciotto, C. A., and Morell, J. I. (1993). Circulating gonadal steroid hormones regulate estrogen receptor mRNA in the male rat forebrain. *Brain Res.* **20,** 79–90.

Livingston, F., and Mooney, R. (2001). Androgens and isolation from adult tutors differentially affect the development of songbird neurons critical to vocal plasticity. *J. Neurophysiol.* **85,** 34–42.

Lohmann, R., and Gahr, R. (2000). Muscle-dependent and hormone-dependent differentiation of the vocal control premotor nucleus robustus archistriatalis and the motor nucleus hypoglossus pars tracheosyringealis of the zebra finch . *J. Neurobiol.* **42,** 220–231.

London, S. E., Monks, D. A., Wade, J., and Schlinger, B. A. (2006). Widespread capacity for steroid synthesis in the avian brain and song system. *Endocrinology* **147,** 5975–5987.

MacDougall-Shackleton, S. A., and Ball, G. F. (1999). Comparative studies of sex differences in the song control system of songbirds. *Trends Neurosci.* **22,** 432–436.

MacDougall-Shackleton, S. A., Ball, G. F., Edmonds, E., Sul, R., and Hahn, T. P. (2005). Age- and sex-related variation in song-control regions in Cassin's finches, *Carpodacus cassinii. Brain Behav. Evol.* **65,** 262–267.

Madge, S., and McGowan, P. (2002). "Pheassants, Partrigdes and Grouse." Princeton University Press, New Jersey.

Maney, D. L., Bernard, D. J., and Ball, G. F. (2001). Gonadal steroid receptor mRNA in catecholaminergic nuclei of the canary brainstem. *Neurosci. Lett.* **311,** 189–192.

Maney, D. L., Cho, E., and Goode, C. T. (2006). Estrogen-dependent selectivity of genomic responses to birdsong. *Eur. J. Neurosci.* **23,** 1523–1529.

Marler, P. (1991). Song-learning behavior: The interface with neuroethology. *Trends Neurosci.* **14,** 199–206.

Marler, P. (1997). Three models of song learning: Evidence from behavior. *J. Neurobiol.* **33,** 501–516.

Marler, P., Kreith, M., and Willis, E. (1962). An analysis of testosterone-induced crowing in young domestic cockerels. *Anim. Behav.* **10,** 48–54.

McEwen, B. S. (1994). How do sex and stress hormones affect nerve cells? *Ann. N. Y. Acad. Sci.* **743,** 1–18.

Mennill, D. J., and Vehrencamp, S. L. (2005). Sex differences in singing and duetting behavior of neotropical rufous-and-white wrens (*Thryothorus rufalbus*). *Auk* **122,** 175–186.

Metzdorf, R., Gahr, M., and Fusani, L. (1999). Distribution of aromatase, estrogen receptor, and androgen receptor mRNA in the forebrain of songbirds and nonsongbirds. *J. Comp. Neurol.* **407,** 115–129.

Mooney, R. (1992). Synaptic basis for developmental plasticity in a birdsong nucleus. *J. Neurosci.* **12,** 2464–2477.

Mooney, R., and Rao, M. (1994). Waiting periods versus early innervation: The development of axonal connections on the zebra finch song system. *J. Neurosci.* **14,** 6532–6543.

Morton, E. S. (1987). The effects of distance and isolation on song-type sharing in the Carolina wren. *Wilson Bull.* **99,** 601–610.

Nastiuk, K. L., and Clayton, D. F. (1995). The canary androgen receptor mRNA is localized in the song control nuclei of the brain and is rapidly regulated by testosterone. *J. Neurobiol.* **26,** 213–224.

Nealen, P. M. (2005). An interspecific comparison using immunofluorescence reveals that synapse density in the avian song system is related to sex but not to male song repertoire size. *Brain Res.* **1032**(1–2), 50–62.

Nealen, P. M., and Perkel, D. J. (2000). Sexual Dimorphism in the Song System of the Carolina Wren *Thryothorus ludovicianus*. *J. Comp. Neurol.* **418,** 346–360.

Nespor, A. A., Lukazewicz, M. J., Dooling, R. J., and Ball, G. F. (1996). Testosterone induction of male-like vocalizations in female budgerigars (*Melopsittacus undulatus*). *Horm. Behav.* **30,** 162–169.

Noble, G. K., and Wurm, M. (1940). The effect of testosterone propionate on the black-crowned night heron. *Endocrinology* **26,** 837–850.

Nottebohm, F. (1980). Testosterone triggers growth of brain vocal control nuclei in adult female canaries. *Brain Res.* **189,** 429–436.

Nottebohm, F. (1981). A brain for all seasons: Cyclical anatomical changes in song control nuclei of the canary brain. *Science* **214,** 1368–1370.

Nottebohm, F., and Arnold, A. P. (1976). Sexual dimorphism in vocal control areas of the songbird brain. *Science* **194,** 211–213.

Nottebohm, F., Stokes, T. M., and Leonard, C. M. (1976). Central control of song in the canary, *Serinus canaria*. *J. Comp. Neurol.* **165,** 457–486.

Nottebohm, F., Kasparian, S., and Pandazis, C. (1981). Brain space for a learned task. *Brain Res.* **213,** 99–109.

Nottebohm, F., Nottebohm, M. E., and Crane, L. A. (1986). Development and seasonal changes in canary song and their relation to changes in the anatomy of song control nuclei. *Behav. Neural. Biol.* **46,** 445–471.

Panzica, G., Viglietti-Panzica, C., and Balthazart, J. (2001). Sexual dimorphism in the neuronal circuits of the quail preoptic and limbic regions. *Microsc. Res. Tech.* **54,** 364–374.

Perlman, W. R., Ramachandran, B., and Arnold, A. P. (2003). Expression of androgen receptor mRNA in the late embryonic and early posthatch zebra finch brain. *J. Comp. Neurol.* **455,** 513–530.

Pesch, A., and Güttinger, H. R. (1985). Der Gesang des weiblichen Kanarienvogels. *J. Ornithol.* **126,** 108–110.

Peterson, R. S., Yarram, L., Schlinger, B. A., and Saldanha, C. J. (2005). Aromatase is pre-synaptic and sexually dimorphic in the adult zebra finch brain. *Proc. Biol. Sci.* **272,** 2089–2096.

Pfaff, D. W. (1980). "Estrogens and Brain Function." Springer Verlag, New York.

Pinaud, R., Fortes, A. F., Lovell, P., and Mello, C. V. (2006). Calbindin-positive neurons reveal a sexual dimorphism within the songbird analogue of the mammalian auditory cortex. *J. Neurobiol.* **66,** 182–195.

Podos, J., Huber, S. K., and Taft, B. (2004). Bird song: The interface of evolution and mechanism. *Ann. Rev. Ecol. Syst.* **35,** 55–87.

Pohl-Apel, G. (1985). The correlation between the degree of brain masculinization and song quality in estradiol treated female zebra finches. *Brain Res.* **336,** 381–383.

Rasika, S., Alvarez-Buylla, A., and Nottebohm, F. (1999). BDNF mediates the effects of testosterone on the survival of new neurons in an adult brain. *Neuron* **22,** 53–62.

Reeves, B. J., Beecher, M. D., and Brenowitz, E. A. (2003). Seasonal changes in avian song control circuits do not cause seasonal changes in song discrimination in song sparrows. *J. Neurobiol.* **57,** 119–129.

Ritchison, G. (1983). Variation in the songs of female black-headed grosbeaks. *Wilson Bull.* **97,** 47–56.

Riters, L. V., and Ball, G. F. (1999). Lesions to the medial preoptic area affect singing in the male European starling (*Sturnus vulgaris*). *Horm. Behav.* **36,** 276–286.

Robisson, P. (1992). Vocalizations in *Aptenodytes* penguins: Application of the two-voice theory. *Auk* **109,** 654–658.

Rowe, L., and Houle, D. (1996). The lek paradox and the capture of genetic variance by condition dependent traits. *Proc. R. Soc. Lond. B Biol. Sci.* **263,** 1215–1421.

Rybak, F., and Gahr, M. (2004). Modulation by steroid hormones of a "sexy" acoustic signal in an Oscine species, the Common Canary *Serinus canaria*. *An. Acad. Bras. Cienc.* **76,** 365–367.

Sakaguchi, H. (1996). Sex differences in the developmental changes of GABergic neurons in zebra finch song control nuclei. *Exp. Brain Res.* **108,** 62–68.

Sakaguchi, H., Li, R., and Taniguchi, I. (2000). Sex differences in the ventral paleostriatum of the zebra finch: Origin of the cholinergic innervation of the song control nuclei. *Neuroreport* **11,** 2727–2731.

Saldanha, C. J., Schultz, J. D., London, S. E., and Schlinger, B. A. (2000a). Telencephalic aromatase but not a song circuit in a suboscine passerine, the golden collared manakin (*Manacus vitellinus*). *Brain Behav. Evol.* **56,** 29–37.

Saldanha, C. J., Tuerek, M. J., Kim, Y.-H., Fernandes, A. O., Arnold, A. P., and Schlinger, B. A. (2000b). Distribution and regulation of telencephalic aromatase expression in the zebra finch revealed with a specific antibody. *J. Comp. Neurol.* **423,** 619–630.

Saino, N., and Moller, A. P. (1995). Testosterone correlates of mate guarding, singing and aggressive behavior in male barn swallows, *Hirundo rustica. Anim. Behav.* **49,** 465–472.

Scharff, C., and Nottebohm, F. (1991). A comparative study of the behavioral deficits following lesions of various parts of the zebra finch song system: Implications for vocal learning. *J. Neurosci.* **11,** 2896–2913.

Scharff, C., Kirn, J. R., Grossman, M., Macklis, J. D., and Nottebohm, F. (2000). Targeted neuronal death affects neuronal replacement and vocal behavior in adult songbirds. *Neuron* **25,** 481–492.

Schlinger, B. A., and Arnold, A. P. (1991). Brain is the major site of estrogen synthesis in a male songbird. *Proc. Natl. Acad. Sci. USA* **88,** 2191–2194.

Schlinger, B. A., and Arnold, A. P. (1992). Plasma sex steroids and tissue aromatization in hatchling zebra finches: Implications for the sexual differentiation of singing behavior. *Endocrinology* **130,** 289–299.

Scholz, B., Kultima, K., Mattsson, A., Axelson, J., Brunström, B., Halldin, K., Stigson, M., and Dencker, L. (2006). Sex-dependent gene expression in early brain development of chicken embryos. *BMC Neuroscience* **7,** 12 doi:10. 1186/1471-2202-7-12.

Searcy, W. A., and Yasukawa, K. (1996). Song and female choice. *In* "Ecology and Evolution of Acoustic Communication in Birds" (D. E. Kroodsma and E. H. Miller, eds.), pp. 454–473. Cornell University Press, New York.

Seibt, U., Wickler, W., Kleindienst, H.-R., and Sonnenschein, E. (2002). Structure, geography and origin of dialects in the traditive song of the forest weaver Ploceus bicolor sclateri in Natal, *S. Africa. Behaviour* **139,** 1237–1265.

Shoemaker, H. H. (1939). Effect of testosterone prorionate on behavior of the female canary. *Proc. Soc. Exp. Biol. Med.* **41,** 299–302.

Simpson, H. B., and Vicario, D. S. (1990). Brain pathways for learned and unlearned vocalizations differ in zebra finches. *J. Neurosci.* **10,** 1541–1556.

Simpson, H. B., and Vicario, D. S. (1991). Early estrogen treatment of female zebra finches masculinizes the brain pathway for learned vocalizations. *J. Neurobiol.* **22,** 777–793.

Soma, M., Takahashi, H., Hasegawa, T., and Okanoya, K. (2006). Trade-offs and correlations among multiple song features in the Bengalese finch. *Ornithol. Sci.* **5,** 77–84.

Striedter, G. F. (1994). The vocal control pathways in budgerigars differ from those in songbirds. *J. Comp. Neurol.* **343,** 35–36.

Takahashi, M. M., and Noumura, T. (1987). Sexually dimorphic and laterally asymmetric development of the embryonic duck syrinx: Effect of estrogen on *in vitro* cell proliferation and chrondrogenesis. *Dev. Biol.* **12,** 417–422.

Taoka, M., Sata, T., Kamada, T., and Okumara, H. (1989). Sexual dimorphism of chatter-calls and vocal recognition in Leach's Storm Petrels (*Oeanodroma leucorhoa*). *Auk* **106,** 498–501.

Terkel, A. S., Moore, C. L., and Beer, C. G. (1976). The Effects of Testosterone and Estrogen on the rate of long-calling vocalization in juvenile laughing gulls, *Larus atricilla*. *Horm. Behav.* **7,** 49–57.

Tobari, Y., Nakamura, K. Z., and Okanoya, K. (2005). Sex differences in the telencephalic song control circuitry in Bengalese Finches (*Lonchura striata var. domestica*). *Zool. Sci.* **22,** 1089–1094.

Todt, D., Hultsch, H., and Duvall, E. P., IInd (1981). Behavioural significance and social function of vocal and non-vocal displays in the monogamous duet-singer Cossypha heuglini H. *Zool. Beitr.* **27,** 426–428.

Tramontin, A. D., and Brenowitz, E. A. (2000). Seasonal plasticity in the adult brain. *Trends Neurosci.* **23,** 251–258.

Tramontin, A. D., Wingfield, J. C., and Brenowitz, E. A. (1999). Contributions of social cues and photoperiod to seasonal plasticity in the adult avian song control system. *J. Neurosci.* **19,** 476–483.

Tramontin, A. D., Wingfield, J. C., and Brenowitz, E. A. (2003). Androgens and estrogens induce seasonal-like growth of song nuclei in the adult songbird brain. *J. Neurobiol.* **57,** 130–140.

Vates, G. E., Vicario, D. S., and Nottebohm, F. (1997). Reafferent thalamo-cortical loops in the song system of ocince songbirds. *J. Comp. Neurol.* **380,** 275–290.

Vicario, D. S., Naqvi, N. H., and Raksin, J. N. (2001). Behavioral discrimination of sexually dimorphic calls by male zebra finches requires an intact vocal motor pathway. *J. Neurobiol.* **47,** 109–120.

Voigt, C., Leitner, S., and Gahr, M. (2006). Repertoire and structure of duet and solo songs in cooperatively breeding white-browed sparrow weavers. *Behaviour* **143,** 159–182.

Vu, E., Mazurek, E., and Kuo, K. (1994). Identification of a forebrain motor programming network for the learned song of zebra finches. *J. Neurosci.* **14,** 6924–6934.

Wade, J., Tang, Y. P., Peabody, C., and Templeman, R. J. (2005). Enhanced gene expression in the forebrain of hatchling and juvenile male zebra finches. *J. Neurobiol.* **64,** 224–238.

Ward, B. C., Nordeen, E. J., and Nordeen, K. W. (1998). Individual variation in neuron number predicts differences in the propensity for avian vocal imitation. *Proc. Natl. Acad. Sci. USA* **95,** 1277–1282.

Weatherhead, P. J., Metz, K. J., Bennett, G. F., and Irwin, R. E. (1993). Parasite faunas, testosterone and secondary sexual traits in male red-winged blackbirds. *Behav. Ecol. Sociobiol.* **33,** 13–23.

Weichel, K., Heid, P., and Güttinger, H.-R. (1989). 17β-Estradiolbenzoate-dependent song induction in juvenile female canaries (*Serinus canaria*). *Ethology* **80,** 55–70.

Wennstrom, K. L., Reeves, B. J., and Brenowitz, E. A. (2001). Testosterone treatment increases the metabolic capacity of adult avian song control nuclei. *J. Neurobiol.* **48,** 256–264.

Wickler, W., and Seibt, U. (1980). Vocal dueting and the pair bond. II. Unisono dueting of the African forest weaver, Symplectes bicolor. *Z. Tierpsychol.* **52,** 217–226.

Wickler, W., and Seibt, U. (1988). Gender dialects: A new phenomenon from birdsong epigenesis. *Naturwissenschaften* **75,** 51–52.

Wickler, W., and Sonnenschein, E. (1989). Ontogeny of song in captive duet-singing slate-coloured boubous (*Lanarius funebris*). *Behaviour* **111,** 220–233.

Wild, J. M. (1997). Neural pathways for the control of birdsong production. *J. Neurobiol.* **33,** 653–670.

Wild, J. M., Li, D., and Eagleton, C. (1997). Projections of the dorsomedial nucleus of the intercollicular complex (DM) in relation to respiratory-vocal nuclei in the brainstem of the pigeon (*Columba livia*) and zebra finch (*Taeniopygia guttata*). *J. Comp. Neurol.* **377,** 392–423.

Williams, L., and Mac Roberts, M. K. (1978). Song variation in dark-eyed juncos in Nova Scotia. *Condor* **80,** 237–240.

Wilson, J. A., and Glick, B. (1970). Ontogeny of mating behavior in the chicken. *Am. J. Physiol.* **218,** 951–955.

Yamaguchi, A. (2001). Sex differences in vocal learning birds. *Nature* **411,** 257–258.

Yu, A. C., Dave, A. S., and Margoliash, D. (1996). Temporal hierarchical control of singing in birds. *Science* **273,** 1871–1875.

4

Gene Regulation as a Modulator of Social Preference in Voles

Elizabeth A. D. Hammock
Department of Pharmacology, Vanderbilt Kennedy Center for Research on Human Development, Vanderbilt University, Nashville, Tennessee 37232

Advances in Genetics, Vol. 59
Copyright 2007, Elsevier Inc. All rights reserved.
0065-2660/07 $35.00
DOI: 10.1016/S0065-2660(07)59004-8

ABSTRACT

Most mammalian species are nonmonogamous: the female alone cares for the young and males and females do not share nest sites. Within the genus *Microtus*, there exists ample diversity in social structure for neuroethological and neurobiological investigation. Prairie voles (*Microtus ochrogaster*) are socially monogamous: both the males and females contribute to care of the young within a shared nest site as a breeding pair through multiple breeding seasons. Closely related species such as the montane (*M. montanus*) and meadow (*M. pennsylvanicus*) voles do not typically show these behaviors. Over a decade of research has demonstrated that species differences in neuropeptide systems play significant roles in the behavioral divergence of these species. In particular, species differences in regional gene expression patterns of neuropeptide receptors in the brain mediate some of the behavioral traits associated with the divergence in social structure. Differences in gene expression patterns of a key gene in mediating social behavior, the arginine vasopressin 1a receptor (*avpr1a*), appear to be due to species divergence in a repeat locus in the 5′ regulatory region of *avpr1a*. This highly repetitive locus is prone to expansion and contraction over relatively short evolutionary timescales and may give rise to the rapid evolution of sociobehavioral traits.

© 2007, Elsevier Inc.

"The key to the sociobiology of mammals is milk."
—E. O. Wilson, 1975

I. INTRODUCTION

A. Monogamy in mammals

The conserved "nuclear unit" (Wilson, 1975) of the mammalian social group is the maternal–infant interaction, which is a defining mammalian trait. Because mammalian neonates require mother's milk for survival, females (and neonates) face substantial selection pressure to maintain evolved neurobiological mechanisms that support maternal–infant interaction. Even though the mammalian nuclear unit is highly conserved, the quantity of maternal care and the quality of the interaction are not conserved. The presence of social bonding outside of the maternal–infant interaction is also not conserved. Additionally, females are the rate-limiting resource in sexual selection in mammals, so polygamy is standard. Male mammals face very little selection pressure to evolve or maintain neurobiological mechanisms of social bonding. This is reflected in the rarity of monogamous social structures among mammals, which is estimated at 3–5% (Kleiman, 1977). Those rare cases of monogamous social structure among mammals appear to reflect harsher environmental conditions where pair bonding and paternal care increase reproductive fitness (Emlen and Oring, 1977). Therefore, for a species to be monogamous, something in the neurobiology of social behavior has to change dramatically. Monogamy, even though rare, has emerged multiple times across diverse mammalian taxa. The repeated appearance of monogamous social structure in distantly related taxa and the diversity of social structure among closely related species suggest that these dramatic changes in underlying neurobiology must happen rapidly, independently, and perhaps reversibly.

B. Historical perspective

From an ecological perspective, lactation certainly is a driving force of mammalian evolution, and the neuropeptide oxytocin is a key player in the biology of lactation. Wilson's (1975) insightful thoughts on the role of milk in the sociobiology of mammals preceded the experimental evidence for the role of oxytocin in the neurobiology of mammalian social behavior. In addition to its role in lactation, oxytocin is co-opted in the brain and influences the neurobiology of maternal–infant interaction. Oxytocin was discovered in 1909 by Sir Henry Dale (1909) and first synthesized in 1953 by Vincent du Vigneaud (du Vigneaud *et al.*, 1954), for which he received the 1955 Nobel Prize in chemistry. Oxytocin is the product of a gene duplication event that also produced the paralogous gene encoding the neuropeptide arginine vasopressin (also known as antidiuretic hormone). Both oxytocin and vasopressin are 9-amino acid peptides that are produced and released within the brain as well as into general circulation via the posterior pituitary. The genes encoding oxytocin and vasopressin face each other on the same chromosome in the mammalian genome (Burbach *et al.*, 2001) and share common gene regulatory elements in the noncoding region between the two coding loci (Fields *et al.*, 2003), even though they are not expressed in the same neurons (Mohr *et al.*, 1988). The two genes derived from a phylogenetically ancient family of peptides which also includes nonmammalian vasotocin, mesotocin, isotocin, conopressin, and even annetocin from annelida. Homologues of oxytocin and vasopressin are evident in extant invertebrates indicating that the precursor to the oxytocin/vasopressin superfamily was present at least 500 million years ago (Satake *et al.*, 1999; Van Kesteren *et al.*, 1995). Peripherally, these peptides serve to regulate physiological homeostasis, especially water and salt balance. With an expanding nervous system across evolutionary timescales, these peptide systems have been put to use in the central nervous system to regulate behavioral aspects of homeostatic control.

In this role, these peptide systems act within the brain to modulate approach/avoid responses to various stimuli, including the approach/avoid dichotomy in social behavior.

II. OXYTOCIN AND PAIR BONDING IN VOLES

A. Oxytocin

Oxytocin and vasopressin, acting within the brain, appear to play key roles in mammalian sociobehavioral strategies. In the early 1990s, drawing on decades of work on the role of oxytocin in maternal–infant interactions (Kendrick *et al.*, 1987; Pedersen and Prange, 1979), Carter *et al.* (1992) postulated that oxytocin may play a role in the neurobiology of adult bonding. This hypothesis was tested in the prairie vole (*Microtus ochrogaster*), which is a socially monogamous rodent species (Getz *et al.*, 1981; Thomas and Birney, 1979). Prior sexual or cohousing experience increased side-by-side resting affiliative behavior in adult prairie voles (Carter *et al.*, 1988). Furthermore, female prairie voles developed a social preference for a male cage-mate after the pair has been housed together for 24 h. The formation of such a "partner preference" was rapidly facilitated if the pair mated during the cohousing period (Williams *et al.*, 1992b). Apparently, these behavioral manipulations (i.e., prolonged cohabitation and/or mating) can be mimicked with a single injection of oxytocin into the brain prior to pairing the male and female. For example, the time required for partner preference formation was truncated with central administration of oxytocin just prior to the cohabitation period (Williams *et al.*, 1992a). Furthermore, this effect was established without the need for estrogen priming, which is required to observe induction of maternal behavior by oxytocin in rats (Pedersen and Prange, 1979) and sheep (Kendrick *et al.*, 1987). Furthermore, a selective oxytocin receptor antagonist eliminated the effect of exogenous oxytocin on partner preference formation (Williams *et al.*, 1994). Therefore, the direct manipulation of the brain oxytocin system in female prairie voles modulates social bonding behavior.

B. Comparative analysis

It is clear that oxytocin plays a role in maternal–infant interactions, and that it also plays a role in adult social interactions in a manner independent of estrogen priming, at least in monogamous prairie voles. Even though the presence of a mother–infant interaction is a conserved feature of mammals, female bonding with other conspecifics is not universal. Diversity in social behavior among closely related species can be utilized to investigate the neural and genetic mechanisms that contribute to social bonding. Specifically, microtine rodents (voles), as a genus, display high levels of diversity in social behavior (Getz *et al.*, 1981; Jannett, 1982; Shapiro and Dewsbury, 1990; Thomas and Birney, 1979). For example, while the prairie vole is socially monogamous with males contributing to care of the young, the closely related montane vole (*M. montanus*) lives in isolation (Jannett, 1982). In field studies, prairie vole males and females share nest sites and can be captured together repeatedly in the same field traps. In contrast, montane voles do not share nest sites and the female alone cares for the offspring. In the laboratory, other behavioral traits also reflect divergent social structure between these two species. In the partner preference test as described above, both male and female prairie voles prefer to spend time with the familiar partner. In contrast, the montane vole shows no preference for either animal, and can even spend most of the test time in isolation. In a quest for an explanation of the species differences in partner preference behavior, Insel and Shapiro (1992) mapped the oxytocin receptor (OTR) in four species of voles. They mapped OTR in two monogamous (*M. ochrogaster* and *M. pinetorum*) and two nonmonogamous (*M. pennsylvanicus* and *M. montanus*) species and noted that patterns of receptor expression were highly variable across species. Further, during the postpartum period in the montane vole (*M. montanus*), OTR levels increased to match the constitutively high prairie vole levels in the lateral amygdala, suggesting that dynamic regulation of OTR may facilitate maternal–infant bonding in the nonmonogamous species. Also, environmental variables, such as day length, can influence the probability of monogamy-related behaviors in the nonmonogamous meadow vole. Short-day length was associated with increased oxytocin receptor density in the extended amygdala and increased probability of partner preference formation (Parker *et al.*, 2001). These data indicate that OTR distribution patterns, whether innate or environmentally regulated, are associated with social bonding behavior.

III. VASOPRESSIN AND PAIR BONDING IN VOLES

A. Vasopressin

While the effects of oxytocin on social behavior were being studied in females, the related neuropeptide, vasopressin, was found to modulate pair bonding in male prairie voles. Prior to the studies in voles, there were early indications that vasopressin might play a role in the brain as a modulator of social behavior. For example, central injection of vasopressin into the medial preoptic area resulted in the onset of stereotypic flank-marking behavior in male and female golden hamsters (Ferris *et al.*, 1984). Further, vasopressin modulates social recognition behavior in rats (Engelmann and Landgraf, 1994; Popik and Van Ree, 1992). In male prairie voles, central administration of vasopressin can induce aggressive behavior and partner preference behavior even with short-duration cohabitation without mating (Winslow *et al.*, 1993). These effects of vasopressin appear to act through the vasopressin 1a receptor (V1aR), as central injection of V1aR antagonist, but not an oxytocin receptor antagonist, abolished both mating-induced aggression toward novel intruders and formation of partner preferences after 24 h of cohabitation with mating (Winslow *et al.*, 1993), which are hallmark behaviors of monogamous social structure. Apparently, the antagonist had to be present during the initial mating/cohabitation bout, since the already high levels of territorial aggression of established breeder males were unaffected by antagonist treatment. In contrast to the effects of exogenous administration of vasopressin on social behavior in prairie voles, central administration of vasopressin into the lateral ventricles of nonmonogamous montane voles had no effect on affiliative behavior (Young *et al.*, 1999). Differential responsivity to vasopressin is likely due to species differences in the brain.

B. Comparative analysis

This species difference in response to vasopressin appears to be explained in part by species differences in receptor distribution patterns for V1aR. As with the oxytocin receptor, V1aR distribution patterns are highly variable across species. Further, V1aR distribution patterns seem to correlate with species differences in social structure. For example, comparisons of distribution patterns among vole species revealed high levels of V1aR in the ventral pallidum in the prairie vole, but very little binding in this brain region in the nonmonogamous montane vole (Insel *et al.*, 1994). The ventral pallidum is a key component of ventral forebrain reward pathways. In similar comparisons in other mammalian taxa, the same relationship exists between V1aR levels in the ventral pallidum and monogamous social structure (Young, 1999). For example, the monogamous mouse, *Peromyscus californicus*, has much higher levels of V1aR in the ventral pallidum compared to its nonmonogamous sister species, *P. leucopus*. Similarly, the monogamous marmoset also has more V1aR in the ventral pallidum than the nonmonogamous rhesus monkey. This comparative analysis points to V1aR in the ventral pallidum as a potential neurobiological correlate of social structure.

C. V1aR in the ventral pallidum

Genetic and pharmacological tools have indeed implicated V1aR in the ventral pallidum as a modulator of social bonding behavior. Increasing the numbers of receptors in the ventral pallidum of the male prairie vole by viral vector gene transfer facilitated partner preference formation in the male prairie vole (Pitkow *et al.*, 2001). A site-specific antagonist for the V1aR administered to the ventral pallidum prior to mating and cohabiting with a female eliminated partner preference formation in male prairie voles (Lim and Young, 2004). These functional studies demonstrate a role for V1aR in the ventral pallidum in the modulation of social behavior in a monogamous species, and suggest that V1aR in the ventral pallidum is an important difference between monogamous and nonmonogamous voles.

Using viral vector gene transfer technology, it is possible to address whether V1aR levels in the ventral pallidum significantly contribute to social behavior. An adeno-associated viral vector expressing the prairie vole V1aR was placed in the ventral pallidum of nonmonogamous meadow voles (*M. pennsylvanicus*) (Lim *et al.*, 2004). The viral vector increased levels of V1aR in the ventral pallidum and induced partner preference formation in male meadow voles. Parental care, another monogamy-related social behavior, was not influenced by this treatment, indicating that V1aR in ventral pallidum modulates the partner preference aspect of monogamy, but not other important behaviors.

D. Ventral forebrain reward pathways

Because the ventral pallidum is a key component of ventral forebrain reward circuitry, and because dopamine is a key neurotransmitter of this circuitry, the role of dopamine in partner preference formation was assessed. Pharmacological manipulation of dopamine in this circuitry modulates partner preference in monogamous prairie voles (Aragona *et al.*, 2003; Gingrich *et al.*, 2000; Wang *et al.*, 1999). Further, in the V1aR viral vector-treated meadow vole, pharmacological blockade of D2 dopamine receptors with eticlopride eliminated the V1aR vector-induced partner preferences. Therefore, similar to the effect of endogenous V1aR in monogamous prairie voles, these exogenous V1aRs in the meadow vole appear to modulate highly conserved ventral forebrain reward circuits (Lim *et al.*, 2004).

E. Note of caution

While these data implicate V1aR in the ventral pallidum in the regulation of partner preference formation, it is important to keep in mind that the V1aR is present elsewhere in the brain. Since hypothalamic neuropeptides like vasopressin diffuse through neural tissue (Ludwig and Leng, 2006), and V1aR is expressed in multiple brain regions, then it is likely that other brain regions also participate in modulating these important and complex suites of behaviors. For example, vasopressin injection into the lateral septum of male prairie voles facilitated partner preference behavior (Liu *et al.*, 2001) and increased paternal behavior (Wang *et al.*, 1994), clearly emphasizing the role of V1aR in other brain regions. Furthermore, activation of V1aR occurs in a context of many other proteins and large circuits. Therefore, V1aR activation in the ventral pallidum is integrated within a larger circuitry. This activity modifies a circuit—it is not a "behavioral switch."

IV. GENE REGULATION IN MALE SPECIES-TYPICAL BEHAVIOR: EVOLUTIONARY TUNING KNOBS

A. Comparative genetics

Caveats aside, receptor distribution patterns do appear to play a role in the modulation of species-typical social behavior. This raises the obvious question of what regulates receptor distribution patterns. Gene sequence comparisons between montane and prairie voles of the gene encoding V1aR (*avpr1a*) have implicated species differences in gene regulatory mechanisms.

The coding regions between montane vole and prairie vole *avpr1a* share 99% identity. There are four amino acid changes out of 420 between these two sister species at this locus, but these differences do not appear to affect quantitative receptor–ligand interaction or affect qualitative second messenger coupling (Insel *et al.*, 1994). Further, there is a second copy of *avpr1a* in the prairie vole which contains a truncating mutation, the effects of which have not been explored (Young *et al.*, 1999). Considering that the V1aR of each species binds vasopressin equally, but that the two species show differences in the *pattern* of receptors in the brain, perhaps differences in gene regulatory mechanisms at the full-length *avpr1a* locus may explain these findings. Sequence comparisons in the 5′ region between these two species also show high levels of identity, with the striking exception of a highly expanded repetitive region in the 5′ regulatory region (~500 base pairs upstream of the transcription start site) in the prairie vole that is minimal in the montane vole (Young *et al.*, 1999). Furthermore, a similar gene:behavior correlation exists in two other closely related North American vole species. The highly expanded microsatellite is present in the monogamous pine vole (*M. pinetorum*) and is very small in the nonmonogamous meadow vole (*M. pennsylvanicus*) (Young *et al.*, 1999).

B. Transgenic mouse

These data suggest that species differences in the 5′ regulatory region of *avpr1a* may play a role in receptor distribution patterns, which are correlated with species differences in behavior. To test this, the prairie vole *avpr1a* locus was inserted into the mouse genome. The inserted transgene included 2.2 kb of the 5′ flanking region, which fully encompassed the prairie vole-specific microsatellite, the entire coding region with the 2.5-kb intron and 2.4 kb of 3′ flanking sequence. These transgenic mice were assessed for their V1aR distribution patterns as well as social behavior (Young *et al.*, 1999). Compared to wild-type litter mates, the transgenic mice displayed a remarkably different receptor distribution pattern that was arguably more similar to a prairie vole than mouse. Further, on injection with vasopressin, the transgenic mice displayed increased olfactory investigation and social grooming of a tethered novel female. These results clearly demonstrate that the prairie vole *avpr1a* locus modifies gene expression and alters affiliative behavior. However, this transgene consisted of over 7 kb of prairie vole DNA, so it is unclear which part of the locus was most responsible for the effect.

C. Functional microsatellite

Luciferase assays in cell culture are routinely performed to assist in delineating the mechanisms of regulation of gene expression. This approach was utilized to address which part(s) of the prairie vole 5′ regulatory region might be most responsible for modifying gene expression (Hammock and Young, 2004). Initial experiments using serial "nested" deletions through 3.5 kb of prairie vole *avpr1a* 5′ region indicated that the microsatellite region contained regulatory elements, but these elements only modified luciferase reporter activity in some cell lines. Further inquiry, utilizing targeted deletion of the microsatellite, revealed that the specific deletion of the microsatellite also modified gene expression in a cell type-dependent manner. Finally, exchanging the large prairie vole microsatellite for the small montane microsatellite, in the surrounding 3.5 kb of prairie vole *avpr1a* 5′ region, also resulted in altered gene expression. These results demonstrate that the repetitive microsatellite DNA in the 5′ region of the prairie vole *avpr1a* alters gene expression in a cell type-specific manner and that species differences at this locus are sufficient to drive gene expression. Therefore, it is plausible that species differences in the microsatellite could in fact modulate gene expression in the vole brain. These data, in combination with the pharmacological and comparative neuroethological data described above, implicate the microsatellite as a significant modulator of species differences in social behavior.

Microsatellites are a common feature of eukaryotic genomes (reviewed in Li *et al.*, 2002, 2004), including mammals. They are appreciated for their length polymorphism and high mutation rates (Lai and Sun, 2003). The repetitive nature of the microsatellite increases mutation through slipped-strand mispairing, which can be slowed by DNA repair mechanisms (reviewed in Tautz and Schlotterer, 1994). Further, microsatellite expansion and contraction mutation can be slowed by single-nucleotide polymorphisms (SNPs) that break up the repeats (Kruglyak *et al.*, 2000).

Polymorphic microsatellite DNA has been implicated in gene regulatory mechanisms in numerous genes such as PIG3, CD30, and the genes encoding tyrosine hydroxylase, and the dopamine transporter, to name just a few (Contente *et al.*, 2002; Croager *et al.*, 2000; Meloni *et al.*, 1998; Michelhaugh *et al.*, 2001; Nakamura *et al.*, 1998). Further, it has been hypothesized that since microsatellites are functional, highly polymorphic, and hypermutable, perhaps they contribute to the rapid evolution of phenotypic traits (Kashi *et al.*, 1997;

King, 1994). Therefore, microsatellites offer a mechanism of continuous morphological variation, a sort of *evolutionary tuning knob* (Kashi *et al.*, 1997; King, 1994).

D. Intraspecific variation

In his *Origin of Species*, Darwin (Darwin, 1859) observed that "...those characters which are mainly distinctive of each breed are in each eminently variable...." If microsatellites offer a mechanism of *evolution* of social behavior, then they will likely show intraspecific variation wherever there is large variation between species. This intriguing idea was explored in the context of the microsatellite in the prairie vole *avpr1a*. It was hypothesized that intraspecific variation in microsatellite length would be associated with intraspecific variation in V1aR distribution patterns and sociobehavioral traits.

As with most other behaviors, there is ample variation within the prairie vole species in social behavior (Getz and Hofman, 1986; Getz *et al.*, 1993; Roberts *et al.*, 1998). Furthermore, both captive (Hammock *et al.*, 2005; Insel *et al.*, 1994) and field-caught (Phelps and Young, 2003) prairie voles show individual variation in V1aR distribution patterns. Finally, the microsatellite, divergent between prairie and montane voles, exhibits significant intraspecific length polymorphism within the prairie vole species (Hammock *et al.*, 2005; Hammock and Young, 2002). There are multiple alleles, showing a roughly normal distribution of lengths within a range of 50 base pairs in the laboratory (Hammock and Young, 2002; Hammock *et al.*, 2005), and a range of ~100 base pairs in a field-caught sample of prairie voles (Phelps, Hammock, and Young, unpublished data). This intraspecific variation, while significant, does not approach the species divergence between prairie and montane voles.

Even though the variation in allele size among prairie voles is less than the observed length differences between prairie and montane voles, the intraspecific variation in microsatellite length modifies gene expression in cell culture. Two different alleles were tested in cell culture with dramatically different effects on gene expression. A longer than average "long" allele drove gene expression of a luciferase reporter vector 2.5 times more than a shorter than average "short" allele (Hammock and Young, 2005). Therefore, these data indicate, that like the species differences in microsatellite length, intraspecific variation in the microsatellite could also modulate gene expression—perhaps explaining some of the intraspecific variation in V1aR distribution patterns and sociobehavioral traits.

To further probe potential function of intraspecific variation in the microsatellite *in vivo*, breeding pairs of naïve prairie voles were mated based on microsatellite length, and their F1 offspring assessed for brain V1aR distribution patterns and social behavior (Fig. 4.1) (Table 4.1) (Hammock and Young, 2005).

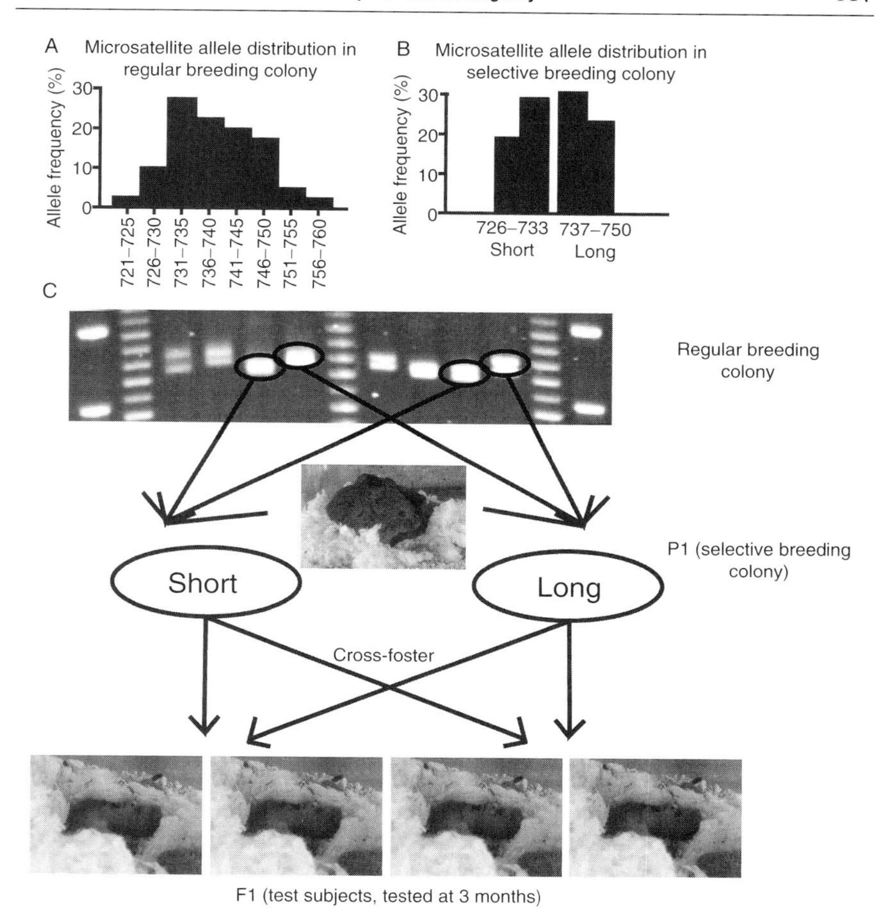

Figure 4.1. Population genotype data and vole microsatellite breeding scheme. (A) *avpr1a* microsatellites are highly polymorphic and relatively normally distributed in this captive breeding population of prairie voles. (B) Genetic selection for breeder pairs based on microsatellite length reduces the range of alleles for selective breeding. (C) Breeding scheme used in Hammock and Young (2005). Adult animals from the regular breeding colony were genotyped then paired based on microsatellite length. Males and females with longer or shorter than average microsatellite alleles were paired. Between 6 and 8 h after the pups were born, the litters were cross-fostered randomly such that short and long pairs received and reared offspring of either long or short breeders. Breeder parental behavior was recorded as was adult behavior of the F1 males. F1 female behavior was not assessed.

Table 4.1. Summary of Findings from Selective Genetic Breeding Strategy from Hammock and Young (2005)

Breeder behavior	**Long compared to short**
Pup mortality	↓
Time on nest	↔
Female nursing postures	↔
Female licking and grooming	↔
Male licking and grooming	↑
F1 behavior	**Long compared to short**
Social olfactory investigation	↑
Nonsocial olfactory investigation	↔
Juvenile investigation	↑, see text
Partner preference probability	↑
Elevated plus maze	↔, see text
Open field test	↔, see text
F1 V1aR density in selected brain regions	**Long compared to short**
Olfactory bulb	↑
Lateral septum	↑
Ventral pallidum	↔
Dorsal bed nucleus of the stria terminalis	↓
Hypothalamus	↓
Posterior cingulate cortex	↓
Basolateral amygdala	↓
Choroid plexus	↓

The arrows represent the direction of the difference of long allele compared to short allele animals.

Males and females homozygous for longer than average microsatellites were paired ("long") and males and females with microsatellites shorter than average were paired ("short"). The offspring from these matings were randomly cross-fostered between 6 and 8 h after birth to minimize postnatal environmental influence. Parental care of the cross-fostered neonates by the breeder pairs was recorded, as were adult social and anxiety-like behaviors of the male F1 offspring.

The first breeding attempt resulted in high pup mortality, which was significantly more prevalent in the "short" allele breeder group. Neither genotype group showed significant pup mortality on subsequent births. The reasons for the high rates of mortality with the first litter are unknown, although it is enticing to hypothesize a role for V1aR in the transition to parenting. The amount of time spent on the nest by either the male or the female, or the amount of time spent huddling were not different between the two genotype groups, nor were nursing rates and postures by the female. However, the male behavior of the breeder pairs was different between the two groups. Specifically, breeder males in the long genotype group displayed significantly higher rates of licking and grooming of the pups than the males of the short allele group. Interestingly, the rates of licking and grooming by the females were not different between the two groups. Further, the differences in licking and grooming rates between males of the long and short groups are of the same magnitude as the observed differences between "high" and "low" licking and grooming rat dams (Francis *et al.*, 1999). In rats, these differences in licking and grooming rates have been shown to have significant consequences on the developmental trajectory of the brain, especially regions supporting socioemotional development. For example, high rates of licking and grooming increase glucocorticoid receptor expression in the hippocampus, which permits tighter inhibitory feedback control on stress responses (Francis *et al.*, 1999; Liu *et al.*, 1997, 2000; Meaney, 2001). This suggests that the group differences in licking and grooming rates provided by the male may be biologically meaningful to the receiving offspring.

Prairie voles, like most rodents, are highly olfactory. In fact, removal of the olfactory bulbs reduce social behavior and eliminate partner preferences in prairie voles (Kirkpatrick *et al.*, 1994; Williams *et al.*, 1992c). Social odors like soiled bedding, in the absence of a live stimulus animal, can be used to probe social approach behavior in rodents. When the F1 males of the above breeding scheme were exposed to 5 min of the odor of an unfamiliar, unrelated female, the males in the long allele group were quicker to approach the odor, spent more time investigating the odor, and had a higher number of approaches toward the social odor than the short allele males. As a control, the same males were exposed to a banana-like odor, *n*-amyl acetate. There were no genotype differences in investigation of this nonsocial odor, suggesting specificity of the effect within the social domain.

To determine if the genotype differences in social approach actually translated to social interaction with a conspecific, a novel juvenile prairie vole was placed in the home cage of the F1 males. While there were no genotype differences in the total amount of social contact time or number of approaches toward the juvenile in the 10-min test, the long allele males were quicker than the short allele males to approach the social object (the juvenile). This difference in latency to approach a conspecific likely has consequences in ethologically relevant contexts. For example, the observed 1.5-min difference in mean latency between the two genotypes would determine whether a vole might interact with a passing conspecific. Furthermore, these behavioral differences, if present early enough, might alter the developmental maturation of experience-dependent neural circuits underlying social behavior.

Because partner preference behavior is dramatically different between montane and prairie voles, this behavior was assessed in the F1 males of the prairie vole breeding scheme. Each male was paired with an unrelated estrogen-primed female. Mating was allowed, but in order to increase the variability in the behavior, the overnight cohabitation period was truncated to 18 h (from 24 h). Under these suboptimal conditions, only the long allele group displayed significant partner preference behavior.

In sum, the long allele males were more likely to engage in social interaction and more likely to form partner preferences in the given conditions. These effects could be due to specific effects on social behavior or more general genotype differences in emotional behavior. This is a considerable confound given that vasopressin has a well-known role in modulation of stress reactivity and anxiety-like behavior (Bielsky *et al.*, 2004; Koolhaas *et al.*, 1998; Landgraf *et al.*, 1995; Murgatroyd *et al.*, 2004; Wigger *et al.*, 2004). This issue was addressed by assessing the anxiety-like phenotype of the F1 males in the elevated plus maze and the open field test. While there were genotype differences in some measures of the test, such as auto grooming, there were no differences in the classic measures of anxiety-like behaviors in these tasks, namely time in the open arms and unprotected area of the open field. The absence of genotype differences in these tasks suggests that the observed genotype differences in the social tasks are not due to underlying differences in trait anxiety.

Finally, the F1 males were assessed for their V1aR distribution patterns. There were significant genotype differences in V1aR densities with long allele males displaying more V1aR in the olfactory bulb and lateral septum, and less in several other brain regions, including the bed nucleus of the stria terminalis, posterior cingulate cortex, and the hypothalamus. In addition to the differences in receptor distribution, there was ample variation in many brain regions that was not associated with genotype. This suggests other mechanisms, in addition to microsatellite length, in the regulation of V1aR patterns in prairie voles.

E. Knobs and switches

In summary (Fig. 4.2), the evidence derived from comparative, genetic, pharmacological, behavioral, and neuroanatomical investigation in two North American vole species indicate that the microsatellite sequence, which can expand and contract over short evolutionary timescales, can modulate the expression patterns of the V1aR, which in turn modulate the influence of the neuropeptide vasopressin on social behavior. If vasopressin is released within the brain in a given context (e.g., during mating or prolonged exposure to a conspecific), and diffuses to multiple receptor sites within different neural circuits, then individual differences in V1aR patterns will result in altered modulation of circuits by vasopressin. This differential circuit activation will likely influence behavior.

This general gene-to-brain-to-behavior model can guide future studies and serves as a powerful explanatory mechanism across several levels of analysis, including genetics, neural circuits, behavior, and evolution. It is possible that

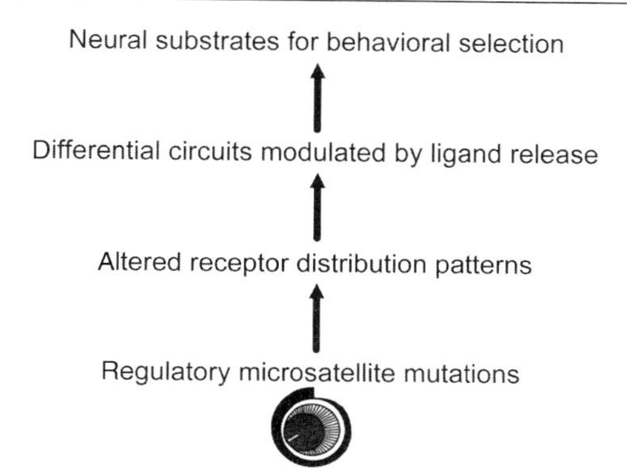

Figure 4.2. General model of microsatellite polymorphism modulating sociobehavioral traits. Like an evolutionary tuning knob (Kashi *et al.* 1997; King, 1994), expansion and contraction of regulatory microsatellites can "dial" variation in receptor distribution patterns. A diffusible ligand like vasopressin will no doubt have different effects on behavior depending on where its receptor is located. This differential response to endogenous ligand will alter neural substrates, thereby producing quantitatively or qualitatively different behaviors, on which natural selection may act.

there are other neuropeptide receptor genes where such a mechanism currently exists. Variation in other gene loci is highly probable, as there are many proteins in the elaborate neural circuitry supporting complex social behavior. It is unlikely that microsatellite variation in just a single gene, like the *avpr1a*, acts as a genetic switch, turning monogamy on and off as environmental niches demand. Indeed, phylogenetic evidence from many vole species indicate that the mere *presence* or *absence* of a microsatellite in the *avpr1a* cannot, in fact, explain the variation in social structure in vole species outside of North America (Fink *et al.*, 2006). However, it is important to clarify that more subtle variation in the microsatellite length and/or sequence, in species that have a microsatellite, can modulate V1aR expression in a "tuning knob" fashion. The presence of a microsatellite confers the potential for length mutation, which can happen rapidly, repeatedly (i.e., independently), and reversibly. This kind of hypermutable locus, whether in *avpr1a* or other modulators of the social brain, has the potential to contribute to the rapid evolution of sociobehavioral traits.

V. SEXUAL DIMORPHISM

This chapter has presented a view of the role of oxytocin and vasopressin systems in female and male species-typical social behavior, respectively. This sexually dimorphic presentation is quite simplistic, and it is therefore important to point out that oxytocin systems can modulate male sociosexual behaviors and, as well, vasopressin can modulate female sociosexual behaviors (Cho *et al.*, 1999). Further, as mentioned previously, the behavioral influence of the neuropeptides oxytocin and vasopressin is not limited to sociosexual behavior, as they each modulate other behaviors, including learning and memory and stress and anxiety.

While there is significant peptide interaction in both sexes and cross talk among peptide and receptor families, there is a significant sexual dimorphism that appears to be relatively conserved in mammals, including prairie voles, and other vertebrates (reviewed in De Vries and Panzica, 2006). In males, higher vasopressin levels (both cell number and projections) in the medial amygdala and bed nucleus of the stria terminalis have repeatedly been observed. This sexual dimorphism in "extrahypothalamic" vasopressin is regulated by gonadal steroids both in the developing and adult brain.

Gonadal steroids not only regulate extrahypothalamic vasopressin content, but also other components of vasopressin and oxytocin systems. In adult mice, estrogen receptors (α and β), oxytocin, and oxytocin receptors all participate in the regulation of social recognition behavior (Choleris *et al.*, 2003). In rats, developmental exposure to steroids regulates oxytocin receptor levels (Uhl-Bronner *et al.*, 2005) and conversely, developmental exposure to oxytocin regulates estrogen receptor α in female, but not male prairie voles (Yamamoto *et al.*, 2006). Such reciprocal molecular interactions throughout development could potentiate sexually dimorphic developmental trajectories. Furthermore, the sex-specific roles of oxytocin and vasopressin systems are likely to be highly species-specific. This species-specificity will be determined not only by species-specific variation in the magnitude of sexual dimorphism in hormonal environments, but also by the species-specific *capacity* for gonadal hormones to influence oxytocin and vasopressin system components.

Future progress in understanding the neurobiology of sexual dimorphism of oxytocin and vasopressin systems will depend on two key factors. First, progress requires an ongoing appreciation of the evolutionary lability of these systems and a utilization of comparative approaches, as described with the vole studies. Second, mechanistic analyses of developing circuitry should continue to be incorporated into "before" and "after" developmental snapshots, which have already provided insight into the effects of gonadal steroids on sexual dimorphism in oxytocin and vasopressin systems.

VI. CODA

Finally, the exploration of the neurobiology of social bonding began with an extension of an idea about the importance of milk in the neurobiology of mammals. Chronologically, these investigations first involved oxytocin and later vasopressin.

However, in the evolutionary history of mammals, both oxytocin and vasopressin were present from the start. The conserved brain distribution of these two neuropeptides, their structural similarities, and cross talk among receptors, albeit with varying affinities, evoke a notion of a highly interwoven evolutionary trajectory of these two systems in the sexual dimorphism of social behavior.

Definitions

Microsatellite: Highly repetitive DNA termed "microsatellite" because it is a small-scale version of a kind of DNA that forms "satellite" bands in cesium chloride centrifugation gradients. Microsatellite DNA is also known as simple sequence repeat (SSR), simple sequence length polymorphism (SSLP), and variable number of tandem repeats (VNTR).

Monogamy: Monogamy refers to a suite of behaviors that involves at least some of the following: (1) nest sharing by a male and a female, (2) shared efforts of taking care of the young, (3) separation-induced vocalizations in the neonate, and (4) mating-induced territorial aggression toward unfamiliar intruders. It is important to note that this definition of monogamy does not include sexual exclusivity. Many monogamous social structures contain ample evidence of extra-pair copulations, including prairie voles, which are a focus of this chapter.

Partner preference test: This test is used as an index of pair bond formation in laboratory rodents, especially voles. This 3-h test is preceded by a 6- to 24-h period of cohabitation (with or without mating). The testing apparatus involves three chambers interconnected with Plexiglas tubing. The test subject has free access to all three chambers, while the partner (from the cohabitation) and a novel stranger (of equal size, sexual receptivity, and so on) are each tethered in separate compartments. A partner preference is defined by how much time the test animal spends in side-by-side contact with the partner versus the stranger animal.

Acknowledgments

I would like to thank Larry J. Young, Ph.D. of Emory University and Pat Levitt, Ph.D. of Vanderbilt University for mentorship, and Drs. Kathie Eagleson, Daniel Campbell and, Barbara Thompson for helpful comments on this chapter. I would also like to acknowledge recent support from NIH F31MH67397 and NIH T32MH65215 and current support from NIH T32MH075883.

References

Aragona, B. J., Liu, Y., Curtis, J. T., Stephan, F. K., and Wang, Z. (2003). A critical role for nucleus accumbens dopamine in partner-preference formation in male prairie voles. *J. Neurosci.* **23**(8), 3483–3490.

Bielsky, I. F., Hu, S. B., Szegda, K. L., Westphal, H., and Young, L. J. (2004). Profound impairment in social recognition and reduction in anxiety-like behavior in vasopressin V1a receptor knockout mice. *Neuropsychopharmacology* **29**(3), 483–493.

Burbach, J. P., Luckman, S. M., Murphy, D., and Gainer, H. (2001). Gene regulation in the magnocellular hypothalamo-neurohypophysial system. *Physiol. Rev.* **81**(3), 1197–1267.

Carter, C. S., Williams, J. R., Witt, D. M., and Insel, T. R. (1992). Oxytocin and social bonding. *Ann. N. Y. Acad. Sci.* **652,** 204–211.

Carter, C. S., Witt, D. M., Thompson, E. G., and Carlstead, K. (1988). Effects of hormonal, sexual, and social history on mating and pair bonding in prairie voles. *Physiol. Behav.* **44**(6), 691–697.

Cho, M. M., DeVries, A. C., Williams, J. R., and Carter, C. S. (1999). The effects of oxytocin and vasopressin on partner preferences in male and female prairie voles (*Microtus ochrogaster*). *Behav. Neurosci.* **113**(5), 1071–1079.

Choleris, E., Gustafsson, J. A., Korach, K. S., Muglia, L. J., Pfaff, D. W., and Ogawa, S. (2003). An estrogen-dependent four-gene micronet regulating social recognition: A study with oxytocin and estrogen receptor-alpha and-beta knockout mice. *Proc. Natl. Acad. Sci. USA* **100**(10), 6192–6197.

Contente, A., Dittmer, A., Koch, M. C., Roth, J., and Dobbelstein, M. (2002). A polymorphic microsatellite that mediates induction of PIG3 by p53. *Nat. Genet.* **30**(3), 315–320.

Croager, E. J., Gout, A. M., and Abraham, L. J. (2000). Involvement of Sp1 and microsatellite repressor sequences in the transcriptional control of the human CD30 gene. *Am. J. Pathol.* **156**(5), 1723–1731.

Dale, H. H. (1909). The action of extracts of the pituitary body. *Biochem. J.* **4**(9), 427–447.

Darwin, C. R. (1859). "On the Origin of Species by Means of Natural Selection." J. Murray, London.

De Vries, G. J., and Panzica, G. C. (2006). Sexual differentiation of central vasopressin and vasotocin systems in vertebrates: Different mechanisms, similar endpoints. *Neuroscience* **138**(3), 947–955.

du Vigneaud, V., Ressler, C., Swan, J. M., Roberts, C. W., Katsoyannis, P. G., and Gordon, S. (1954). The synthesis of an octapeptide amide with the hormonal activity of oxytocin. *J. Am. Chem. Soc.* **75,** 4879–4880.

Emlen, S. T., and Oring, L. W. (1977). Ecology, sexual selection, and the evolution of mating systems. *Science* **197**(4300), 215–223.

Engelmann, M., and Landgraf, R. (1994). Microdialysis administration of vasopressin into the septum improves social recognition in Brattleboro rats. *Physiol. Behav.* **55**(1), 145–149.

Ferris, C. F., Albers, H. E., Wesolowski, S. M., Goldman, B. D., and Luman, S. E. (1984). Vasopressin injected into the hypothalamus triggers a stereotypic behavior in golden hamsters. *Science* **224**(4648), 521–523.

Fields, R. L., House, S. B., and Gainer, H. (2003). Regulatory domains in the intergenic region of the oxytocin and vasopressin genes that control their hypothalamus-specific expression *in vitro*. *J. Neurosci.* **23**(21), 7801–7809.

Fink, S., Excoffier, L., and Heckel, G. (2006). Mammalian monogamy is not controlled by a single gene. *Proc. Natl. Acad. Sci. USA* **103**(29), 10956–10960.

Francis, D., Diorio, J., Liu, D., and Meaney, M. J. (1999). Nongenomic transmission across generations of maternal behavior and stress responses in the rat. *Science* **286**(5442), 1155–1158.

Getz, L. L., and Hofman, J. E. (1986). Social organization in free-living prairie voles, *Microtus ochrogaster*. *Behav. Ecol. Sociobiol.* **18,** 275–282.

Getz, L. L., Carter, C. S., and Gavish, L. (1981). The mating system of the prairie vole, *Microtus ochrogaster*: Field and laboratory evidence for pair-bonding. *Behav. Ecol. Sociobiol.* **8,** 189–194.

Getz, L. L., McGuire, B., Pizzuto, T., Hofman, J. E., and Frase, B. (1993). Social organization of the prairie vole (*Microtus ochrogaster*). *J. Mammol.* **74**(1), 44–58.

Gingrich, B., Liu, Y., Cascio, C., Wang, Z., and Insel, T. R. (2000). Dopamine D2 receptors in the nucleus accumbens are important for social attachment in female prairie voles (*Microtus ochrogaster*). *Behav. Neurosci.* **114**(1), 173–183.

Hammock, E. A., and Young, L. J. (2002). Variation in the vasopressin V1a receptor promoter and expression: Implications for inter- and intraspecific variation in social behaviour. *Eur. J. Neurosci.* **16**(3), 399–402.

Hammock, E. A., and Young, L. J. (2004). Functional microsatellite polymorphism associated with divergent social structure in vole species. *Mol. Biol. Evol.* **21**(6), 1057–1063.

Hammock, E. A., and Young, L. J. (2005). Microsatellite instability generates diversity in brain and sociobehavioral traits. *Science* **308**(5728), 1630–1634.

Hammock, E. A., Lim, M. M., Nair, H. P., and Young, L. J. (2005). Association of vasopressin 1a receptor levels with a regulatory microsatellite and behavior. *Genes Brain Behav.* **4**(5), 289–301.

Insel, T. R., and Shapiro, L. E. (1992). Oxytocin receptor distribution reflects social organization in monogamous and polygamous voles. *Proc. Natl. Acad. Sci. USA* **89**(13), 5981–5985.

Insel, T. R., Wang, Z. X., and Ferris, C. F. (1994). Patterns of brain vasopressin receptor distribution associated with social organization in microtine rodents. *J. Neurosci.* **14**(9), 5381–5392.

Jannett, F. J. (1982). Nesting patterns of adult voles, *Microtus montanus*, in field populations. *J. Mammol.* **63**(3), 495–498.

Kashi, Y., King, D., and Soller, M. (1997). Simple sequence repeats as a source of quantitative genetic variation. *Trends Genet.* **13**(2), 74–78.

Kendrick, K. M., Keverne, E. B., and Baldwin, B. A. (1987). Intracerebroventricular oxytocin stimulates maternal behaviour in the sheep. *Neuroendocrinology* **46**(1), 56–61.

King, D. G. (1994). Triple repeat DNA as a highly mutable regulatory mechanism. *Science* **263**(5147), 595–596.

Kirkpatrick, B., Williams, J. R., Slotnick, B. M., and Carter, C. S. (1994). Olfactory bulbectomy decreases social behavior in male prairie voles (*M. ochrogaster*). *Physiol. Behav.* **55**(5), 885–889.

Kleiman, D. G. (1977). Monogamy in mammals. *Q. Rev. Biol.* **52**(1), 39–69.

Koolhaas, J. M., Everts, H., de Ruiter, A. J., de Boer, S. F., and Bohus, B. (1998). Coping with stress in rats and mice: Differential peptidergic modulation of the amygdala-lateral septum complex. *Prog. Brain Res.* **119,** 437–448.

Kruglyak, S., Durrett, R., Schug, M. D., and Aquadro, C. F. (2000). Distribution and abundance of microsatellites in the yeast genome can be explained by a balance between slippage events and point mutations. *Mol. Biol. Evol.* **17**(8), 1210–1219.

Lai, Y., and Sun, F. (2003). The relationship between microsatellite slippage mutation rate and the number of repeat units. *Mol. Biol. Evol.* **20**(12), 2123–2131.

Landgraf, R., Gerstberger, R., Montkowski, A., Probst, J. C., Wotjak, C. T., Holsboer, F., and Engelmann, M. (1995). V1 vasopressin receptor antisense oligodeoxynucleotide into septum reduces vasopressin binding, social discrimination abilities, and anxiety-related behavior in rats. *J. Neurosci.* **15**(6), 4250–4258.

Li, B., Xia, Q., Lu, C., Zhou, Z., and Xiang, Z. (2004). Analysis on frequency and density of microsatellites in coding sequences of several eukaryotic genomes. *Genomics Proteomics Bioinformatics* **2**(1), 24–31.

Li, Y. C., Korol, A. B., Fahima, T., Beiles, A., and Nevo, E. (2002). Microsatellites: Genomic distribution, putative functions and mutational mechanisms: A review. *Mol. Ecol.* **11**(12), 2453–2465.

Lim, M. M., and Young, L. J. (2004). Vasopressin-dependent neural circuits underlying pair bond formation in the monogamous prairie vole. *Neuroscience* **125**(1), 35–45.

Lim, M. M., Wang, Z., Olazabal, D. E., Ren, X., Terwilliger, E. F., and Young, L. J. (2004). Enhanced partner preference in a promiscuous species by manipulating the expression of a single gene. *Nature* **429**(6993), 754–757.

Liu, D., Diorio, J., Tannenbaum, B., Caldji, C., Francis, D., Freedman, A., Sharma, S., Pearson, D., Plotsky, P. M., and Meaney, M. J. (1997). Maternal care, hippocampal glucocorticoid receptors, and hypothalamic-pituitary-adrenal responses to stress. *Science* **277**(5332), 1659–1662.

Liu, D., Diorio, J., Day, J. C., Francis, D. D., and Meaney, M. J. (2000). Maternal care, hippocampal synaptogenesis and cognitive development in rats. *Nat. Neurosci.* **3**(8), 799–806.

Liu, Y., Curtis, J. T., and Wang, Z. (2001). Vasopressin in the lateral septum regulates pair bond formation in male prairie voles (*Microtus ochrogaster*). *Behav. Neurosci.* **115**(4), 910–919.

Ludwig, M., and Leng, G. (2006). Dendritic peptide release and peptide-dependent behaviours. *Nat. Rev. Neurosci.* **7**(2), 126–136.

Meaney, M. J. (2001). Maternal care, gene expression, and the transmission of individual differences in stress reactivity across generations. *Annu. Rev. Neurosci.* **24,** 1161–1192.

Meloni, R., Albanese, V., Ravassard, P., Treilhou, F., and Mallet, J. (1998). A tetranucleotide polymorphic microsatellite, located in the first intron of the tyrosine hydroxylase gene, acts as a transcription regulatory element *in vitro*. *Hum. Mol. Genet.* **7**(3), 423–428.

Michelhaugh, S. K., Fiskerstrand, C., Lovejoy, E., Bannon, M. J., and Quinn, J. P. (2001). The dopamine transporter gene (SLC6A3) variable number of tandem repeats domain enhances transcription in dopamine neurons. *J. Neurochem.* **79**(5), 1033–1038.

Mohr, E., Bahnsen, U., Kiessling, C., and Richter, D. (1988). Expression of the vasopressin and oxytocin genes in rats occurs in mutually exclusive sets of hypothalamic neurons. *FEBS Lett.* **242**(1), 144–148.

Murgatroyd, C., Wigger, A., Frank, E., Singewald, N., Bunck, M., Holsboer, F., Landgraf, R., and Spengler, D. (2004). Impaired repression at a vasopressin promoter polymorphism underlies overexpression of vasopressin in a rat model of trait anxiety. *J. Neurosci.* **24**(35), 7762–7770.

Nakamura, Y., Koyama, K., and Matsushima, M. (1998). VNTR (variable number of tandem repeat) sequences as transcriptional, translational, or functional regulators. *J. Hum. Genet.* **43**(3), 149–152.

Parker, K. J., Phillips, K. M., Kinney, L. F., and Lee, T. M. (2001). Day length and sociosexual cohabitation alter central oxytocin receptor binding in female meadow voles (*Microtus pennsylvanicus*). *Behav. Neurosci.* **115**(6), 1349–1356.

Pedersen, C. A., and Prange, A. J., Jr. (1979). Induction of maternal behavior in virgin rats after intracerebroventricular administration of oxytocin. *Proc. Natl. Acad. Sci. USA* **76**(12), 6661–6665.

Phelps, S. M., and Young, L. J. (2003). Extraordinary diversity in vasopressin (V1a) receptor distributions among wild prairie voles (*Microtus ochrogaster*): Patterns of variation and covariation. *J. Comp. Neurol.* **466**(4), 564–576.

Pitkow, L. J., Sharer, C. A., Ren, X., Insel, T. R., Terwilliger, E. F., and Young, L. J. (2001). Facilitation of affiliation and pair-bond formation by vasopressin receptor gene transfer into the ventral forebrain of a monogamous vole. *J. Neurosci.* **21**(18), 7392–7396.

Popik, P., and Van Ree, J. M. (1992). Long-term facilitation of social recognition in rats by vasopressin related peptides: A structure-activity study. *Life Sci.* **50**(8), 567–572.

Roberts, R. L., Williams, J. R., Wang, A. K., and Carter, C. S. (1998). Cooperative breeding and monogamy in prairie voles: Influence of the sire and geographical variation. *Anim. Behav.* **55**(5), 1131–1140.

Satake, H., Takuwa, K., Minakata, H., and Matsushima, O. (1999). Evidence for conservation of the vasopressin/oxytocin superfamily in Annelida. *J. Biol. Chem.* **274**(9), 5605–5611.

Shapiro, L. E., and Dewsbury, D. A. (1990). Differences in affiliative behavior, pair bonding, and vaginal cytology in two species of vole (*Microtus ochrogaster* and *M. montanus*). *J. Comparat. Psychol.* **104**(3), 268–274.

Tautz, D., and Schlotterer, C. (1994). Simple sequences. *Curr. Opin. Genet. Dev.* **4**(6), 832–837.

Thomas, J. A., and Birney, E. C. (1979). Parental care and mating system in the prairie vole, *Microtus ochrogaster*. *Behav. Ecol. Sociobiol.* **5,** 171–186.

Uhl-Bronner, S., Waltisperger, E., Martinez-Lorenzana, G., Lara, M. Condes, and Freund-Mercier, M. J. (2005). Sexually dimorphic expression of oxytocin binding sites in forebrain and spinal cord of the rat. *Neuroscience* **135**(1), 147–154.

Van Kesteren, R. E., Smit, A. B., De Lange, R. P., Kits, K. S., Van Golen, F. A., Van Der Schors, N. D., De With, N. D., Burke, J. F., and Geraerts, W. P. (1995). Structural and functional evolution of the vasopressin/oxytocin superfamily: Vasopressin-related conopressin is the only member present in Lymnaea, and is involved in the control of sexual behavior. *J. Neurosci.* **15**(9), 5989–5998.

Wang, Z., Ferris, C. F., and De Vries, G. J. (1994). Role of septal vasopressin innervation in paternal behavior in prairie voles (*Microtus ochrogaster*). *Proc. Natl. Acad. Sci. USA* **91**(1), 400–404.

Wang, Z., Yu, G., Cascio, C., Liu, Y., Gingrich, B., and Insel, T. R. (1999). Dopamine D2 receptor-mediated regulation of partner preferences in female prairie voles (*Microtus ochrogaster*): A mechanism for pair bonding? *Behav. Neurosci.* **113**(3), 602–611.

Wigger, A., Sanchez, M. M., Mathys, K. C., Ebner, K., Frank, E., Liu, D., Kresse, A., Neumann, I. D., Holsboer, F., Plotsky, P. M., and Landgraf, R. (2004). Alterations in central neuropeptide expression, release, and receptor binding in rats bred for high anxiety: Critical role of vasopressin. *Neuropsychopharmacology* **29**(1), 1–14.

Williams, J. R., Carter, C. S., and Insel, T. (1992a). Partner preference development in female prairie voles is facilitated by mating or the central infusion of oxytocin. *Ann. N. Y. Acad. Sci.* **652,** 487–489.

Williams, J. R., Catania, K. C., and Carter, C. S. (1992b). Development of partner preferences in female prairie voles (*Microtus ochrogaster*): The role of social and sexual experience. *Horm. Behav.* **26**(3), 339–349.

Williams, J. R., Slotnick, B. M., Kirkpatrick, B. W., and Carter, C. S. (1992c). Olfactory bulb removal affects partner preference development and estrus induction in female prairie voles. *Physiol. Behav.* **52**(4), 635–639.

Williams, J. R., Insel, T. R., Harbaugh, C. R., and Carter, C. S. (1994). Oxytocin administered centrally facilitates formation of a partner preference in female prairie voles (*Microtus ochrogaster*). *J. Neuroendocrinol.* **6**(3), 247–250.

Wilson, E. O. (1975). "Sociobiology: A New Synthesis." Belknap Press of Harvard University Press, Cambridge, Massachusetts.

Winslow, J. T., Hastings, N., Carter, C. S., Harbaugh, C. R., and Insel, T. R. (1993). A role for central vasopressin in pair bonding in monogamous prairie voles. *Nature* **365**(6446), 545–548.

Yamamoto, Y., Carter, C. S., and Cushing, B. S. (2006). Neonatal manipulation of oxytocin affects expression of estrogen receptor alpha. *Neuroscience* **137**(1), 157–164.

Young, L. J. (1999). Frank A. Beach Award. Oxytocin and vasopressin receptors and species-typical social behaviors. *Horm. Behav.* **36**(3), 212–221.

Young, L. J., Nilsen, R., Waymire, K. G., MacGregor, G. R., and Insel, T. R. (1999). Increased affiliative response to vasopressin in mice expressing the V1a receptor from a monogamous vole. *Nature* **400**(6746), 766–768.

5

Genetic Basis for MHC-Dependent Mate Choice

Kunio Yamazaki and Gary K. Beauchamp
Monell Chemical Senses Center, Philadelphia, Pennsylvania 19104

ABSTRACT

Genes in the major histocompatibility complex (MHC), best known for their role in immune recognition and transplantation success, are also involved in modulating mate choice in mice. Early studies with inbred, congenic mouse lines showed that mate choice tended to favor nonself MHC types. A similar phenomenon was

Advances in Genetics, Vol. 59
0065-2660/07 $35.00
DOI: 10.1016/S0065-2660(07)59005-X
Copyright 2007, Elsevier Inc. All rights reserved.

demonstrated with semi-wild mice as well. Subsequent studies showed that, rather than nonself choices, it was more accurate to say that mice chose nonparental MHC types for mates since preferences for nonself could be reversed if mice were fostered from birth on parents with nonself MHC types. Other studies have demonstrated that parent–offspring recognition is also regulated by MHC-determined signals suggesting that this system is one of general importance for mouse behavior. Many studies have now demonstrated that volatile mouse body odors are regulated by MHC genes and it is presumably these odor differences that underlie mate choice and familial recognition. Recent studies have shown that many odorants are controlled by the MHC but the mechanism by which MHC genes exert their influence has not been identified. Surprisingly, not only are volatile body odors influenced by MHC genes but so too are nonvolatile signals. Peptides bound to the MHC protein may also function in individual recognition. The extent to which this system is involved in mate choice of other species is unclear although there are some suggestive studies. Indeed, there is tentative evidence that MHC differences, presumably acting via odor changes, may influence human partner selection. Further studies should clarify both the mechanism underlying MHC influence on body odors as well as the generality of their importance in mate selection. © 2007, Elsevier Inc.

I. INTRODUCTION

For kin recognition and mate selection to exist, animals must be able to discriminate among individuals. In the case of kin recognition, phenotypic differences that distinguish kin from nonkin allow the animal to treat these two classes of individuals differently, generally to the benefit of kin. For mate choice, it is generally assumed that this is based on phenotypic differences that somehow indicate "good" genes in the selected mate. We and others have found that the major histocompatibility complex (MHC) of genes, which is critical for immune response, is also one source of chemosensory information that enables mice to identify one another as individuals.

II. THE MHC

The importance of this group of linked genes can be gauged from the fact that a similar set of genes probably exists in all vertebrates (Klein, 1986). The MHC of the mouse, called H-2, comprises many linked genes and is divided into regions, the main ones being H-2K (K), H-2D (D), and Tla. The mouse's "MHC type" of "H-2 type" is the total set of variable alleles of all genes in the MHC region. The set of MHC alleles (in the mouse on chromosome 17) is called a haplotype, and a vast number of haplotypes are possible. This family of ~50 genes is characterized in many species by their extreme diversity. In fact, the number of potential MHC types, comprising two MHC sets in each diploid individual, might theoretically exceed the population of a given species. There are several sources of this diversity. As regards class I MHC genes, H-2K and H-2D each has at least 50 and perhaps many more alleles, generated in part by a mechanism by which one class I gene confers part of its sequence on another. (Ivanyi, 1995; Kelley *et al.*, 2005).

The MHC is best known from studies on tissue transplantations because incompatibility of MHC types causes rapid rejection of grafts. The fate of organ transplants depends mainly on MHC compatibility. Throughout the MHC region there are also genes that determine the degree of response to particular antigens, other genes expressed selectively in lymphocytes and other apparently unrelated genes. MHC gene-encoded proteins play a critical role during immune recognition by serving as antigen receptors that bind peptide fragments for cell surface presentation to T lymphocytes (Germain, 1994; Parham and Ohta, 1996; Van Kaer, 2002). Thus, the MHC is concerned in many aspects of how immune cells, lymphocytes equipped with specific receptors for antigen, handle chemical information from the environment.

The diversity of MHC genes makes them excellent candidates for marking each individual of a species with a unique odortype (genetically programmed body odors that distinguish one individual from another–Boyse *et al.*, 1987, 1991a), as originally suggested by Lewis Thomas and described below in more detail. Since this suggestion, many laboratories have verified Thomas's remarkable predictions in mice (Beauchamp and Yamazaki, 2003; Beauchamp *et al.*, 1985; Boyse *et al.*, 1991a,b; Brown and Eklund, 1994), rats (Brown *et al.*, 1987; Singh *et al.*, 1987, 1988), and perhaps humans (Ferstl *et al.*, 1992).

The use of inbred congenic mice has been essential in work on MHC-associated chemosensory communication. An H-2 congenic strain is produced by crossing two selected inbred strains of different H-2 types and by backcrossing to one of these strains, serologically selecting for the donor H-2 type in each generation. Consequently, the final inbred congenic strain is genetically identical to the base strain, except for a segment of chromosome 17 bearing the H-2 haplotype, introduced from the donor strain. Any difference between the base strain and its congenic partner strain, if genetic, must be due to genes in the H-2 region because this is the only genetic difference between the inbred base strain and its congenic partner.

III. MOUSE MATING PREFERENCES

Thomas (1975) was the first to suggest that MHC genes might be involved in individual recognition. Building on theories that histocompatibility arises from genes intended originally to protect the integrity of organisms by molecular cellular identity, he suggested that histocompatibility genes might also impart to each individual a characteristic scent. Dogs, he surmised, might be able to distinguish different human MHC types by the sense of smell. This speculation received support from the laboratory of Edward A. Boyse who informally noticed that the social behavior of mice seemed to be influenced by their H-2 types. To evaluate this formally, a test system was set up where breeding trios were studied in which an inbred male was caged with two females, a male was caged with two H-2 congenic females, both in estrus, and the trio was observed continuously until successful copulation, verified by a vaginal plug, had occurred with one of the females, after which the receptivity of the other female was confirmed by successful copulation with a second male. The various genetic and other controls are described in reports of the time. These studies confirmed that mice do indeed exhibit natural mating preferences related to their H-2 types, and that this bias generally favors matings between males and females of different rather than similar H-2 types (Andrews and Boyse, 1978; Yamaguchi *et al.*, 1978; Yamazaki *et al.*, 1976, 1978).

Other investigators have also reported mating bias based on MHC type (Coopersmith and Lenington, 1990; Egid and Brown, 1989; Eklund *et al.*, 1991). In particular, females also seem to be biased to mate with males of different H-2 types (Egid and Brown, 1989). Importantly, Potts *et al.* (1991) reported a negative assortative mating bias according to H-2 type in penned mice held under semi-natural conditions, lending support to the hypothesis that such a bias serves to maintain heterozygosity at this locus and to avoid the deleterious effects of inbreeding.

Manning *et al.* (1992) demonstrated another context in which H-2-determined odortypes are most likely involved: Females nest with other females that are similar at the MHC. These studies, using partially outbred animals, strongly suggest that MHC odortypes represent salient cues in the context of varied genetic (non-MHC) backgrounds. This has been demonstrated experimentally by showing that mice can be trained to discriminate MHC odortypes in spite of a varying set of background genes (Yamazaki *et al.*, 1994) and they can recognize MHC differences on a novel genetic background (Willse *et al.*, 2006). Roberts and Gosling (2003) showed that female mice are attracted to the scent marks of MHC-dissimilar males. However, there is still no evidence for MHC-based mating or odor preferences in completely wild house mice.

IV. FUNCTIONS OF MHC-MEDIATED MATING PREFERENCE

MHC-determined mate preference should be an effective way of avoiding inbreeding, and enhancing offspring variability and hence viability (Apanius *et al.*, 1997; Grob *et al.* 1998; Penn, 2002; Penn and Potts, 1999). MHC diversity should be advantageous for immunological reasons. For instance, the hypothesis of viral mimicry supposes that a virus may use its capacity for genetic variation to alter the constitution of its glycoprotein, thus simulating self (host) and avoiding immune recognition and response (Potts *et al.*, 1994). Thus, negative associative mating according to MHC odortype can serve to decrease the genetic dangers associated with inbreeding. Structural diversification of MHC glycoproteins may be viewed as a counter to this viral gambit calculated to deny access of such a mutant virus to the population at large. Moreover, since heterozygous offspring can protect themselves against a wider range of parasites than homozygous ones (Penn *et al.*, 2002), MHC-negative assortative mating might provide progeny with an enhanced immunological surveillance and reduced levels of either infectious or genetic disease. (Doherty and Zinkernagel, 1975; Wakeland *et al.*, 1990).

V. FAMILIAL IMPRINTING DETERMINES H-2 SELECTIVE MATING PREFERENCES

Inbred male mice typically prefer to mate with females of a different, nonself H-2 haplotype. To determine whether this natural preference is influenced by rearing events early in life, a test system was used which relied on previous observations that B6 females (H-2^b) mate preferentially with congenic B6-H-2^k rather than B6 females, and B6-H-2^k males with B6 females. Within 16 h of birth, entire litters were removed from their natural parents and transferred to foster parents whose own litters, born at approximately the same time, were simultaneously removed. At 21 days of age, the fostered mice were weaned and the males maintained in stock cages containing only males of the same genotype and fostering history until sexual maturity (3.5 months of age, minimum) when tests of mating preference began.

Fostering completely reversed the preference for nonself MHC type. Thus, H-2 selective mating preference is acquired by early learning or imprinting on familial H-2 types (Beauchamp *et al.*, 1988; Yamazaki *et al.*, 1988). It should be noted that influences of non-MHC genes on mating preference are not excluded by this study because the genomes of the participating mice were identical except at the MHC. The terms self and nonself with regard to mating preference are thus misleading: preference was determined not by the animal's own MHC type but by the parental MHC type.

A. Choice by the male

Particularly in light of the conclusion from our studies described above that familial imprinting determines mating preferences of males, it is important to be sure that the choice of MHC odortype in our mating preference test is in fact exercised primarily by the male and not by the female. This latter alternative interpretation is seemingly excluded by the reversal of preference of fostered males, since the congenic females in the test system were normally reared. However, there remains a possibility that the unaltered females were responding to a male odortype acquired from the foster mother. This possibility is unlikely: when the odortype of B6 males fostered by B6-H-2^k parents was compared in the Y-maze with the odortype of control B6 males fostered by B6 parents, no distinction was possible in this nor in an identical study with B6-H-2^k males. Therefore, it is highly likely that choice in the mating preference test is the prerogative of the male.

B. Choice by the female

Most theories of evolutionary biology suggest that female choice, in most cases, is more important than male choice. Thus, it is important to indicate that there is evidence for female choice based on MHC type. Roberts and Gosling (2003) show in mice that although MHC dissimilarity is a "good genes" indicator (investment in scent-marking), both have a role in determining female preference, and their relative influence can vary depending on the degree of variability in each trait among available males. Such interactions between condition-dependent and disassortative mate choice criteria suggest a mechanism by which female choice can contribute to maintenance of additive genetic variance in both the MHC and condition-dependent traits. Ziegler *et al.* (2005) explore the diverse ways females might influence MHC-associated mating patterns (especially postcopulatory mechanisms) and cast their observations into an immunological perspective.

VI. PREGNANCY BLOCK

Another system from which to view MHC-associated communication concerns the phenomenon known as pregnancy block, or the Bruce effect (Bruce, 1960). If a female mouse is separated from her mate shortly after mating and is then exposed to a male of a strain different from that of her first mate, or to the urine of such a male, there is an increased probability that pregnancy or pseudopregnancy will be terminated and she will return to normal estrous cycling. We found that the incidence of pregnancy or pseudopregnancy block was substantially greater when the blocking male differed from the studmale at the MHC locus than when the blocking male had the same MHC type as the stud (Yamazaki *et al.*, 1983b). Even when MHC disparity of the two males was limited to a single class I gene, H-2^b versus the mutant H-2^{bm1}, a significant rise in incidence of pregnancy block was evident (Yamazaki *et al.*, 1986). Pregnancy block likely represents a neuroendocrine response to the genetic (individual) identity of the blocking male generally associated with a male signal. Thus, sensory recognition according to genotype may operate in the broader context of neuroendocrine responses linked to reproduction.

VII. PARENT–INFANT RECOGNITION

Parent–infant recognition based on individual-specific odors is well documented among many genera (Beauchamp *et al.*, 1976; Halpin, 1980; Johnston *et al.*, 1999). We have found (Yamazaki *et al.*, 1992) that H-2 odortypes are evident in mice as young as 1 day of age, raising the possibility that a dam might thereby identify her offspring. Moreover, since olfactory function is well developed in mouse pups as young as a few days of age (Hepper, 1983), they might reciprocally recognize and prefer their mothers' MHC type.

Our studies were designed to test these reciprocal hypotheses, and determine whether the MHC plays a role in early behavioral development. We found that mothers recognize and preferentially retrieve syngeneic (genetically identical) pups from otherwise identical pups differing only for MHC (Yamazaki *et al.*, 2000). Reciprocally, pups move toward parental odors according to parental MHC type in which the pups are fostered on congenic parents, they then choose congenic over syngeneic odor (Yamazaki *et al.*, 2000). As described above, mating preferences according to MHC type favor dissimilar choice, presumably to enhance outbreeding and H-2 heterozygosity, promote diversity of H-2 genes, and edit spontaneous mutations. In contrast, familial choice favors H-2 similarity, thus ensuring appropriate mother–young attachment, and hence survival. However, both selective mating and parent–pup preference are influenced by perinatal acclimatization, which plays a key role in both behaviors. In the former, early learning promotes selection of the unfamiliar in subsequent mate choice; in the latter, the biologically familiar is favored. The fact that these behaviors are learned responses to MHC type is supported by the fact that both mate choice and pup-preference are amenable to manipulation by foster nursing (Yamazaki *et al.*, 1988, 2000).

We have also found (Beauchamp *et al.*, 1994) that mice can be trained to discriminate among genetically identical pregnant females that are carrying fetuses that differ only according to H-2 type. Evidently, the odortype of a pregnant female is made up of the combination of her own odortype and those of her fetuses; this appears to be the case for humans as well (Beauchamp *et al.*, 1995). Fetal odortypes may thus play a functional role in social interactions among mice (Beauchamp *et al.*, 1994, 2000; Yamazaki and Beauchamp, 2005).

VIII. CHEMOSENSORY DISCRIMINATION

A. MHC variation and body odor

What cues do the mice use to identify MHC type? Since mice are known to be highly oriented to odors, it is likely that behavioral and neuroendocrine effects are based on MHC-determined body odor variation. To test whether MHC genes influence body odor, we have extensively used a very simple apparatus, a Y-maze. In the Y-maze, air is drawn through two odor boxes, containing urine of H-2 congenic mice. The air is then conducted to the left and right arms of the maze, which are thereby scented differentially by urines of mice whose only genetic difference is H-2. Some mice are trained to run for water reward toward the odor of one H-2 congenic type, whereas others are trained to run to the other, the test subject mice having been deprived of water for 23 h beforehand. These studies demonstrated that trained mice distinguished the scents of urine from mice of dissimilar H-2 types (Yamaguchi *et al.*, 1981), even differences as slight as a subdivision or single-gene mutation of the H-2 complex. Bard *et al.* (2000), Yamazaki *et al.* (1982, 1983a) and Carroll *et al.* (2002), used different techniques to also show mutant discrimination. Rats are also capable of making these distinctions (Beauchamp *et al.*, 1990), as are some humans (Gilbert *et al.*, 1986).

That odortype information permeates voided urine led to the question of whether bacteria might be involved in odorant specification (Singh *et al.*, 1990). We showed that urine drawn directly from the bladder and urine of germ-free mice are adequate sources of MHC odortype. Clearly, bacteria are not essential (Yamazaki *et al.*, 1990) although they may serve to provide substrates or to increase the amounts of odor available (Schellinck and Brown, 1992). Urine is a complex mixture of hundreds or even thousands of low molecular weight compounds and peptides. Most of the protein content of rodent urine is made up of major urinary protein (MUP, up to 70 mg ml^{-1}) that are thought to contribute to chemical signaling of individual identity by acting as carriers for certain volatile compounds (Hurst and Beynon, 2004; Hurst *et al.*, 2001). Therefore, mouse urine is likely to contain both MHC-dependent and MUP-dependent signals of individual identity, which may be used in different behavioral contexts.

We have shown that serum, following treatment with a protease, contains MHC-determined odorants that have some of the same odor qualities as found in urine (Yamazaki *et al.*, 1999). Because the pattern of odorants characterizing the MHC-determined odortype is sufficiently similar in serum and urine, it follows that the odorant pattern is established prerenally; one proven source is the hematopoietic system (Yamazaki *et al.*, 1985). A likely mechanism for odortype specification may be that soluble MHC gene products themselves bind circulating odorants selectively, presumably after they have lost their bound peptide, and then release them mainly during the course of renal processing and excretion. Alternatively, the bound peptides may be broken down to volatiles and excreted in the urine.

B. The chemical code

Most of our studies have focused on the volatile odorants underlying MHC recognition. We have reported that congenic mice differ in a number of volatile metabolites but none have been specifically tied to odor discriminations (Singer *et al.* 1993, 1997; Willse *et al.*, 2005). However, the system is more complex than originally believed. We have found (Willse *et al.*, 2006) that the identity of the odorants that distinguish congenic mouse strains are strongly influenced by the background genotype of the strains. That is, there is an interaction between background genotype and MHC genotype. In spite of this, mice trained to discriminate MHC differences on one genetic background can recognize the same MHC difference on a different background (Willse *et al.*, 2006), demonstrating that some MHC-regulated odorants are common across different genetic backgrounds.

In addition to volatile signals that we have documented, new studies strongly suggest that MHC peptides may act as an additional, nonvolatile chemosignal of MHC identity (Boehm and Zufall, 2005; Leinders-Zufall *et al.*, 2004; Milinski *et al.*, 2005; Restrepo *et al.*, 2006; Slev *et al.*, 2006; Spehr *et al.*, 2006). These peptides are 8–10 amino acids long. MHC peptides convey information on MHC haplotype because they have amino acid anchor residues that fit within the binding groove of the MHC molecule. Depending on the anchor residues a particular MHC peptide would bind to MHC molecules of a particular haplotype but not to others.

The production of both volatile and nonvolatile signals might confer unique advantages to the animal over the use of either chemosignal alone. Thus, the presence of both the odortype (volatile) and the MHC peptide (nonvolatile) as chemosignal of MHC haplotype provides a distinct advantage for the individual (Restrepo *et al.*, 2006).

IX. MHC-DEPENDENT MATING PREFERENCES IN OTHER TAXA

It appears that a variety of species other than rodents use MHC-regulated odors in organizing their social and reproductive behavior. For example, fish apparently use MHC-mediated odors for recognizing kin and choosing mates. Arctic charr (*Salvelinus alpinus*) can discriminate MHC-similar and MHC-dissimilar siblings (Olsen *et al.*, 1998), and some indirect evidence suggests that Atlantic salmon (*Salmo salar*) choose their mates in order to increase the heterozygosity of their offspring at the MHC and, more specifically, at the peptide-binding region, presumably in order to provide them with better defense against parasites and pathogens (Landry *et al.*, 2001). Milinski and his colleagues found that female three-spined sticklebacks (*Gasterosteus aculeatus*) are attracted to the scent of males that have many MHC alleles, but only if they have few alleles themselves (Aeschlimann *et al.*, 2003; Reusch *et al.*, 2001). When females have many alleles, then they prefer males with few alleles. To understand these preferences, Milinski and his colleague examined how MHC influences resistance to infection by parasites. They found that individuals with intermediate levels of MHC heterozygosity are most resistant to three common parasites (Wegner *et al.*, 2003). It would be interesting to know whether these mating preferences produce offspring with intermediate MHC diversity. Increasing the number of MHC molecules expressed during development is probably beneficial, but the expression of too many molecules may have detrimental effects on immune function.

Olsson *et al.* (2003) found that female Swedish sand lizards (*Lacerta agilis*) bias paternity by selecting the sperm of genetically compatible males. Recently, they found that, in the laboratory, females prefer to associate with MHC-dissimilar males, whereas in the wild, body size and MHC similarity seem to interact in mate choice.

No evidence for MHC disassortative mating preferences was found in Soay sheep (*Ovis aries*), this may be because of male–male competition masking the effect of female choice in this system (Paterson and Pemberton, 1997). Similarly, mate choice was determined to be independent of MHC type in female great reed warblers (*Acrocephalus arundinaceus*) (Westerdahl, 2004), rhesus macques (*Macaca mulatta*) (Sauermann *et al.*, 2001), and great snipes (*Gallinago media*) (Ekblom *et al.*, 2004). In birds, the relevance of MHC genes in mating and social systems is still largely unknown. There is some suggestion that MHC-based mate choice occurs in the ring-necked pheasant. MHC genotype is correlated with viability and spur length, a character females use in mate choice (Von Schantz *et al.*, 1989, 1996). A study on Savannah sparrows (*Passerculus sandwichensis*) by Freeman-Gallant *et al.* (2003) found that females are able to discriminate between males carrying different MHC haplotypes, and avoid mating with MHC-similar males.

X. HLA: ODORTYPES AND MATE SELECTION IN HUMANS

Several studies have tested whether MHC dissimilarities influence odor and mating preferences in humans. Wedekind *et al.* (1995), Wedekind and Furi (1997) showed that women preferred the odors of T-shirts worn by men with dissimilar MHC-genotypes, and that the scent of these men reminded the women of their own present or former partners, although this study did not measure mating partner preferences, as is implied by the chapter's title. Thornhill *et al.* (2003) found that men preferred the odor of t-shirts worn by MHC-dissimilar women. And Santos *et al.* (2005) found a significant correlation only when female smellers evaluated male sweat odors.

Ober *et al.* (1997) and Hedrick and Black (1997) have addressed the hypothesis that HLA haplotype affects the choice of mates in human populations. Hedrick and Black examined 194 couples selected from 11 South Amerindian tribes and found no evidence for negative assortative mate choice. But this group had a small sample size and significantly more HLA types, a situation that does not provide the power for detecting a nonrandom pattern. In contrast, Ober *et al.* found evidence for negative assortative mate choice according to HLA type. Taking a somewhat similar approach to that of Hedrick and Black, Ober and colleagues evaluated 411 Hutterite couples (Hutterites are a North American, reproductively isolated, cultural, and religious group of European ancestry) and found fewer matches of HLA haplotypes between spouses than were expected. Caution that this may not be a very easily observed phenomenon and it is possible that in most human societies cultural and biological factors overwhelm it. Nevertheless, it has been shown that woman single out body odors from alternative sensory attributes as the most important variable for mate choice (Herz and Cahill, 1997). It is interesting to speculate whether negative assortative mating according to MHC type might have played a larger role earlier in human or primate evolution, perhaps when female choice, unencumbered by many cultural factors, was more easily expressed.

Jacob *et al.* (2002) studied odor preferences of women as they relate to the body odors of males. They found that women prefer the smell of donors that have an intermediated level of HLA difference when compared with the sequence of their own HLA genes. When they compared the HLA sequences of the odor donor with those of the woman's parents, they found a significant correlation between the HLA sequences from the preferred odor donors and those she inherited from her father. Indeed, the function of HLA odor choice may be more broadly social, including kin recognition and choice of social networks. Interestingly, Milinski and Wedekind (2001) found that an individual's MHC is also associated with his or her preferences for perfumes. Finally, Balseiro and Correia (2006) reported that volatile organic compounds produced by tumors, which are products of MHC genes, are detected by dogs.

XI. CONCLUSIONS AND IMPLICATIONS FOR FUTURE WORK

MHC genes (as well as yet unknown other genes) contribute to a mouse's individual body odor and these odors are important regulators of behavioral and neuroendocrine functions, including mating choices. In addition to mice, there are evidence that (1) human body odors also may be influenced by HLA genetic variation and (2) these odors may mediate behavioral responses to other people. These findings raise many questions. First, we still do not know how variation in these genes leads to variation in odorant concentrations. We initially thought that by identifying the odorants that are regulated by MHC genetic variation, we could gain insight into these mechanisms but that hope has not yet been realized. Additional biochemical studies are needed. Second, the generality of our findings (i.e., which other species show MHC-regulated mating preference based on odortype differences) remains unclear. Although as documented above this is seen in several species, it is important to see if, for example, this occurs for other primates.

The original suggestions for an MHC odortype connection that were made by Thomas stemmed in part from his interests in how dogs track people. The implication was that each person had a unique body odor that the dogs exploited. Presuming that this is true, then it should be possible to replace the dog with an instrument that can identify individual people much as can be done with fingerprints. Our mouse studies implicating MHC genes and genes at other locations in the genome in odortype specification imply that the odortype of an individual may be extremely complex. This is consistent with the necessity, at least so far, for the sophisticated olfactory apparatus of the dog or the mouse to discriminate such odors. Nevertheless, it seems reasonable to guess that it may not be so long before we can take an odor sample from an individual and determine the identity of the person. It may indeed be that people are already doing this at a subconscious level and that odor plays a role in sexual attraction and human "mate choice."

References

Aeschlimann, P. B., Haberli, M. A., Reusch, T. B., Boehm, T., and Milinski, M. (2003). Female sticklebacks Gasterosteus aculeatus use self-reference to optimize MHC allele number during mate selection. *Behav. Ecol.* **54,** 119–126.

Andrews, P. W., and Boyse, E. A. (1978). Mapping of an H-2-linked gene that influences mating preferences in mice. *Immnogenetics* **6,** 265–268.

Apanius, V., Penn, D., Slev, P., Ruff, L. R., and Potts, W. K. (1997). The nature of selection on the major histocompatibility complex. *Crit. Rev. Immunol.* **17,** 179–224.

Balseiro, S. C., and Correia, H. R. (2006). Is olfactory detection of human cancer by dogs based on major histocompatibility complex-dependent odor component? A possible cure and a precocious diagnosis of cancer. *Med. Hypotheses.* **66,** 270–272.

Bard, J., Yamazaki, K., Curran, M., Boyse, E. A., and Beauchamp, G. K. (2000). Effect of ß2 m gene disruption on MHC-determined odortypes. *Immunogenetics* **51**(7), 514–518.

Beauchamp, G. K., and Yamazaki, K. (2003). Chemical signalling in mice. *Biochem. Soc. Trans.* **31,** 147–151.

Beauchamp, G. K., Doty, R. L., Moulton, D. G., and Mugford, R. A. (1976). The pheromone concept in mammalian chemical communication: A critique. *In* "Mammalian Olfaction, Reproductive Processes, and Behavior" (R. L. Doty, ed.), pp. 143–160. Academic Press, New York.

Beauchamp, G. K., Yamazaki, K., and Boyse, E. A. (1985). The chemosensory recognition of genetic individuality. *Sci. Am.* **253,** 86–92.

Beauchamp, G. K., Yamazaki, K., Bard, J., and Boyse, E. A. (1988). Pre-weaning experience in the control of mating preferences by genes in the major histocompatibility complex of the mouse. *Behav. Genet.* **18,** 537–547.

Beauchamp, G. K., Yamazaki, K., Duncan, H., Bard, J., and Boyse, E. A. (1990). Genetic determination of individual mouse odor. *In* "Chemical Signals in Vertebrates" (D. W. Macdonald, D. Müller-Schwarze, and S. E. Natynczuk, eds.), Vol. 5, pp. 244–254. Oxford University Press, New York.

Beauchamp, G. K., Yamazaki, K., Curran, M., Bard, J., and Boyse, E. A. (1994). Fetal H-2 odortypes are evident in the urine of pregnant female mice. *Immunogenetics* **39**(2), 109–113.

Beauchamp, G. K., Katahira, K., Yamazaki, K., Mennella, J. A., Bard, J., and Boyse, E. A. (1995). Evidence suggesting that the odortypes of pregnant women are a compound of maternal and fetal odortypes. *Proc. Natl. Acad. Sci. USA* **92,** 2617–2621.

Beauchamp, G. K., Curran, M., and Yamazaki, K. (2000). MHC-mediated fetal odortypes expressed by pregnant females influence male associative behaviour. *Anim. Behav.* **60,** 289–295.

Boehm, T., and Zufall, F. (2005). MHC peptides and the sensory evaluation of genotype. *Trends Neurosci.* **29,** 100–107.

Boyse, E. A., Beauchamp, G. K., and Yamazaki, K. (1987). The genetics of body scent. *Trends Genet.* **3,** 97–102.

Boyse, E. A., Beauchamp, G. K., Bard, J., and Yamazaki, K. (1991a). Behavior and the major histocompatibility complex (MHC), H-2, of the mouse. *In* "Psychoneuroimmunology-II" (R. Ader, D. L. Felter, and Cohen, eds.), pp. 831–846. Academic Press, San Diego.

Boyse, E. A., Beauchamp, G. K., Yamazaki, K., and Bard, J. (1991b). Genetic components of kin recognition in mammals. *In* "Kin Recognition" (P. G. Hepper, ed.), pp. 148–161. Cambridge University Press, New York.

Brown, J. L., and Eklund, A. (1994). Kin recognition and the major histocompatibility complex: An integrative review. *Am. Nat.* **143,** 435–461.

Brown, R. E., Singh, P. B., and Roser, B. (1987). The major histocompatibility complex and the chemosensory recognition of individuality in rats. *Physiol. Behav.* **40,** 65–73.

Bruce, H. M. (1960). A block to pregnancy in the mouse caused by proximity of strange males. *J. Reprod. Fertil.* **1,** 96–103.

Carroll, L. S., Penn, D. J., and Potts, W. K. (2002). Discrimination of MHC-derived odors by untrained mice is consistent with divergence in peptide-binding region residues. *Proc. Natl. Acad. Sci. USA* **99,** 11260–11264.

Coopersmith, C. B., and Lenington, S. (1990). Preferences of female mice for males whose t-haplotype differs from their own. *Anim. Behav.* **40,** 1179–1181.

Doherty, P. C., and Zinkernagel, R. M. (1975). Enhanced immunological surveillance in mice heterozygous at the H-2 gene complex. *Nature* **256,** 50–52.

Egid, K., and Brown, J. L. (1989). The major histocompatibility complex and female mating preferences in mice. *Anim. Behav.* **38**(3), 548–549.

Ekblom, R., Saether, S. A., Grahn, M., Fiske, P., Kalas, J. A., and Hoglund, J. (2004). Major histocompatibility complex variation and mate choice in a lekking bird, the great snipe (*Gallinago media*). *Mol. Ecol.* **13,** 3821–3828.

Eklund, A., Egid, K., and Brown, J. L. (1991). The major histocompatibility complex and mating preferences of male mice. *Anim. Behav.* **42,** 693–694.

Ferstl, R., Eggert, F., Westphal, E., Zavazava, N., and Muller-Ruchholtz, W. (1992). MHC-related odors in humans. *In* "Chemical Signals in Vertebrates 6" (R. L. Doty and D. Müller-Schwarze, eds.), pp. 205–211. Plenum Publishing, New York.

Freeman-Gallant, C. R., Meguerdichian, M., Wheelwright, N. T., and Sollecito, S. V. (2003). Social pairing and female mating fidelity predicted by restriction fragment length polymorphism similarity at the major histocompatibility complex in a songbird. *Mol. Ecol.* **12,** 3077–3083.

Germain, R. N. (1994). MHC-dependent antigen processing and peptide presentation: Providing ligands for T lymphocyte activation. *Cell* **76,** 287–299.

Gilbert, A. N., Yamazaki, K., Beauchamp, G. K., and Thomas, L. (1986). Olfactory discrimination of mouse strains (*Mus musculus*) and major histocompatibility types by humans. *J. Comp. Psychol.* **100,** 262–265.

Grob, B., Knapp, L. A., Martin, R. D., and Anzenberger, G. (1998). The major histocompatibility complex and mate choice: Inbreeding avoidance and selection of good genes. *Exp. Clin. Immunogenet.* **15,** 119–129.

Halpin, Z. T. (1980). Individual odors and individual recognition. *Biol. Behav.* **5,** 233–248.

Hepper, P. G. (1983). Sibling recognition in the rat. *Anim. Behav.* **31,** 1177–1191.

Herz, R. S., and Cahill, E. D. (1997). Differential use of sensory information in sexual behavior as a function of gender. *Hum. Nat.* **8,** 275–289.

Hedrick, P. W., and Black, F. L. (1997). HLA and mate selection: No evidence in South Amerindians. *Am. J. Hum. Genet.* **61,** 505–511.

Hurst, J. L., and Beynon, R. J. (2004). Scent wars: The chemobiology of competitive signaling in mice. *Bioessays* **26,** 1288–1298.

Hurst, J. L., Thom, M. D., Nevison, C. M., Humphries, R. E., and Beynon, R. J. (2001). Individual recognition in mice mediated by major urinary proteins. *Nature* **414,** 631–634.

Ivanyi, P. (1995). Biological meaning of the MHC. *Folia Biologica (Praha)* **41,** 178–189.

Jacob, S., McClintock, M. K., Zelano, B., and Ober, C. (2002). Paternally inherited HLA alleles are associated with woman's choice of male odor. *Nat. Genet.* **30,** 175–179.

Johnston, R. E., Muller-Schwarze, D., and Sorensen, P. W. (1999). "Advances in Chemical Signals in Vertebrates." Kluwer Academic/Plenum Publishers, New York.

Kelley, J., Walter, L., and Trowsdale, J. (2005). Comparative genomics of major histocompatibility complexes. *Immunogenetics* **56,** 683–695.

Klein, J. (1986). "Natural History of the Major Histocompatibility Complex." J. Wiley & Sons, New York.

Landry, C., Garant, D., Duchesne, P., and Bernatchez, L. (2001). "Good genes as heterozygosity": The major histocompatibility complex and mate choice in Atlantic salmon (Salmo salar). *Proc. R. Soc. Lond. B* **268,** 1279–1285.

Leinders-Zufall, T., Brennan, P., Widmayer, P., Maul-pavicic, A., Jager, M., Li, X. H., Breer, H., Zufall, F., and Boehm, T. (2004). MHC classI peptides as chemosensory signals in the vomeronasal organ. *Science* **306,** 1033–1037.

Manning, C. J., Wakeland, E. K., and Potts, W. K. (1992). Communal nesting patterns in mice implicate MHC genes in kin recognition. *Nature* **360,** 581–583.

Milinski, M., and Wedekind, C. (2001). Evidence for MHC-correlated perfume preferences in humans. *Behav. Ecol.* **12,** 140–149.

Milinski, M., Griffiths, S., Wegner, K. M., Reusch, T. B., Haas-Assenbaum, A., and Boehm, T. (2005). Mate choice decisions of stickleback females predictably modified by MHC peptide ligands. *Proc. Natl. Acad. Sci. USA* **102,** 4414–4418.

Ober, C., Weitkamp, L. R., Cox, N., Dytch, H., Kostyyu, D., and Elias, S. (1997). HLA and mate choice in humans. *Am. J. Hum. Genet.* **61,** 497–504.

Olsen, K. H., Grahn, M., Lohm, J., and Langefors, A. (1998). MHC and kin discrimination in juvenile Arctic charr (*Salvelinus alpinus*). *Anim. Behav.* **56,** 319–327.

Olsson, M., Madsen, T., and Nordby, N. (2003). Major histocompatibility complex and mate choice in sand lizards. *Proc. R. Soc. Lond. B* **270,** 254–256.

Parham, P., and Ohta, T. (1996). Population biology of antigen presentation by MHC class I molecules. *Science* **272,** 67–74.

Paterson, S., and Pemberton, J. M. (1997). No evidence for major histocompatibility complex-dependent mating patterns in a free living ruminant population. *Proc. R. Soc. Lond. B.* **264,** 1813–1819.

Penn, D. J. (2002). The Scent of genetic compatibility: Sexual selection and the major histocompatibility complex. *Ethology* **108,** 1–21.

Penn, D. J., and Potts, W. K. (1999). The evolution of mating preferences and major histocompatibility complex genes. *Am. Nat.* **153,** 145–164.

Penn, D. J., Damjanovich, K., and Potts, W. K. (2002). MHC heterozygosity confers a selective advantage against multiple-strain infections. *Proc. Natl. Acad. Sci. USA* **99,** 11260–11264.

Potts, W. K., Manning, C. J., and Wakeland, E. K. (1991). MHC genotype influences mating patterns in semi-natural populations of Mus. *Nature* **352,** 619–621.

Potts, W. K., Manning, C. J., and Wakeland, E. K. (1994). The role of infectious disease, inbreeding and mating preferences in maintaining MHC genetic diversity: An experimental test. *Proc. R. Soc. Soc. Lond. B* **260,** 369–378.

Restrepo, D., Lin, W., Salcedo, E., Yamazaki, K., and Beauchamp, G. (2006). Odortypes and MHC peptides: Complementary chemosignals of MHC haplotype? *Trends Neurosci.* **29,** 604–609.

Reusch, T. B., Haberli, M. A., Aeschlimann, P. B., and Milinski, M. (2001). Female sticklebacks count alleles in a strategy of sexual selection explaining MHC polymorphism. *Nature* **414,** 300–302.

Roberts, S. G., and Gosling, L. M. (2003). Genetic similarity and quality interact in mate choice decisions by female mice. *Nat. Genet.* **35,** 103–106.

Santos, P. S., Schinemann, J. A., Gabardo, J., and Bicalho, M. G. (2005). New evidence that the MHC influences odor perception in humans: A study with 58 Southern Brazilian students. *Horm. Behav.* **47,** 384–388.

Sauermann, U., Nurnberg, P., Bercovitch, F. B., Berard, J. D., Trefilov, A., Widdig, A., Kessler, M., Schmidtke, J., and Krawczak, M. (2001). Increased reproductive success of MHC class II heterozygous males among free-ranging rhesus macaques. *Hum. Genet.* **108,** 249–254.

Schellinck, H. M., and Brown, R. E. (1992). Why does germfree rearing eliminate the odors of individuality in rats but not in mice. *In* "Chemical Signals in Vertebrates VI" (R. L. Doty and D. Muller-Schwarze, eds.), pp. 237–241. Plenum Press, New York.

Singer, A. G., Tsuchiya, H., Wellington, J. L., Beauchamp, G. K., and Yamazaki, K. (1993). Chemistry of odortypes in mice: Fractionation and bioassay. *J. Chem. Ecol.* **19,** 569–579.

Singer, A. G., Beauchamp, G. K., and Yamazaki, K. (1997). Volatile signals of the major histocompatibility complex in male mouse urine. *Proc. Natl. Acad. Sci. USA* **94,** 2210–2214.

Singh, P. B., Brown, R. E., and Roser, B. (1987). MHC antigens in urine as olfactory recognition cues. *Nature* **327,** 161–164.

Singh, P. B., Brown, R. E., and Roser, B. (1988). Class I transplantation antigens in solution in body fluids and in the urine: Individuality signals to the environment. *J. Exp. Med.* **168,** 195–211.

Singh, P. B., Herbert, J., Roser, B., Arnott, L., Tucker, D. K., and Brown, R. E. (1990). Rearing rats in germ-free environment eliminates their odors of individuality. *J. Chem. Ecol.* **16,** 1667–1682.

Slev, P. R., Nelson, A. C., and Potts, W. K. (2006). Sensory neurons with MHC-like peptide binding properties: Disease consequences. *Curr. Opin. Immunol.* **18,** 1–9.

Spehr, M., Kelliher, K. R., Li, X. H., Boehm, T., Leinders-Zufall, T., and Zufall, F. (2006). Essential role of the main olfactory system in social recognition of major histocompatibility complex peptide ligands. *J.Neurosci.* **26**(7), 1961–1970.

Thomas, L. (1975). Symbiosis as an immunologic problem: The immune system and infectious diseases. *In* "Fourth International Congress of Immunology" (E. Neter and F. Milgrom, eds.), p. 2. Karger, Basel.

Thornhill, R., Gangestad, S. W., and Miller, R. (2003). Major histocompatibility complex genes, symmetry, and body scent attractiveness in men and women. *Behav. Ecol.* **14,** 668–678.

Van Kaer, L. (2002). Major histocompatibility complex class I-restricted antigen processing and presentation. *Tissue Antigens* **60,** 1–9.

Von Schantz, T., Goransson, G., Andersson, G., Froberg, I., Grahn, M., and Helgee, M. (1989). Female choice selects for viability-based male trait in pheasants. *Nature* **337,** 166–169.

Von Schantz, T., Wittzell, H., Goransson, G., Grahn, M., and Persson, K. (1996). MHC genotype and male ornamentation: Genetic evidence for the Hamilton-Zuk model. *Proc. R. Soc. Lond.* **263,** 265–271.

Wakeland, E. K., Bohhme, S., She, J. X., Lu, C. C., and Eckert, C. G. (1990). Ancestral polymorphisms of MHC class II genes: Divergent allele advantage. *Immunol. Res.* **9,** 115–122.

Wedekind, C., and Furi, S. (1997). Body odour preferences in men and women: Do they aim for specific MHC combinations or simply heterozygosity? *Proc. R. Soc. Lond. B* **264,** 1471–1479.

Wedekind, C., Seebeck, T., Bettens, F., and Paepke, A. J. (1995). MHC-dependent mate preferences in humans. *Proc. R. Soc. Lond. B* **260,** 245–249.

Wegner, K. M., Kalbe, M., Kurtz, J., Reusch, T. B. H., and Milinski, M. (2003). Parasite selection for immunogenetic optimality. *Science* **301,** 1343.

Westerdahl, H. (2004). No evidence of an MHC-based female mating preference in great reed warblers. *Mol. Ecol.* **13,** 2465–2470.

Willse, A., Belcher, A. M., Preti, G., Wahl, J. H., Thresher, M., Yang, P., Yamazaki, K., and Beauchamp, G. K. (2005). Identification of MHC-regulated body odorants by statistical analysis of a comparative gas chromatography-mass spectrometry experiment. *Anal. Chem.* **77,** 2348–2361.

Willse, A., Kwak, J., Yamazaki, K., Preti, G., Wahl, J. H., and Beauchamp, G. K. (2006). Individual odortypes: Interaction of MHC and background genes. *Immunogenetics* **58**(12), 967–982.

Yamaguchi, M., Yamazaki, K., and Boyse, E. A. (1978). Mating tests with the recombinant congenic strain BALB. HTG. *Immunogenetics* **6,** 261–264.

Yamaguchi, M., Yamazaki, K., Beauchamp, G. K., Bard, J., Thomas, L., and Boyse, E. A. (1981). Distinctive urinary odors governed by the major histocompatibility locus of the mouse. *Proc. Natl. Acad. Sci. USA* **78,** 5817–5820.

Yamazaki, K., and Beauchamp, G. K. (2005). Chemosensory recognition of olfactory individuality. *Chem. Senses* Suppl. **1,** i142–i143.

Yamazaki, K., Boyse, E. A., Mike, V., Thaler, H. T., Mathieson, B. J., Abbott, J., Boyse, J., Zayas, Z. A., and Thomas, L. (1976). Control of mating preferences in mice by genes in the major histocompatibility complex. *J. Exp. Med.* **144,** 1324–1335.

Yamazaki, K., Yamaguchi, M., Andrews, P. W., Peake, B., and Boyse, E. A. (1978). Mating preferences of F_2 segregants of crosses between MHC-congenic mouse strains. *Immunogenetics* **6,** 253–259.

Yamazaki, K., Beauchamp, G. K., Bard, J., Thomas, L., and Boyse, E. A. (1982). Chemosensory recognition of phenotypes determined by the Tla and H-2K regions of chromosome 17 of the mouse. *Proc. Natl. Acad. Sci. USA* **79,** 7828–7831.

Yamazaki, K., Beauchamp, G. K., Egorov, I. K., Bard, J., Thomas, L., and Boyse, E. A. (1983a). Sensory distinction between H-2^{b} and H-2^{bm1} mutant mice. *Proc. Natl. Acad. Sci USA* **80,** 5685–5688.

Yamazaki, K., Beauchamp, G. K., Wysocki, C. J., Bard, J., Thomas, L., and Boyse, E. A. (1983b). Recognition of H-2 types in relation to the blocking of pregnancy in mice. *Science* **221,** 186–188.

Yamazaki, K., Beauchamp, G. K., Thomas, L., and Boyse, E. A. (1985). The hematopoietic system is a source of odorants that distinguish major histocompatibility types. *J. Exp. Med.* **162,** 1377–1380.

Yamazaki, K., Beauchamp, G. K., Matsuzaki, O., Kupniewski, D., Bard, J., Thomas, L., and Boyse, E. A. (1986). Influence of a genetic difference confined to mutation of H-2K on the incidence of pregnancy block in mice. *Proc. Natl. Acad. Sci. USA* **83,** 740–741.

Yamazaki, K., Beauchamp, G. K., Kupniewski, D., Bard, J., Thomas, L., and Boyse, E. A. (1988). Familial imprinting determines H-2 selective mating preferences. *Science* **240,** 1331–1332.

Yamazaki, K., Beauchamp, G. K., Imai, Y., Bard, J., Phelan, S. P., Thomas, L., and Boyse, E. A. (1990). Odortypes determined by the major histocompatibility complex in germ-free mice. *Proc. Natl. Acad. Sci. USA* **87,** 8413–8416.

Yamazaki, K., Beauchamp, G. K., Imai, Y., and Boyse, E. A. (1992). Expression of urinary H-2 odortypes by infant mice. *Proc. Natl. Acad. Sci. USA* **89,** 2756–2758.

Yamazaki, K., Beauchamp, G. K., Shen, F. W., Bard, J., and Boyse, E. A. (1994). Discrimination of odortypes determined by the major histocompatibility complex among outbred mice. *Proc. Natl. Acad. Sci. USA* **91,** 3735–3738.

Yamazaki, K., Beauchamp, G. K., Singer, A., Bard, J., and Boyse, E. A. (1999). Odortypes: Their origin and composition. *Proc. Natl. Acad. Sci. USA* **96,** 1522–1525.

Yamazaki, K., Beauchamp, G. K., Curran, M., Bard, J., and Boyse, E. A. (2000). Parent-progeny recognition as a function of MHC odortype identity. *Proc. Natl. Acad. Sci. USA* **97,** 10500–10502.

Ziegler, A., Kentenich, H., and Uchanska-Ziegler, B. (2005). Female choice and the MHC. *Trends Immunol.* **26,** 496–502.

6 Molecular Biology of Peptide Pheromone Production and Reception in Mice

Kazushige Touhara
Department of Integrated Biosciences, The University of Tokyo
Chiba 277–8562, Japan

ABSTRACT

Intraspecies communication via pheromones plays an important role in social and sexual behaviors, which are critical for survival and reproduction in many animal species. In mice, pheromonal signals are processed by the parallel action

Advances in Genetics, Vol. 59
Copyright 2007, Elsevier Inc. All rights reserved.
0065-2660/07 $35.00
DOI: 10.1016/S0065-2660(07)59006-1

of two olfactory systems: the main olfactory system and the vomeronasal pathway. Pheromones are recognized by chemosensory receptors expressed in the main olfactory epithelium and by V1R- and V2R-type receptors expressed in the vomeronasal organ (VNO). Mice take advantage of the chemical properties of both types of pheromones (i.e., volatile/nonvolatile) to precisely control the spatial and temporal transmission of their individual signals. The recent discovery of the exocrine gland-secreting peptide (ESP) family, which appears to encode a VNO-specific ligand repertoire, should open a new avenue to understanding peptide pheromone-mediated communication via the vomeronasal pathway in mice. In this chapter, I will review the current knowledge on genetic and molecular aspects of peptide pheromones and their receptors, by focusing primarily on the mouse VNO system. It is also an intriguing aspect to discuss peptide pheromones in the context of the evolutionary importance of species-specific chemical communication. © 2007, Elsevier Inc.

I. INTRODUCTION

Karlson and Luscher (1959) defined pheromones as "substances secreted to the outside of an individual and received by a second individual of the same species in which they release a specific reaction, for example, a definite behavior or developmental process." This new term was created based on identification of a volatile sex attractant, bombykol, that is released by the female silk moth *Bombyx mori* and elicits the full sexual behavior of male moths (Butenandt *et al.*, 1959). In contrast to such releaser pheromones that evoke an immediate innate behavioral reaction in the conspecifics, primer pheromones induce relatively long-term endocrine changes that affect developmental processes, including sexual maturation (Vandenbergh, 1994).

This definition for pheromones is restrictive, so whether certain isolated substances are pheromones has been controversial (Stowers and Marton, 2005). The recent discovery of new candidate pheromones, however, suggests that pheromones play much more diverse functional roles than included in the previous definition. Thus, a pheromone should be more broadly defined as a substance that is utilized for intraspecies communication. The key criterion is that pheromones must be released to the outside of an individual and received by conspecifics, and the pheromones themselves can send information about sex, strain, and species to the receiver (Meredith, 2001). Based on this revised definition, many candidate pheromones that I will discuss in this chapter can indeed be considered pheromones.

This definition does not contain any words such as "odor" or "odorant" because a pheromone does not have to be odorous or volatile as long as the signal is a chemical substance that is transferred between conspecifics. This means that

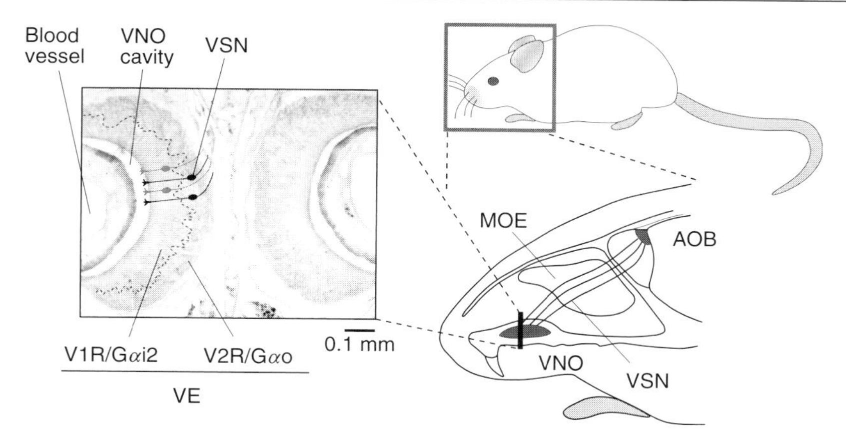

Figure 6.1. Molecular and anatomical organization of the mouse vomeronasal system: coronal view of the vomeronasal epithelium and sagittal view of the nasal cavity and forebrain. The vomeronasal epithelium (VE) in the vomeronasal organ (VNO), located below the main olfactory epithelium (MOE), can be divided into apical and basal layers, which express V1R/Gαi2 and V2R/Gαo, respectively. Pheromones are taken up into the VNO cavity by an active vascular pumping mechanism of the blood vessel, and recognized by vomeronasal sensory neurons (VSNs) that send their axons to the accessory olfactory bulb (AOB).

pheromones can be nonvolatile substances with a molecular weight larger than a few hundred, including relatively large organic compounds, peptides, and proteins. In the aqueous environment, peptides and proteins can be easily transferred to other individuals, so how terrestrial animals transmit and receive nonvolatile pheromones is an important question. Animals have addressed this by acquiring a specific organ called the vomeronasal organ (VNO) during their evolution from an aquatic to a terrestrial environment (Halpern and Martinez-Marcos, 2003). In rodent, the VNO lies within the nasal cavity and is filled with chemosensory neurons (Fig. 6.1). This organ can take up nonvolatile stimuli via an active vascular pumping mechanism during direct nasal contact or specific behaviors such as facial grooming (Meredith, 1994). Pheromonal signals, therefore, are likely processed by the parallel action of the main olfactory system, which primarily senses volatile pheromones, and the vomeronasal pathway, which detects stimuli with other chemical properties such as nonvolatile peptides (Baxi *et al.*, 2006; Spehr *et al.*, 2006b).

In this chapter, I will primarily focus on nonvolatile pheromonal cues (i.e., peptide pheromones) that are encoded by the genome, which ensure the stable expression of fixed information about the species, sex, and identity of an individual. Peptide pheromones are widely distributed throughout the prokaryotic and eukaryotic kingdoms, but in this chapter, I discuss only peptide pheromones produced by mice in the context of the evolutionary importance of species-specific chemical communication. Finally, I summarize what is currently known about the mechanisms underlying pheromone reception.

II. INTRASPECIES EFFECTS OF VOLATILE PHEROMONES IN MICE

In the mid-1950s, a few years before the term pheromone was introduced, several studies demonstrated that odors derived from mice affected the physiological status of conspecifics (Bruce, 1959; Van Der Lee and Boot, 1955; Whitten, 1956). Evidence for pheromone communication in mice has continued to accumulate since then. For example, odors from male mice accelerate the onset of puberty (Vandenbergh, 1969), synchronization of estrous cycles (Whitten, 1956), and pregnancy block in female mice (Bruce, 1959). In contrast, female odors postpone puberty of female mice and lengthen the estrous cycle (Van Der Lee and Boot, 1955). Female odors also influence developmental processes in male mice, for example, affecting the concentration of androgens (Macrides *et al.*, 1975) and the maturation of sperm (Koyama and Kamimura, 2000). The possession of such intraspecies effects satisfies the definition of a primer pheromone, which is a substance that primes changes in the physiological conditions of conspecifics (Koyama, 2004).

Studies on the molecular basis for the primer effects led to identification of physiologically active compounds from various tissues such as the bladder and preputial gland (Fig. 6.2). For example, 2,5-dimethylpyrazine, a compound in female urine, delays the onset of puberty and induces longer estrous cycles in female mice (Jemiolo and Novotny, 1993; Ma *et al.*, 1998), while it decreases the level of testosterone in male mice (see discussion in Koyama, 2004). In addition, 3,4-dehydro-*exo*-brevicomin, 2-*sec*-butyl-4,5-dihydrothiazole, 6-hydroxy-6-methyl-3-heptanone, and α- and β-farnesene from male urine or the preputial

	Compound	MW	Origin	Effect	Reception
Volatile	(Methylthio)methanethiol	94	Male urine	Attract female	MOE
	2,5-Dimethylpyrazine	108	Female urine	Postpone puberty Smaller testis	MOE/VNO
	2-*sec*-Butyl-4,5-dihydrothiazole	143	Male urine	Induce puberty	MOE/VNO
	6-Hydroxy-6-methyl-3-heptanone	144	Male urine	Induce puberty	MOE/VNO
	3,4-Dehydro-*exo*-brevicomin	154	Male urine	Induce puberty	MOE/VNO
	α,β-Farnesene	204	Preputial gland	Induce estrus Induce puberty	MOE/VNO
Non volatile	MHC peptide	~1 kDa	?	Pregnancy block	MOE/VNO
	ESP1	~7 kDa	ELG	Male signal?	VNO
	MUP	~15 kDa	Urine	Individual recognition?	VNO?

Abbreviations: MOE, main olfactory epithelium; VNO, vomeronasal organ; ELG, extraorbital lacrimal gland
ESP, exocrine gland-secreting peptide; MHC, major histocompatibility complex; MUP, major urinary protein

Figure 6.2. Chemical property, molecular weight, origin, and effect of mouse pheromones, and sensory systems mediating the responses.

gland accelerate the onset of puberty and induce estrous in female mice (Jemiolo *et al.*, 1986; Ma *et al.*, 1999; Novotny *et al.*, 1999). These male odors also induce aggressive behavior in male mice (Novotny *et al.*, 1985). Finally, a study showed that female mice are attracted by (methylthio)methanethiol (MTMT), a highly volatile compound in male urine (Lin da *et al.*, 2005). These studies suggest that the urine-derived odors not only cause primer effects, but they also attract the opposite sex and therefore act as a kind of releaser pheromone.

III. PEPTIDES AND PROTEINS AS SOCIAL SIGNALS IN MICE

Volatile signals are used to immediately transmit the current status of an animal to other animals in the vicinity, but they are transient. Thus, nonvolatile peptides and proteins have the advantage that they remain at scent marks for an extended time. Because volatile pheromones are metabolically produced, they vary according to the reproductive condition and health of the individual, whereas peptide pheromones are genetically encoded, so that the information remains constant throughout the lifespan of the animal. Because of their stability, peptide pheromones appear to have evolved to carry information about the species, sex, and identity of the individual (Altstein, 2004).

The first evidence for involvement of peptides in pheromonal action in mice came from research on the major urinary proteins (MUPs), which are found in large quantities in the urine of sexually mature male mice (Beynon and Hurst, 2004; Hurst and Beynon, 2004). MUPs form a group in the lipocalin family of peptides, and they bind the volatile male pheromones, 3,4-dehydro-*exo*-breviomin and 2-*sec*-butyl-4,5-dihydrothiazole (Cavaggioni and Mucignat-Caretta, 2000; Timm *et al.*, 2001). In this way, MUPs extend the duration of pheromone release from deposited urine to many hours. In addition, MUPs consist of over 30 homologous genes, and further, MUPs are highly polymorphic in wild populations, suggesting that the pattern of MUPs differs between individuals (Beynon *et al.*, 2002). The extreme polymorphism and sexual dimorphism of MUPs indicate that, in addition to the function as a pheromone reservoir, they convey individual identity (Beynon and Hurst, 2004). Indeed, MUPs play an important role in the individual recognition mechanism (Hurst *et al.*, 2001). Whether MUPs themselves function as a chemosignal in mice awaits further studies on whether the olfactory or vomeronasal system responds to ligandless MUPs and transmits a signal to the brain.

Major histocompatibility complex (MHC) peptides comprise a second class of peptide pheromones (Boehm and Zufall, 2006). MHC peptides are ligands for MHC molecules and are encoded by genes with a high degree of polymorphism (Rammenses *et al.*, 1997) (see also the section by Yamazaki, this volume). This polymorphism leads to differences in the type of MHC peptides presented by MHC molecules (Trombetta and Mellman, 2005). Peptide–MHC complexes on the cell surface are recognized by T cells and send information about self or nonself, which is important for the immune defense system (Boehm, 2006). It is also possible that the presented MHC peptides can eventually be secreted to the outside in bodily fluids and therefore utilized for interindividual communication (Boehm and Zufall, 2006). This possibility has recently been examined in the context of pregnancy blocks (Bruce's effects). A MHC peptide from an unfamiliar male caused pregnancy block when added to the familiar male's urine, which otherwise had no effect (Leinders-Zufall *et al.*, 2004). It is not yet clear whether a MHC peptide alone is sufficient to cause this effect because the experiments were done in the presence of urine, nor it is clear whether the peptide is indeed secreted outside body. Nonetheless, MHC peptides appear to function as individual recognition cues that cause neuroendorinological changes.

IV. THE EXOCRINE GLAND-SECRETING PEPTIDE FAMILY IN MICE

A. The search for a VNO ligand

Initially, it was thought that pheromonal information is processed by the VNO system and odorants by the main olfactory system. Increasing evidence, however, has suggested that both systems can detect pheromones (Baxi *et al.*, 2006; Spehr *et al.*, 2006b; Stowers and Marton, 2005). For example, imaging and electrophysiological techniques have revealed that the volatile urinary pheromones implicated in various intraspecies responses activate both olfactory and vomeronasal sensory neurons (Leinders-Zufall *et al.*, 2000; Liberles and Buck, 2006; Spehr *et al.*, 2006a; Xu *et al.*, 2005). The male-specific urinary pheromone, MTMT, has been shown to attract female mice by activating the main olfactory neuronal pathway (Lin da *et al.*, 2005). Also, MHC peptides appear to stimulate both olfactory and vomeronasal neurons, leading to distinct physiological and behavioral outcomes (Spehr *et al.*, 2006a). Finally, *in vitro* evidence that the VNO can detect nonpheromonal odorants suggests that the VNO has more diverse functions than previously imagined (Sam *et al.*, 2001).

To examine the function of the VNO alone, we searched for substances from mice that activate vomeronasal sensory neurons *in vivo*. At the time when we began this search, no single compound was known that could clearly activate the VNO pathway *in vivo*. Two experiments were thought to monitor neuronal responses in behaving mice: electrophysiological recording of single cells in the accessory olfactory bulb (AOB) and immunohistological monitoring of the expression of immediate early genes in the VNO. Although the former technique has been successfully performed by Luo *et al.* (2003), we did not expect it to be suitable for screening of active compounds because the signals would likely be detected by narrowly tuned receptor neurons connected to only a few mitral cells in the AOB (Del Punta *et al.*, 2002b; Wagner *et al.*, 2006) and therefore difficult to find the target neurons. In contrast, the latter strategy using c-Fos or egr-1 as a marker in the VNO and AOB has been applied to identify responsive neurons (Brennan *et al.*, 1992, 1999; Halem *et al.*, 1999, 2001). In fact, we successfully detected c-Fos-induced vomeronasal sensory neurons in BALB/c female mice that respond to male-soiled bedding but not to female-soiled bedding, indicating the presence of a sex-specific substance(s) in BALB/c male-soiled bedding (Kimoto and Touhara, 2005) (Fig. 6.3A).

Unexpectedly, we found little or no c-Fos-inducing activity in urine; however, we identified an activity in fur shaved from male mice, suggesting that the active compounds originate in an exocrine gland of adult male mice rather than from urine (Kimoto *et al.*, 2005). Surprisingly, the activity was observed in the extraorbital lacrimal gland, a tear gland below the ear, suggesting that the activity was released in male tear fluid (Kimoto *et al.*, 2005). Purification of the active compound by DEAE anionexchange and C4 reverse-phase chromatography revealed that it is a 7-kDa peptide encoded by a previously uncharacterized gene on chromosome 17 (Fig. 6.4). We named this new peptide ESP1, for *e*xocrine gland-*s*ecreting *p*eptide (ESP) (Kimoto *et al.*, 2005). ESP1 was exclusively expressed in the extraorbital lacrimal gland of sexually mature male mice,

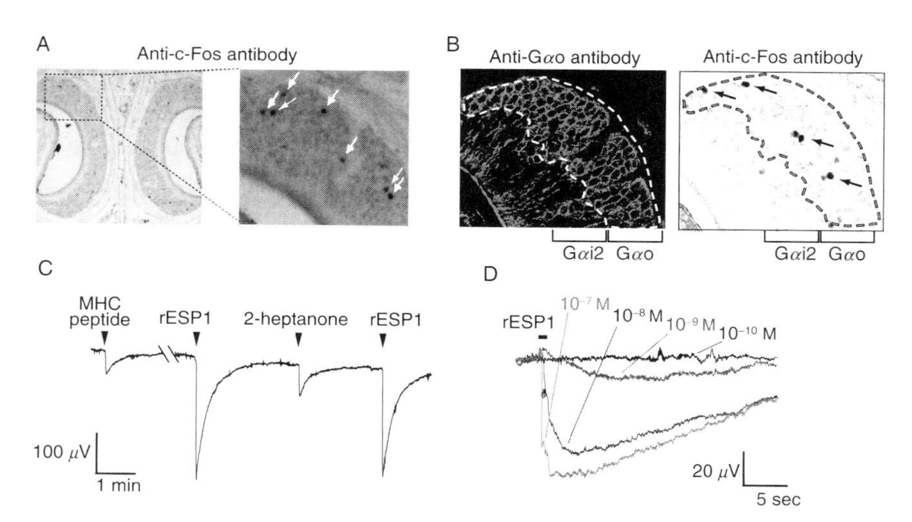

Figure 6.3. Induction of c-Fos expression in a subset of BALB/c female VSNs in the basal Gαo-expressing layer of the VNO by male-soiled bedding (A, B) and electrical responses of the female VNO to recombinant ESP1 and other related pheromones (C, D) (adapted from Kimoto *et al.*, 2005, 2006).

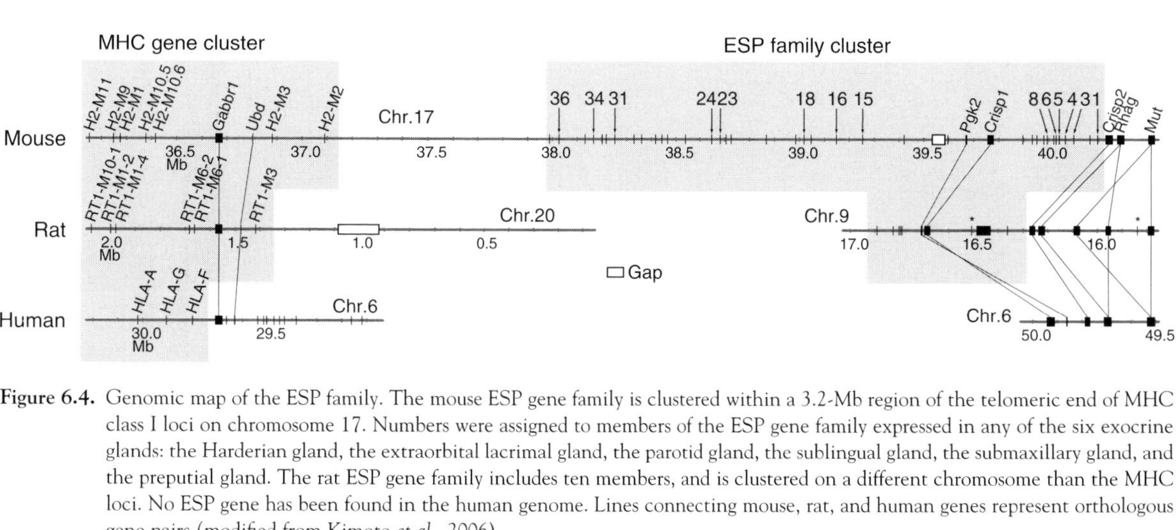

Figure 6.4. Genomic map of the ESP family. The mouse ESP gene family is clustered within a 3.2-Mb region of the telomeric end of MHC class I loci on chromosome 17. Numbers were assigned to members of the ESP gene family expressed in any of the six exocrine glands: the Harderian gland, the extraorbital lacrimal gland, the parotid gland, the sublingual gland, the submaxillary gland, and the preputial gland. The rat ESP gene family includes ten members, and is clustered on a different chromosome than the MHC loci. No ESP gene has been found in the human genome. Lines connecting mouse, rat, and human genes represent orthologous gene pairs (modified from Kimoto *et al.*, 2006).

suggesting that it is a sex pheromone (Kimoto *et al.*, 2005). Recombinant ESP1 induced c-Fos expression in vomeronasal sensory neurons of female mice, and importantly, elicited an electrical response in the VNO (Kimoto *et al.*, 2005) (Fig. 6.3C and D). ESP1 did not induce c-Fos expression in BALB/c male mice, apparently because the ability to express c-Fos is lost in vomeronasal sensory neurons when they are continuously exposed to their own pheromones (see discussion in Kimoto *et al.*, 2005). These results provided direct evidence that a sex-specific substance released by male mice is taken up by the VNO of female mice *in vivo*, where it stimulates vomeronasal sensory neurons.

B. Genomic organization of the ESP family

ESP1 turned out to be a member of a large multigene family that was unknown when we reported it in October 2005 (Kimoto *et al.*, 2005). This novel family has been named the ESP family (Kimoto *et al.*, 2005). In October 2005, the ESP family included 24 members, but by May 2006, the number had increased to 38 because there had been many gaps in the region of the ESP gene cluster (Fig. 6.4) (Kimoto *et al.*, 2006). Of the 38 ESP genes, 15 are expressed in an exocrine gland around the face area, supporting the idea that the ESP gene family encodes ESP (Fig. 6.5A).

Thus, although analysis of the mouse genome is largely thought to be complete, many important genes remain to be discovered. This may be due to the fact that comprehensive proteomic studies usually target proteins larger than 10 kDa, while peptidome studies have targeted peptides smaller than 5 kDa. As a result, proteins between 5 and 10 kDa tend to be missed. Furthermore, ESP1 consists of two introns and three exons, making it difficult to predict the mature protein (Kimoto *et al.*, 2005). Therefore, the discovery of the ESP1 has not only shed light on sexual communication via the vomeronasal system in mice but also led to a reconsideration of the comprehensiveness of so-called comprehensive bioinformatics.

The ESP gene family is clustered within a 3.2-Mb region at the telomeric end of MHC class I loci on mouse chromosome 17 (Kimoto *et al.*, 2005) (Fig. 6.4). This is of particular interest because of two previous observations: (1) polymorphic MHC molecules have been shown to influence behavioral decisions in the context of social recognition in mice (Beauchamp and Yamazaki, 2003; Yamazaki *et al.*, 1999) (see also the section by Yamazaki, this volume), and (2) the M10 family of MHC class I molecules is expressed in mouse vomeronasal sensory neurons (Ishii *et al.*, 2003; Loconto *et al.*, 2003). The genetic proximity to the MHC cluster suggests an evolutionary relationship, but the fact that the rat ESP gene family is clustered on a different chromosome than the MHC loci excludes the possibility of coevolution. In rat, there are only

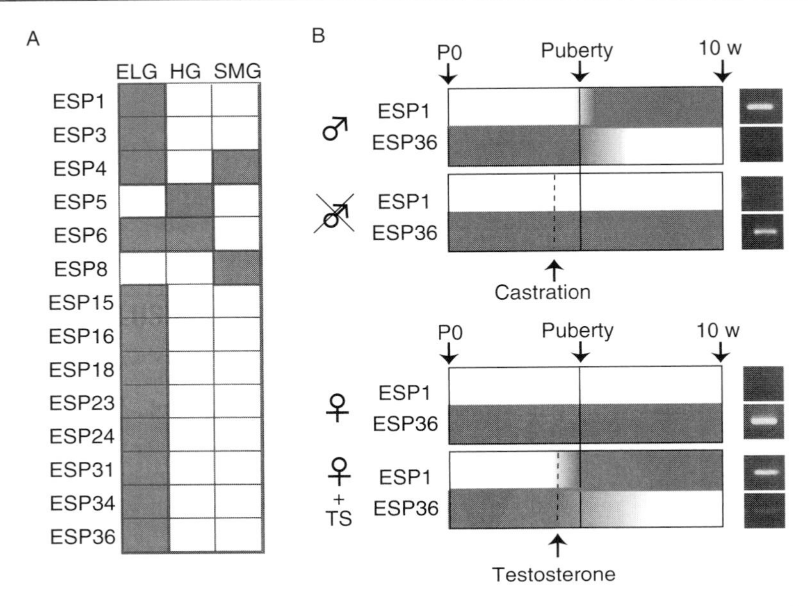

Figure 6.5. Expression patterns of the ESP family members and sexual dimorphism in BALB/c mice. (A) Expression patterns of the ESP family in the extraorbital lacrimal gland (ELG), the Harderian gland (HG), and the submaxillary gland (SMG). (B) Sexual dimorphic expression of ESP1 and ESP36 and testosterone dependency (see details in Kimoto *et al.*, 2006).

ten ESPs, and homologous genes have not been found in the human genome, suggesting that the ESP gene family has undergone rapid modification during the process of evolution (Fig. 6.4) (Kimoto *et al.*, 2006).

Sequence motifs have not been found in the ESP family, and, therefore, the origin of the ESP family remains unknown. ESP peptides range from 60 to 160 amino acids in length, and most possess a putative signal sequence motif, suggesting that they are secreted. Indeed, Western blot analysis using an anti-ESP1 antibody demonstrated that ESP1 is secreted in male tear fluid (Kimoto *et al.*, 2005). The mature ESP1 is secreted after cleavage of the signal sequence, processing of several N-terminal amino acids, and truncation of 1–3 amino acids at the C-terminus (Kimoto *et al.*, 2005). Thus, each ESP gene appears to encode a single polypeptide from which several isoforms are produced. Although apparent polymorphism between individuals has not been found, the possibility of differences between strains cannot be excluded. Preliminary results suggest that there are some differences in the 5′-upstream region of ESPs between BALB/c and B6 strains (Haga and Touhara, unpublished data).

C. Sexual dimorphism of the ESP family

The expression of ESP1 in the extraorbital lacrimal gland is male-specific and is observed after 4 weeks of age in BALB/c mice (Kimoto *et al.*, 2005) (Fig. 6.5B, upper panel). The onset of the expression appears to be hormonally regulated, and mice castrated at 3 weeks fail to express ESP1 in the adult stages (Fig. 6.5B, upper panel). Daily administrations of testosterone after castration rescues ESP1 expression, suggesting that the expression of ESP1 is androgen dependent (Kimoto *et al.*, 2006).

A female-specific ESP, ESP36, has been found, and its expression is negatively regulated by androgen (Kimoto *et al.*, 2006). Both male and female mice express ESP36 before puberty, but in male mice, the expression stops after puberty, and it is not observed after 5 weeks of age (Fig. 6.5B, lower panel). Treatment of adult female mice with testosterone downregulates the expression of ESP36 and induces the expression of male-specific ESP1 (Fig. 6.5B, lower panel). These observations are of particular interest in terms of the epigenetic occurrence of sexual dimorphism.

D. Implications in the vomeronasal system

What is the function of the ESP family? One obvious possibility is that the ESP family encodes a repertoire of VNO ligands. Fifteen out of the 38 mouse ESPs appear to be expressed in the extraorbital lacrimal gland, the Harderian gland, and/or the submaxillary gland. Therefore, these expressed ESPs are expected to be secreted in tear fluid or saliva. Secreted ESPs could be taken up by the VNO as observed for ESP1. Electrophysiological testing of recombinant proteins shows that all 15 ESPs evoke a negative field potential in the VNO, suggesting that the ESPs comprise a ligand repertoire for vomeronasal sensory neurons (Kimoto *et al.*, 2006). Importantly, electro-olfactograms do not show a response in the olfactory epithelium, indicating that the effects of ESPs are specific to the VNO.

V. MOLECULAR BIOLOGY OF PHEROMONE RECEPTION

In mice, pheromones are detected by two anatomically distinct olfactory systems: the main olfactory system and the vomeronasal (or accessory olfactory) system. The olfactory neurons express a repertoire of olfactory receptors (ORs), which is responsible for detecting volatile odorants and pheromones. The vomeronasal sensory epithelium in the VNO can be divided into apical and basal layers, which express V1R- and V2R-type receptors, respectively (Berghard and Buck, 1996; Dulac and Axel, 1995; Herrada and Dulac, 1997; Matsunami and Buck, 1997; Ryba and Tirindelli, 1997; Tirindelli *et al.*, 1998) (Fig. 6.1). V1Rs and V2Rs begin to be expressed after birth (Dulac and Axel, 1995), whereas ORs

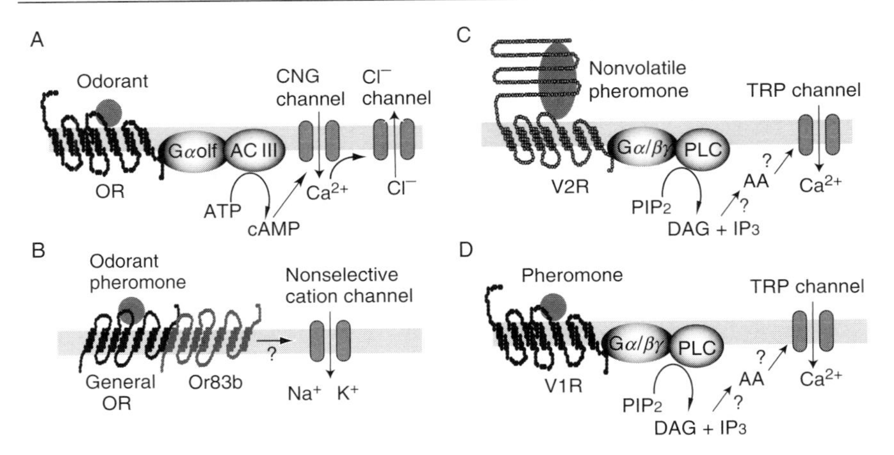

Figure 6.6. Signal transduction mechanisms mediated by olfactory receptor in olfactory neurons in mice (A) and insect (B), and vomeronasal receptor in vomeronasal sensory neurons in mice (C, D). OR, olfactory receptor; AC, adenylate cyclase; CNG, cyclic nucleotide gated; VR, vomeronasal receptor; PLC, phospholipase C; TRP, transient receptor potential; PIP2, phosphoinositide-4,5-bisphosphate; DAG, diacylglycerol; IP3, inositol 1,4,5-triphosphate; AA, arachidonic acid.

begin to be expressed during embryogenesis (Sullivan *et al.*, 1995). The segregation of the vomeronasal epithelium into two layers occurs during the postnatal period (Berghard and Buck, 1996), consistent with the development of the VNO as an organ that functions in adult communication. It has been suggested that ORs, V1Rs, and V2Rs mediate both odorant and pheromone signals by coupling to G-proteins, specifically, Gαolf, Gαi2, and Gαo, respectively (Fig. 6.6). The pheromone signals transduced by these receptors are eventually converted to electronic activity via second messenger-mediated channel opening (Fig. 6.6). In this section, I summarize the current knowledge about pheromone–receptor interactions.

A. V1R-type receptors

The V1R family consists of a total of 352 sequences in mice, 187 of which appear to encode full-length open reading frames (Grus *et al.*, 2005) (Fig. 6.7). Most of the gene family is located on chromosomes 6, 7, and 13 and, like OR genes, they are encoded by a single exon (Shi *et al.*, 2005; Zhang *et al.*, 2004). In humans, over 90% of the V1R genes have been converted to pseudogenes so that only five V1R genes appear to be intact (Rodriguez and Mombaerts, 2002; Rodriguez *et al.*, 2000). V1Rs do not show significant homology with the ORs, but they are weakly related to the T2R family of bitter receptors (Dulac and Axel, 1995).

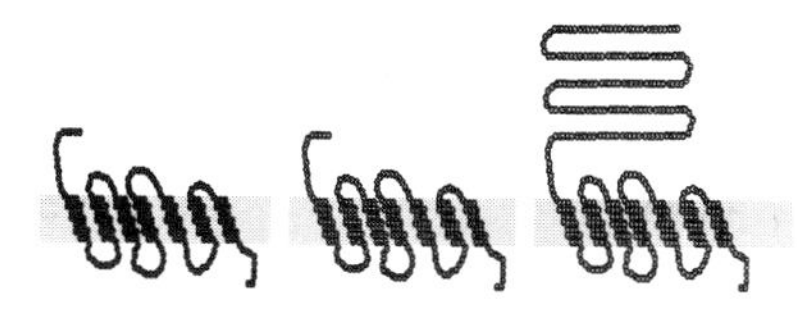

Organism	OR	V1R	V2R
Human	388 (802)	5 (~100)	0 (??)
Mouse	1037 (1391)	187 (352)	61 (209)
Zebra fish	98 (133)	1 (1)	88 (106)
Puffer fish	40 (94)	1 (1)	~15 (~30)

Figure 6.7. Topology and numbers of olfactory receptor (OR), V1R-type vomeronasal receptor (V1R), and V2R-type vomeronasal receptor (V2R) in various organisms. The numbers in parentheses represent numbers of total genes including pseudogenes (Grus *et al.*, 2005; Yang *et al.*, 2005; Niimura and Nei, 2005).

The expression of V1Rs is restricted to the apical neuroepithelium in mice (Dulac and Torello, 2003), whereas in humans, which lack a functional VNO, V1Rs are expressed in the olfactory epithelium (Rodriguez *et al.*, 2000). The genomic structure and expression patterns of V1Rs appear to have undergone rapid changes during the process of the evolution (Grus *et al.*, 2005).

Evidence for V1Rs as pheromone receptors has emerged (Mombaerts, 2004; Rodriguez, 2004). First, a V1R expressed ectopically in vomeronasal sensory neurons was shown to respond to urine (Hagino-Yamagishi *et al.*, 2001). Second, mutant mice lacking a cluster of V1Rs fail to respond to some volatile pheromones and show decreased aggressiveness and sexual behavior (Del Punta *et al.*, 2002a). Third, V1Rb2 expressed in vomeronasal neurons responds to 2-heptanone, one of the components that cause extension of the estrous cycle (Boschat *et al.*, 2002). Together with evidence from Ca^{2+} imaging of intact epithelial slices showing that vomeronasal neurons in the apical layer respond to various volatile pheromones (Leinders-Zufall *et al.*, 2000), these findings suggest that V1Rs serve as sensors for volatile pheromones. Vomeronasal neurons, however, can also detect nonpheromonal odorants (Sam *et al.*, 2001), implying that V1R-expressing neurons may not be restricted to detection of species-specific volatile pheromones for mating but also may perceive nonspecies-related signals such as those from prey, predators, and the physical environment.

B. V2R-type receptors

In mice, the V2R family comprises 61 intact genes out of a total of 209 genes including pseudogenes (Yang *et al.*, 2005) (Fig. 6.7). Several years were needed for the comprehensive genomic analysis of the V2R family because of their complex intron/exon structures (Herrada and Dulac, 1997; Matsunami and Buck, 1997; Ryba and Tirindelli, 1997). V2Rs possess a long extracellular N-terminus that is common to Ca^{2+}-sensing and metabotropic glutamate receptors and the T1R family of sweet and umami receptors (Mombaerts, 2004). Intact V2R genes have not been found in the human genome; there appear to be only ~10 V2R pseudogenes, suggesting that V2R genes have changed more dramatically than V1R genes. The expression of V2Rs is restricted to the basal neuroepithelium in mice (Herrada and Dulac, 1997; Matsunami and Buck, 1997; Ryba and Tirindelli, 1997), and, as in the case of ORs and V1Rs, the one cell–one receptor rule seems to be conserved (Del Punta *et al.*, 2002b), with the exception that one V2R, called V2R2, appears to be coexpressed with other V2Rs (Martini *et al.*, 2001).

The potential ligands for V2Rs include peptide pheromones such as MUPs, MHC peptides, and ESPs. Biochemical studies in rats have demonstrated that some MUPs activate vomeronasal neurons in the basal layer expressing Gαo and V2Rs (Krieger *et al.*, 1999). Ca^{2+} imaging studies in mice indicate that vomeronasal sensory neurons responding to MHC peptides are located in the basal neuroepithelium where V2Rs are expressed (Leinders-Zufall *et al.*, 2004). Also, male-specific ESP1 induces c-Fos expression in Gαo/V2R-expressing neurons in female mice (Kimoto *et al.*, 2005) (Fig. 6.3B). Although these studies do not provide direct evidence for pheromone–V2R interactions, the results support the idea that V2R-expressing neurons respond to nonvolatile pheromones.

To determine whether this is correct, further studies were performed to identify which V2R(s) is expressed in vomeronasal neurons that recognize ESP1. Vomeronasal neurons were double-labeled with an anti-c-Fos antibody and an *in situ* RNA probe for V2Rs. Of 12 different RNA probes, 1 of them, V2Rp, which was designed to hybridize with 5 highly homologous V2Rp genes, clearly recognized c-Fos-positive neurons responding to ESP1 (Kimoto *et al.*, 2005). In contrast, none of other probes overlapped with c-Fos. Further studies using specific probes to discriminate the five V2Rp revealed that V2Rp5 was expressed in all of the c-Fos-induced neurons (Haga *et al.*, 2007). The results provide evidence that single V2Rs are responsible for the detection of peptide pheromones, which results in a narrow ligand spectrum for individual vomeronasal sensory neurons. V2Rp5, which recognizes male-specific ESP1, however, is expressed in both male and female mice after birth, suggesting that the peptide pheromone but not the receptor shows an apparent sexually dimorphic pattern of expression. V2Rs for other ESPs remain to be identified.

C. Olfactory receptors for pheromones

The main olfactory system detects not only general odorants but also volatile pheromones (Liberles and Buck, 2006; Restrepo *et al.*, 2004; Spehr *et al.*, 2006a; Xu *et al.*, 2005). Electrical recording and Ca^{2+} imaging techniques have shown that the mouse urinary pheromones 2,5-dimethylpyrazine and 2-heptanone stimulate olfactory neurons (Spehr *et al.*, 2006b). Also, the female attractant MTMT was originally discovered in male urine by screening for volatile chemicals that induced a neural response in the main olfactory bulb (Lin da *et al.*, 2005). Although cognate ORs for these volatile pheromones have not been fully identified, the results strongly suggest that ORs can function as pheromone sensors. A second family of ORs, called "trace amine-associated receptors" (TAARs), is found to be expressed in the olfactory epithelium and to recognize volatile pheromonal amine compounds present in urine, suggesting that volatile pheromones are perceived by a diverse array of chemosensory receptors (Liberles and Buck, 2006). If these receptors participate in pheromone detection, it is reasonable to expect that these possess a high sensitivity and selectivity for ligands and that the neuronal circuit is constructed so that the sensory inputs from receptor-expressing neurons are connected to anatomical areas controlling reproductive physiology and behavior. Indeed, mouse olfactory neurons detect social chemosignals with extremely high sensitivity ($\sim 10^{-10}$) (Spehr *et al.*, 2006b), and a subset of olfactory neurons have been shown to connect to neurons that connect to hypothalamic neurons regulating endocrine status and, ultimately, sociosexual behavior (Boehm *et al.*, 2005; Yoon *et al.*, 2005).

In other animal species, ORs sometimes function as pheromone detectors. In fish, olfactory neurons in the olfactory epithelium express 1 V1R, 50–100 V2Rs, and 50–100 ORs, each of which is expressed in single neurons in a mutually exclusive manner (Asano-Miyoshi *et al.*, 2000; Cao *et al.*, 1998; Hansen *et al.*, 2004; Ngai *et al.*, 1993; Pfister and Rodriguez, 2005). V2Rs are thought to detect amino acids and small peptides, which includes food odorants for fish, whereas ORs may help recognize pheromones such as bile acids, steroid-derived compounds, and prostaglandins (Friedrich and Korsching, 1998; Luu *et al.*, 2004; Sorensen and Caprio, 1998). Because fish lack a VNO, a discrete population of olfactory neurons has acquired the ability to send information related to odorants or pheromones by innervating a specific neural circuit (Sato *et al.*, 2005).

Insects, the largest terrestrial phyla, express ~60 ORs in olfactory neurons of the antennae and maxillary palps (Clyne *et al.*, 1999; Gao and Chess, 1999; Vosshall *et al.*, 1999). In the silk moth, an OR, BmOR1, acts as a specific receptor for bombykol, a sex pheromone released by female moths that attracts male moths (Nakagawa *et al.*, 2005; Sakurai *et al.*, 2004). Insects appear to have selected pheromone receptors from a repertoire of ORs, rather than creating a new family of receptors.

In humans, which lack a functional VNO, at most five V1Rs are expressed in the nasal mucosa, where they could function as pheromone receptors (Rodriguez and Mombaerts, 2002; Rodriguez *et al.*, 2000). It is also possible that some ORs and TAARs recognize a putative human pheromone, but clarification of this possibility awaits identification of an authentic pheromone in human.

VI. PHEROMONE SIGNALS TO THE BRAIN

VNO-mediated chemosensory signals are sent to the AOB, which then projects to the basal forebrain, regions that are important for reproductive and mating behavior (Dulac and Torello, 2003; Halpern and Martinez-Marcos, 2003; Keverne, 2004; Meredith, 1998). The AOB projects to the medial and posteromedial cortical nuclei of the amygdala from which the information is relayed to the medial preoptic area of the hypothalamus, leading to behavioral and endocrine responses. Because the medial preoptic area contains many gonadotropin-releasing hormone (GnRH) or luteinizing hormone-releasing hormone (LHRH) neurons, the VNO inputs to the amygdala-hypothalamic pathway regulate reproductive function. Expression of a targeted transneuronal tracer in transgenic mice was used to visualize the VNO-AOB neuronal circuit that connects to the GnRH/LHRH neurons (Boehm *et al.*, 2005). Although one study suggests that this input is negligible (Yoon *et al.*, 2005), VNO signaling appears to be processed at the basal forebrain area, eventually leading to gender recognition and aggression.

The main olfactory pathway projects to the main olfactory bulb, which then projects to much wider areas, including the anterior olfactory nucleus, olfactory tubercle, pyriform cortex, entorhinal cortex, and the anterior cortical and posterolateral cortical nuclei of the amygdala, which are distinct from the vomeronasal amygdala [i.e., medial amygdala (MeA) and posteromedial cortical nuclei] (Horowitz *et al.*, 1999; Yoshihara *et al.*, 1999; Zelano and Sobel, 2005; Zou *et al.*, 2001, 2005). Genetic approaches revealed that some GnRH neurons might also receive olfactory input from the main olfactory epithelium, suggesting that the main olfactory system plays an essential role in reproductive and mating behavior (Boehm *et al.*, 2005; Yoon *et al.*, 2005). The signal inputs to the hypothalamus via the olfactory pathway are consistent with the idea that both the olfactory and vomeronasal systems participate in social and sexual behavior and reproduction.

How are the inputs from the two pathways integrated to produce behavioral or endocrine outcomes of pheromonal communication? This can be illustrated using the example of pheromonal communication in mice (Fig. 6.8). When a male mouse is introduced into the cage of a virgin female mouse, the female mouse is attracted to the male mouse by the male-specific volatile pheromone MTMT, which is received by her main olfactory system. Subsequent direct contact allows for detection of nonvolatile cues such as ESP1 and MHC peptides by the VNO, allowing the mouse to recognize the gender and identity of the other mouse. Thus, the processing of pheromones in the two systems appears to trigger a final behavior, in this example, mating or aggression. Generation of the output behavior may require parallel sensing of volatile and nonvolatile pheromones by the olfactory and vomeronasal systems, respectively, but it is

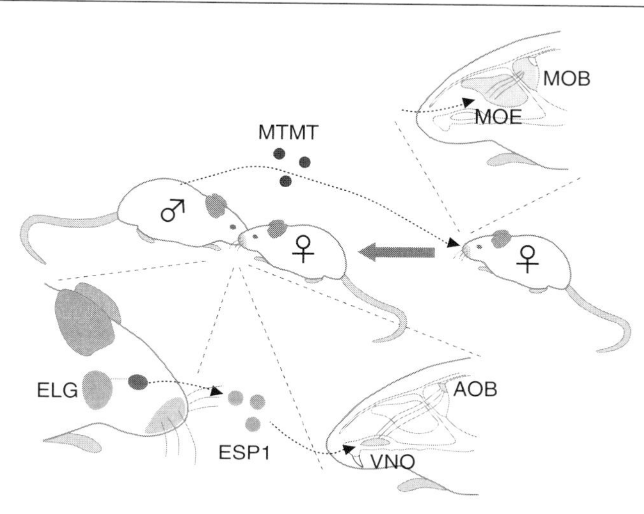

Figure 6.8. A model of sexual communication mediated by volatile and nonvolatile pheromones in mice. MOE, main olfactory epithelium; MOB, main olfactory bulb; VNO, vomeronasal organ; AOB, accessory olfactory bulb; ELG, extraorbital lacrimal gland; ESP, exocrine gland-secreting peptide; MTMT, (methylthio)methanethiol.

also possible that the olfactory detection of volatile pheromones is a prerequisite for the efficient incorporation of nonvolatile pheromones in the VNO cavity (Mandiyan *et al.*, 2005). Further studies are needed to determine how mice utilize two chemically distinct pheromone categories and two anatomically distinct olfactory pathways that connect to neighboring regions of the brain to establish social and sexual communication.

VII. PHEROMONES AND EVOLUTION

Volatile pheromones are a relatively recent development in animal evolution and appear to have arisen when animals underwent a transition to a terrestrial environment during the Cambrian explosion (Fig. 6.9). In contrast, peptide pheromones are widely observed in mice as well as in prokaryotic and eukaryotic species, including bacteria, fungi, worms, mollusks, and amphibians (Altstein, 2004). For example, bacteria produce a mating peptide pheromone that initiates cell–cell interaction and a peptide signal that transmits information about cell density as part of the so-called quorum-sensing system (Waters and Bassler, 2005). Male newts release a female-attracting decapeptide, called sodefrin, from the abdominal glands, and it is received by the female VNO

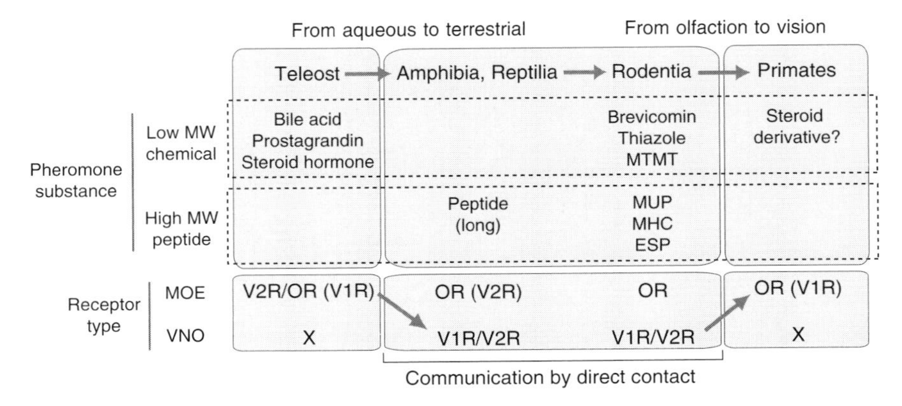

Figure 6.9. Evolutional changes in chemical properties of pheromones and the sensing systems. The appearance of the VNO coincides with the acquisition of the lung respiration system when animals underwent a transition to a terrestrial environment during the Cambrian explosion. Development of the visual system appears to have provided a positive pressure for elimination of the vomeronasal system and peptide pheromones. Peptide pheromones appear to be utilized by animals that make communication through the vomeronasal system by direct contact. MW, molecular weight; MOE, main olfactory epithelium; VNO, vomeronasal organ; OR, olfactory receptor; VR, vomeronasal receptor; ESP, exocrine gland-secreting peptide; MHC, major histocompatibility complex; MUP, major urinary protein; MTMT, (methylthio)methanethiol.

(Kikuyama and Toyoda, 1999). Also, frogs release a 25-amino acid peptide, called splendipherin, that attracts female frogs (Wabnitz *et al.*, 1999), and male salamanders produce a 22-kDa protein from the mental gland, termed plethodontid receptivity factor, that appears to increase female receptivity (Rollmann *et al.*, 1999). This peptide-mediated sexual communication in amphibians is considered to be analogous to the ESP-mediated interaction in mice, although the peptide sequences are not homologous, suggesting that the genes encoding peptide pheromones did not coevolve. Nonetheless, because of the diversity of organisms producing peptide pheromones, comparative studies could help establish the evolutionary importance of peptide-mediated communication.

The appearance of the VNO coincides with the acquisition of the lung respiration system during the Cambrian explosion (Halpern and Martinez-Marcos, 2003) (Fig. 6.9). A change in tissues that express the V1R and V2R family is also observed at this point in evolution: fish express V1Rs and V2Rs in the olfactory epithelium (Eisthen, 2004; Hansen *et al.*, 2004; Pfister and Rodriguez, 2005), whereas, in terrestrial animals, V1Rs and V2Rs are expressed in the VNO, which is thought to have first appeared in amphibians. As described above, peptide pheromones are widely utilized in amphibians and reptiles, and the VNO in these species functions as a peptide pheromone detector.

Although the VNO and peptide pheromones appear to play the most active part in communication between individuals within rodent species, it appears that there was a tendency for a loss of VNO function and the elimination of peptide pheromones at this point in evolution. Indeed, the number of ESP family members quickly decreased within rodent species, and they disappeared completely somewhere prior to the appearance of primates (Kimoto *et al.*, 2006). The loss of VNO function occurred a little later, and a vestigial VNO and expression of intact V1Rs can still be found in humans (Meredith, 2001; Rodriguez and Mombaerts, 2002; Rodriguez *et al.*, 2000) (Fig. 6.9). Nonetheless, development of the visual system appears to have provided a positive pressure for elimination of the vomeronasal system and peptide pheromones during the evolution of terrestrial animals.

Whether peptide recognition in the VNO can be considered part of olfaction remains unclear, but we made an interesting observation in this regard: to be taken up by the VNO and stimulate the vomeronasal sensory neurons, ESP1 must be dry (Kimoto *et al.*, 2005). This suggests that dried ESP1 enters the nasal cavity in a powder phase due to sniffing behavior, which allows it to access the vomeronasal cavity. Similarly, the nonvolatile fluorescent dye rhodamine can access the nasal cavity (Spehr *et al.*, 2006a). Thus, nonvolatile cues have characteristics of airborne volatile compounds for animals that show sniffing behavior such as rodents and even dogs. In addition, the peptide recognition system also appears to be analogous to the somatosensory system, wherein direct contact or touch induces sensation. Further, water-soluble taste molecules are sensed by the T1Rs, which have weak homology with the V2Rs, suggesting the intriguing possibility that the taste system replaced the V2R-mediated detection of water-soluble peptides. Such a mixed concept for the peptide–VNO system in mice might be related to the origins of the multimodal sensing strategy that human and other primates use to sense complex external information.

VIII. CONCLUDING REMARKS

In this chapter, I reviewed the current knowledge on genetic and molecular aspects of peptide pheromones and their receptors, by focusing primarily on the mouse system. The discovery of the ESP family, which appears to encode a VNO-specific ligand repertoire, should open a new avenue to understanding peptide pheromone-mediated communication via the VNO in mice (Kimoto *et al.*, 2005). To fully understand pheromonal behavior in mice, however, both volatile and nonvolatile cues must be considered. Nonvolatile cues are utilized for communication within a short distance, whereas volatile cues are intended to transmit a signal across a relatively long distance. Nonvolatile pheromones are more stable and remain longer where they are deposited, whereas volatile pheromones tend to be diluted within a short time. Mice appear to take advantage of the chemical properties of both types of pheromones to precisely control the spatial and temporal transmission of their individual signals. Thus, each organism has developed systems that use pheromone substances with appropriate physical characteristics for their environment (i.e., volatile/nonvolatile and aqueous/nonaqueous) and an appropriate detection system through which the pheromones control the mating behavior (i.e., main olfactory epithelium and/or VNO), and these systems have been combined to maximize the efficiency and probability of sexual behavior. Pheromonal communication plays an important role in social and sexual behaviors, which are critical for survival and reproduction. Elucidation of pheromonal communication systems in a wide variety of animals should eventually help determine where, when, how, and why we humans evolved.

Acknowledgments

I would like to thank H. Kimoto for tremendous amount of work leading to the discovery of the ESP family, and S. Haga and K. Sato for providing crucial data on the ESP project. I also thank other members of Touhara lab for helpful support. This work was supported by grants from the Japan Society for the Promotion of Science (JSPS) and the Program for Promotion of Basic Research Activities for Innovative Biosciences (PROBRAIN) in Japan.

References

Altstein, M. (2004). Peptide pheromones: An overview. *Peptides* **25,** 1373–1376.

Asano-Miyoshi, M., Suda, T., Yasuoka, A., Osima, S., Yamashita, S., Abe, K., and Emori, Y. (2000). Random expression of main and vomeronasal olfactory receptor genes in immature and mature olfactory epithelia of Fugu rubripes. *J. Biochem. (Tokyo)* **127,** 915–924.

Baxi, K. N., Dorries, K. M., and Eisthen, H. L. (2006). Is the vomeronasal system really specialized for detecting pheromones? *Trends Neurosci.* **29,** 1–7.

Beauchamp, G. K., and Yamazaki, K. (2003). Chemical signalling in mice. *Biochem. Soc. Trans.* **31,** 147–151.

Berghard, A., and Buck, L. B. (1996). Sensory transduction in vomeronasal neurons: Evidence for G alpha o, G alpha i2, and adenylyl cyclase II as major components of a pheromone signaling cascade. *J. Neurosci.* **16,** 909–918.

Beynon, R. J., and Hurst, J. L. (2004). Urinary proteins and the modulation of chemical scents in mice and rats. *Peptides* **25,** 1553–1563.

Beynon, R. J., Veggerby, C., Payne, C. E., Robertson, D. H., Gaskell, S. J., Humphries, R. E., and Hurst, J. L. (2002). Polymorphism in major urinary proteins: Molecular heterogeneity in a wild mouse population. *J. Chem. Ecol.* **28,** 1429–1446.

Boehm, T. (2006). Quality control in self/nonself discrimination. *Cell* **125,** 845–858.

Boehm, T., and Zufall, F. (2006). MHC peptides and the sensory evaluation of genotype. *Trends Neurosci.* **29,** 100–107.

Boehm, U., Zou, Z., and Buck, L. B. (2005). Feedback loops link odor and pheromone signaling with reproduction. *Cell* **123,** 683–695.

Boschat, C., Pelofi, C., Randin, O., Roppolo, D., Luscher, C., Broillet, M. C., and Rodriguez, I. (2002). Pheromone detection mediated by a V1r vomeronasal receptor. *Nat. Neurosci.* **5,** 1261–1262.

Brennan, P. A., Hancock, D., and Keverne, E. B. (1992). The expression of the immediate-early genes c-fos, egr-1 and c-jun in the accessory olfactory bulb during the formation of an olfactory memory in mice. *Neuroscience* **49,** 277–284.

Brennan, P. A., Schellinck, H. M., and Keverne, E. B. (1999). Patterns of expression of the immediate-early gene egr-1 in the accessory olfactory bulb of female mice exposed to pheromonal constituents of male urine. *Neuroscience* **90,** 1463–1470.

Bruce, H. M. (1959). An exteroceptive block to pregnancy in the mouse. *Nature* **184,** 105.

Butenandt, A., Beckmann, R., Stamm, D., and Hecker, E. (1959). Uber den Sexuallockstoff des Seidenspinners *Bombyx mori*, Reindarstellung und Konstitution. *Z Naturforschung* **14b,** 283–284.

Cao, Y., Oh, B. C., and Stryer, L. (1998). Cloning and localization of two multigene receptor families in goldfish olfactory epithelium. *Proc. Natl. Acad. Sci. USA* **95,** 11987–11992.

Cavaggioni, A., and Mucignat-Caretta, C. (2000). Major urinary proteins, alpha(2U)-globulins and aphrodisin. *Biochim. Biophys. Acta* **1482,** 218–228.

Clyne, P. J., Warr, C. G., Freeman, M. R., Lessing, D., Kim, J., and Carlson, J. R. (1999). A novel family of divergent seven-transmembrane proteins: Candidate odorant receptors in *Drosophila*. *Neuron* **22,** 327–338.

Del Punta, K., Leinders-Zufall, T., Rodriguez, I., Jukam, D., Wysocki, C. J., Ogawa, S., Zufall, F., and Mombaerts, P. (2002a). Deficient pheromone responses in mice lacking a cluster of vomeronasal receptor genes. *Nature* **419,** 70–74.

Del Punta, K., Puche, A., Adams, N. C., Rodriguez, I., and Mombaerts, P. (2002b). A divergent pattern of sensory axonal projections is rendered convergent by second-order neurons in the accessory olfactory bulb. *Neuron* **35,** 1057–1066.

Dulac, C., and Axel, R. (1995). A novel family of genes encoding putative pheromone receptors in mammals. *Cell* **83,** 195–206.

Dulac, C., and Torello, A. T. (2003). Molecular detection of pheromone signals in mammals: From genes to behaviour. *Nat. Rev. Neurosci.* **4,** 551–562.

Eisthen, H. L. (2004). The goldfish knows: Olfactory receptor cell morphology predicts receptor gene expression. *J. Comp. Neurol.* **477,** 341–346.

Friedrich, R. W., and Korsching, S. I. (1998). Chemotopic, combinatorial, and noncombinatorial odorant representations in the olfactory bulb revealed using a voltage-sensitive axon tracer. *J. Neurosci.* **18,** 9977–9988.

Gao, Q., and Chess, A. (1999). Identification of candidate *Drosophila* olfactory receptors from genomic DNA sequence. *Genomics* **60,** 31–39.

Grus, W. E., Shi, P., Zhang, Y. P., and Zhang, J. (2005). Dramatic variation of the vomeronasal pheromone receptor gene repertoire among five orders of placental and marsupial mammals. *Proc. Natl. Acad. Sci. USA* **102,** 5767–5772.

Haga, S., Kimoto, H., and Touhara, K. (2007). Molecular characterization of vomeronasal sensory neurons responding to the male-specific peptide: Sex communication in mice. *Pure Appl. Chem.* **79,** 775–783.

Hagino-Yamagishi, K., Matsuoka, M., Ichikawa, M., Wakabayashi, Y., Mori, Y., and Yazaki, K. (2001). The mouse putative pheromone receptor was specifically activated by stimulation with male mouse urine. *J. Biochem. (Tokyo)* **129,** 509–512.

Halem, H. A., Cherry, J. A., and Baum, M. J. (1999). Vomeronasal neuroepithelium and forebrain Fos responses to male pheromones in male and female mice. *J. Neurobiol.* **39,** 249–263.

Halem, H. A., Baum, M. J., and Cherry, J. A. (2001). Sex difference and steroid modulation of pheromone-induced immediate early genes in the two zones of the mouse accessory olfactory system. *J. Neurosci.* **21,** 2474–2480.

Halpern, M., and Martinez-Marcos, A. (2003). Structure and function of the vomeronasal system: An update. *Prog. Neurobiol.* **70,** 245–318.

Hansen, A., Anderson, K. T., and Finger, T. E. (2004). Differential distribution of olfactory receptor neurons in goldfish: Structural and molecular correlates. *J. Comp. Neurol.* **477,** 347–359.

Herrada, G., and Dulac, C. (1997). A novel family of putative pheromone receptors in mammals with a topographically organized and sexually dimorphic distribution. *Cell* **90,** 763–773.

Horowitz, L. F., Montmayeur, J. P., Echelard, Y., and Buck, L. B. (1999). A genetic approach to trace neural circuits. *Proc. Natl. Acad. Sci. USA* **96,** 3194–3199.

Hurst, J. L., and Beynon, R. J. (2004). Scent wars: The chemobiology of competitive signalling in mice. *Bioessays* **26,** 1288–1298.

Hurst, J. L., Payne, C. E., Nevison, C. M., Marie, A. D., Humphries, R. E., Robertson, D. H., Cavaggioni, A., and Beynon, R. J. (2001). Individual recognition in mice mediated by major urinary proteins. *Nature* **414,** 631–634.

Ishii, T., Hirota, J., and Mombaerts, P. (2003). Combinatorial coexpression of neural and immune multigene families in mouse vomeronasal sensory neurons. *Curr. Biol.* **13,** 394–400.

Jemiolo, B., and Novotny, M. (1993). Long-term effect of a urinary chemosignal on reproductive fitness in female mice. *Biol. Reprod.* **48,** 926–929.

Jemiolo, B., Harvey, S., and Novotny, M. (1986). Promotion of the Whitten effect in female mice by synthetic analogs of male urinary constituents. *Proc. Natl. Acad. Sci. USA* **83,** 4576–4579.

Karlson, P., and Luscher, M. (1959). Pheromones: A new term for a class of biologically active substances. *Nature* **183,** 55–56.

Keverne, E. B. (2004). Importance of olfactory and vomeronasal systems for male sexual function. *Physiol. Behav.* **83,** 177–187.

Kikuyama, S., and Toyoda, F. (1999). Sodefrin: A novel sex pheromone in a newt. *Rev. Reprod.* **4,** 1–4.

Kimoto, H., and Touhara, K. (2005). Induction of c-Fos expression in mouse vomeronasal neurons by sex-specific non-volatile pheromone(s). *Chem. Senses* **30**(Suppl. 1), i146–i147.

Kimoto, H., Haga, S., Sato, K., and Touhara, K. (2005). Sex-specific peptides from exocrine glands stimulate mouse vomeronasal sensory neurons. *Nature* **437,** 898–901.

Kimoto, H., Sato, K., Nodari, F., Haga, S., Holy, T. E., and Touhara, K. (2007). The mouse ESP family: Sexual dimorphic expression and implications for the vomeronasal sensory system. Submitted.

Koyama, S. (2004). Primer effects by conspecific odors in house mice: A new perspective in the study of primer effects on reproductive activities. *Horm. Behav.* **46,** 303–310.

Koyama, S., and Kamimura, S. (2000). Influence of social dominance and female odor on the sperm activity of male mice. *Physiol. Behav.* **71,** 415–422.

Krieger, J., Schmitt, A., Lobel, D., Gudermann, T., Schultz, G., Breer, H., and Boekhoff, I. (1999). Selective activation of G protein subtypes in the vomeronasal organ upon stimulation with urine-derived compounds. *J. Biol. Chem.* **274,** 4655–4662.

Leinders-Zufall, T., Lane, A. P., Puche, A. C., Ma, W., Novotny, M. V., Shipley, M. T., and Zufall, F. (2000). Ultrasensitive pheromone detection by mammalian vomeronasal neurons. *Nature* **405,** 792–796.

Leinders-Zufall, T., Brennan, P., Widmayer, P. S. P. C., Maul-Pavicic, A., Jager, M., Li, X. H., Breer, H., Zufall, F., and Boehm, T. (2004). MHC class I peptides as chemosensory signals in the vomeronasal organ. *Science* **306,** 1033–1037.

Liberles, S. D., and Buck, L. B. (2006). A second class of chemosensory receptors in the olfactory epithelium. *Nature* **442,** 645–650.

Lin da, Y., Zhang, S. Z., Block, E., and Katz, L. C. (2005). Encoding social signals in the mouse main olfactory bulb. *Nature* **434,** 470–477.

Loconto, J., Papes, F., Chang, E., Stowers, L., Jones, E. P., Takada, T., Kumanovics, A., Fischer Lindahl, K., and Dulac, C. (2003). Functional expression of murine V2R pheromone receptors involves selective association with the M10 and M1 families of MHC class Ib molecules. *Cell* **112,** 607–618.

Luo, M., Fee, M. S., and Katz, L. C. (2003). Encoding pheromonal signals in the accessory olfactory bulb of behaving mice. *Science* **299,** 1196–1201.

Luu, P., Acher, F., Bertrand, H. O., Fan, J., and Ngai, J. (2004). Molecular determinants of ligand selectivity in a vertebrate odorant receptor. *J. Neurosci.* **24,** 10128–10137.

Ma, W., Miao, Z., and Novotny, M. V. (1998). Role of the adrenal gland and adrenal-mediated chemosignals in suppression of estrus in the house mouse: The Lee-Boot effect revisited. *Biol. Reprod.* **59,** 1317–1320.

Ma, W., Miao, Z., and Novotny, M. V. (1999). Induction of estrus in grouped female mice (*Mus domesticus*) by synthetic analogues of preputial gland constituents. *Chem. Senses* **24,** 289–293.

Macrides, F., Bartke, A., and Dalterio, S. (1975). Strange females increase plasma testosterone levels in male mice. *Science* **189,** 1104–1106.

Mandiyan, V. S., Coats, J. K., and Shah, N. M. (2005). Deficits in sexual and aggressive behaviors in Cnga2 mutant mice. *Nat. Neurosci.* **8,** 1660–1662.

Martini, S., Silvotti, L., Shirazi, A., Ryba, N. J., and Tirindelli, R. (2001). Co-expression of putative pheromone receptors in the sensory neurons of the vomeronasal organ. *J. Neurosci.* **21,** 843–848.

Matsunami, H., and Buck, L. B. (1997). A multigene family encoding a diverse array of putative pheromone receptors in mammals. *Cell* **90,** 775–784.

Meredith, M. (1994). Chronic recording of vomeronasal pump activation in awake behaving hamsters. *Physiol. Behav.* **56,** 345–354.

Meredith, M. (1998). Vomeronasal, olfactory, hormonal convergence in the brain. Cooperation or coincidence? *Ann. N. Y. Acad. Sci.* **855,** 349–361.

Meredith, M. (2001). Human vomeronasal organ function: A critical review of best and worst cases. *Chem. Senses* **26,** 433–445.

Mombaerts, P. (2004). Genes and ligands for odorant, vomeronasal and taste receptors. *Nat. Rev. Neurosci.* **5,** 263–278.

Nakagawa, T., Sakurai, T., Nishioka, T., and Touhara, K. (2005). Insect sex-pheromone signals mediated by specific combinations of olfactory receptors. *Science* **307,** 1638–1642.

Ngai, J., Dowling, M. M., Buck, L., Axel, R., and Chess, A. (1993). The family of genes encoding odorant receptors in the channel catfish. *Cell* **72,** 657–666.

Niimura, Y., and Nei, M. (2005). Evolutionary dynamics of olfactory receptor genes in fishes and tetrapods. *Proc. Natl. Acad. Sci. USA* **102,** 6039–6044.

Novotny, M., Harvey, S., Jemiolo, B., and Alberts, J. (1985). Synthetic pheromones that promote inter-male aggression in mice. *Proc. Natl. Acad. Sci. USA* **82,** 2059–2061.

Novotny, M. V., Jemiolo, B., Wiesler, D., Ma, W., Harvey, S., Xu, F., Xie, T. M., and Carmack, M. (1999). A unique urinary constituent, 6-hydroxy-6-methyl-3-heptanone, is a pheromone that accelerates puberty in female mice. *Chem. Biol.* **6,** 377–383.

Pfister, P., and Rodriguez, I. (2005). Olfactory expression of a single and highly variable V1r pheromone receptor-like gene in fish species. *Proc. Natl. Acad. Sci. USA* **102,** 5489–5494.

Rammenses, H. G., Bachmann, J., and Stefanovic, S. (1997). *In* "MHC Ligands and Peptide Motifs." Landes Bioscience, Georgetown, TX.

Restrepo, D., Arellano, J., Oliva, A. M., Schaefer, M. L., and Lin, W. (2004). Emerging views on the distinct but related roles of the main and accessory olfactory systems in responsiveness to chemosensory signals in mice. *Horm. Behav.* **46,** 247–256.

Rodriguez, I. (2004). Pheromone receptors in mammals. *Horm. Behav.* **46,** 219–230.

Rodriguez, I., and Mombaerts, P. (2002). Novel human vomeronasal receptor-like genes reveal species-specific families. *Curr. Biol.* **12,** R409–R411.

Rodriguez, I., Greer, C. A., Mok, M. Y., and Mombaerts, P. (2000). A putative pheromone receptor gene expressed in human olfactory mucosa. *Nat. Genet.* **26,** 18–19.

Rollmann, S. M., Houck, L. D., and Feldhoff, R. C. (1999). Proteinaceous pheromone affecting female receptivity in a terrestrial salamander. *Science* **285,** 1907–1909.

Ryba, N. J., and Tirindelli, R. (1997). A new multigene family of putative pheromone receptors. *Neuron* **19,** 371–379.

Sakurai, T., Nakagawa, T., Mitsuno, H., Mori, H., Endo, Y., Tanoue, S., Yasukochi, Y., Touhara, K., and Nishioka, T. (2004). Identification and functional characterization of a sex pheromone receptor in the silkmoth Bombyx mori. *Proc. Natl. Acad. Sci. USA* **101,** 16653–16658.

Sam, M., Vora, S., Malnic, B., Ma, W., Novotny, M. V., and Buck, L. B. (2001). Neuropharmacology. Odorants may arouse instinctive behaviours. *Nature* **412,** 142.

Sato, Y., Miyasaka, N., and Yoshihara, Y. (2005). Mutually exclusive glomerular innervation by two distinct types of olfactory sensory neurons revealed in transgenic zebrafish. *J. Neurosci.* **25,** 4889–4897.

Shi, P., Bielawski, J. P., Yang, H., and Zhang, Y. P. (2005). Adaptive diversification of vomeronasal receptor 1 genes in rodents. *J. Mol. Evol.* **60,** 566–576.

Sorensen, P. W., and Caprio, J. (1998). Chemoreception. *In* "The Physiology of Fishes" (D. H. Evans, ed.), pp. 375–405. CRC, Boca Raton, FL.

Spehr, M., Kelliher, K. R., Li, X. H., Boehm, T., Leinders-Zufall, T., and Zufall, F. (2006a). Essential role of the main olfactory system in social recognition of major histocompatibility complex peptide ligands. *J. Neurosci.* **26,** 1961–1970.

Spehr, M., Spehr, J., Ukhanov, K., Kelliher, K. R., Leinders-Zufall, T., and Zufall, F. (2006b). Parallel processing of social signals by the mammalian main and accessory olfactory systems. *Cell. Mol. Life Sci.* **63,** 1476–1484.

Stowers, L., and Marton, T. F. (2005). What is a pheromone? Mammalian pheromones reconsidered. *Neuron* **46,** 699–702.

Sullivan, S. L., Bohm, S., Ressler, K. J., Horowitz, L. F., and Buck, L. B. (1995). Target-independent pattern specification in the olfactory epithelium. *Neuron* **15,** 779–789.

Timm, D. E., Baker, L. J., Mueller, H., Zidek, L., and Novotny, M. V. (2001). Structural basis of pheromone binding to mouse major urinary protein (MUP-I). *Protein Sci.* **10,** 997–1004.

Tirindelli, R., Mucignat-Caretta, C., and Ryba, N. J. (1998). Molecular aspects of pheromonal communication via the vomeronasal organ of mammals. *Trends Neurosci.* **21,** 482–486.

Trombetta, E. S., and Mellman, I. (2005). Cell biology of antigen processing *in vitro* and *in vivo*. *Annu. Rev. Immunol.* **23,** 975–1028.

Van Der Lee, S., and Boot, L. M. (1955). Spontaneous pseudopregnancy in mice. *Acta Physiol. Pharmacol. Neerl.* **4,** 442–444.

Vandenbergh, J. (1994). Pheromones and mammalian reproduction. *In* "The Physiology of Reproduction" (E. Knobil and J. D. Neill, eds.), pp. 343–359. Raven Press, New York.

Vandenbergh, J. G. (1969). Male odor accelerates female sexual maturation in mice. *Endocrinology* **84,** 658–660.

Vosshall, L. B., Amrein, H., Morozov, P. S., Rzhetsky, A., and Axel, R. (1999). A spatial map of olfactory receptor expression in the *Drosophila* antenna. *Cell* **96,** 725–736.

Wabnitz, P. A., Bowie, J. H., Tyler, M. J., Wallace, J. C., and Smith, B. P. (1999). Aquatic sex pheromone from a male tree frog. *Nature* **401,** 444–445.

Wagner, S., Gresser, A. L., Torello, A. T., and Dulac, C. (2006). A multireceptor genetic approach uncovers an ordered integration of VNO sensory inputs in the accessory olfactory bulb. *Neuron* **50,** 697–709.

Waters, C. M., and Bassler, B. L. (2005). Quorum sensing: Cell-to-cell communication in bacteria. *Annu. Rev. Cell Dev. Biol.* **21,** 319–346.

Whitten, W. K. (1956). Modification of the oestrus cycle of the mouse by external stimuli associated with the male. *J. Endocrinol.* **13,** 399–404.

Xu, F., Schaefer, M., Kida, I., Schafer, J., Liu, N., Rothman, D. L., Hyder, F., Restrepo, D., and Shepherd, G. M. (2005). Simultaneous activation of mouse main and accessory olfactory bulbs by odors or pheromones. *J. Comp. Neurol.* **489,** 491–500.

Yamazaki, K., Beauchamp, G. K., Singer, A., Bard, J., and Boyse, E. A. (1999). Odortypes: Their origin and composition. *Proc. Natl. Acad. Sci. USA* **96,** 1522–1525.

Yang, H., Shi, P., Zhang, Y. P., and Zhang, J. (2005). Composition and evolution of the V2r vomeronasal receptor gene repertoire in mice and rats. *Genomics* **86,** 306–315.

Yoon, H., Enquist, L. W., and Dulac, C. (2005). Olfactory inputs to hypothalamic neurons controlling reproduction and fertility. *Cell* **123,** 669–682.

Yoshihara, Y., Mizuno, T., Nakahira, M., Kawasaki, M., Watanabe, Y., Kagamiyama, H., Jishage, K., Ueda, H., Suzuki, H., Tabuchi, K., Sawamoto, K., Okano, H., *et al.* (1999). A genetic approach to visualization of multisynaptic neural pathways using plant lectin transgene. *Neuron* **22,** 33–41.

Zelano, C., and Sobel, N. (2005). Humans as an animal model for systems-level organization of olfaction. *Neuron* **48,** 431–454.

Zhang, X., Rodriguez, I., Mombaerts, P., and Firestein, S. (2004). Odorant and vomeronasal receptor genes in two mouse genome assemblies. *Genomics* **83,** 802–811.

Zou, Z., Horowitz, L. F., Montmayeur, J. P., Snapper, S., and Buck, L. B. (2001). Genetic tracing reveals a stereotyped sensory map in the olfactory cortex. *Nature* **414,** 173–179.

Zou, Z., Li, F., and Buck, L. B. (2005). Odor maps in the olfactory cortex. *Proc. Natl. Acad. Sci. USA* **102,** 7724–7729.

7

Environmental Programming of Phenotypic Diversity in Female Reproductive Strategies

Michael J. Meaney
Developmental Neuroendocrinology Laboratory, Douglas Hospital Research Centre, McGill University, Montreal, QC, Canada H4H 1R3

ABSTRACT

Among invertebrates, certain hermaphroditic species reproduce sexually, but with no process of sexual differentiation. In such cases the brain is bisexual: Each member of the species develops male and female sexual organs and retains

Advances in Genetics, Vol. 59
Copyright 2007, Elsevier Inc. All rights reserved.
0065-2660/07 $35.00
DOI: 10.1016/S0065-2660(07)59007-3

the capacity to express both male and female patterns of reproductive behavior. Members of such species can reproduce socially or alone. Mammals and many other species reproduce both sexually and socially, which requires an active process of sexual differentiation of reproductive organs and brain. The primary theme of this chapter is simply that this process admits to variation and thus individual differences in gender-specific patterns of reproductive function. The focus on this chapter is the often neglected variation in the development of reproductive function in the female mammal. The basic premise is that evolution has not defined any single, optimal reproductive phenotype, but rather encourages plasticity in specific reproductive traits among same sex members of the species that are derived from variations in the quality of the prevailing environment during development that are mediated by alterations in parent–offspring interactions. Thus, the variations in parental care that define the reproductive phenotype of the offspring are influenced by the quality of the environment (i.e., nutrient availability, predation, infection, population density, and so on). © 2007, Elsevier Inc.

I. INTRODUCTION

Among invertebrates, certain hermaphroditic species reproduce sexually, but with no process of sexual differentiation. In such cases the brain is bisexual: Each member of the species develops male and female sexual organs and retains the capacity to express both male and female patterns of reproductive behavior. Members of such species can reproduce socially or alone. Mammals and many other species reproduce both sexually and socially, which requires an active process of sexual differentiation of reproductive organs and brain. The primary theme of this chapter is simply that this process admits to variation and thus individual differences in gender-specific patterns of reproductive function. The focus on this chapter is the often neglected variation in the development of reproductive function in the female mammal. The basic premise is that evolution has not defined any single, optimal reproductive phenotype (Gross, 1996), but rather encourages plasticity in specific reproductive traits among same sex members of the species that are derived from variations in the quality of the prevailing environment during development that are mediated by alterations in parent–offspring interactions. Thus, the variations in parental care that define the reproductive phenotype of the offspring are influenced by the quality of the environment (i.e., nutrient availability, predation, infection, population density, and so on; Chisholm, 1993; Coall and Chisholm, 2003) (Fig. 7.1).

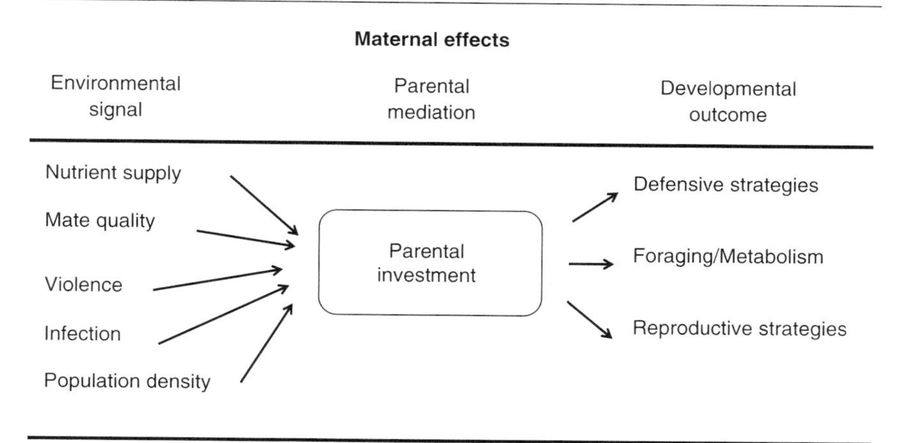

Figure 7.1. Provides a summary of the literature from evolutionary biology on "maternal effects." The consistent theme reflected in these studies is that various environmental signals can alter multiple phenotypic outcomes through effects on parent–offspring interactions, broadly referred to as "Parental Investment." The relevant form of the variation in parent–offspring interaction will vary depending on the species. The principal idea is that of parental mediation and of coordinated effects on multiple phenotypic outcomes.

A. Sexual differentiation

Sexual differentiation in mammals is linked to the composition of the sex chromosomes. In the genetic male, the sexually determining region (*Sry*) gene on the Y chromosome promotes the differentiation of the testis from the indifferent gonad, resulting in the production of two critical hormonal signals, testosterone and anti-Mullerian hormone (Arnold, 2004; Breedlove and Hampson, 2002; Goy and McEwen, 1980; McCarthy, 1994). These hormones initiate the differentiation of the reproductive tracts and genitalia. Anti-Mullerian hormone supports the regression of the Mullerian duct resulting in the initial defeminization of the reproductive system. Testosterone promotes the development of the male-typical Wolffian ducts and the male genitalia resulting in the masculinization of the reproductive system. Thus, genetic sex (XX or XY) determines gonadal phenotype, from which derives the process of sexual differentiation into the male or female phenotype, or gender. Gonadal secretions then form the basis for the sexual differentiation of the reproductive system and the brain.

It is important to note that the process of sexual differentiation is mediated by multiple intracellular pathways and is therefore subject to variation. For example, testosterone is subject to intracellular metabolism by 5α-reductase into 5α-dihydrotestosterone, which binds to androgen receptors with a higher affinity than does testosterone. The formation of 5α-dihydrotestosterone from testosterone within reproductive tissues can enhance the process of virulization of the male genitalia. The differential expression of the androgen receptor or of factors regulating the downstream effect of the androgen receptor on gene expression or protein–protein interactions could likewise modulate the degree of masculinization. A familial reductase deficiency common in the Dominican Republic (Imperato-McGinley, 1979) results in the severely limited levels of 5α-dihydrotestosterone and in incomplete masculinization. It is reasonable to assume that the process of masculinization admits to degrees, even at the more rudimentary level of the reproductive tract and genitalia. Such variation is evident in cases of testicular feminization (or androgen insensitivity) or congenital adrenal hyperplasia, in which varying levels of androgen sensitivity (tfm) or androgen exposure (CAH) result in varying degrees of masculinization. The latter case is interesting since it affects the sexual differentiation of genetic females.

Successful reproduction requires the appropriate genitalia and endocrine organs, as well as the necessary "software," in the form of hypothalamic and pituitary secretions that regulate reproductive function and corticolimbic brain systems that influence sexual motivation and parental care of the offspring. Predictably, gonadal hormones also support the sexual differentiation of hypothalamic-pituitary-gonadal (HPG) axis. The essential gender difference in mammalian HPG function is the positive surge in hypothalamic gonadotrophin-releasing hormone (GnRH) in response to estrogen. This GnRH surge supports the pulse of pituitary luteinizing hormone (LH) essential for ovulation. The capacity for such positive feedback effects of estrogen is lost in the male rodent (although not in all male mammals) as a function of perinatal testosterone exposure (Levine, 1997). In genetic females, perinatal testosterone exposure results in the loss of the LH surge and the capacity for ovulation. Similarly, males castrated in early life retain the capacity for an estrogen-induced LH surge. Thus, HPG function in the adult is subject to the developmental influences of gonadal hormones in early life.

A comparable scenario is apparent in the sexual differentiation of sexual behavior in the rodent. As the female rat transits from diestrus to proestrus, there occurs a prolonged and increasing exposure to estrogen that ultimately triggers an LH surge and an increase in circulating levels of progesterone. These hormonal signals render the female sexually receptive and promote the classic lordosis reflex in response to male mounts (Kow and Pfaff, 1998). The female is unresponsive at other points in the cycle. The female pattern of behavior is coordinated with ovulation and the sexual differentiation of female sexual behavior follows the same pattern of influences as that observed for the HPG axis. Thus, as first suggested by Pheonix *et al.* (1959) in studies with guinea pigs, androgens act during perinatal development to "organize" the sexual differentiation of sexual behaviors. Females exposed to increased levels of perinatal androgens fail to exhibit lordosis in response to male mounting even when primed appropriately with estrogen and progesterone. Males castrated at birth retain the capacity to exhibit lordosis in response to estrogen and progesterone priming. This chapter refers largely to studies with rodents, and particularly the rat. Nevertheless, similar hormone-dependent mechanisms of neural and behavioral differentiation are observed in amphibians (Kelley, 1997), birds (Balthazart and Ball, 1995), and reptiles (Godwin and Crews, 1997).

Sex differences in phenotype are also analyzed at the level of the supporting neuroanatomy (Madeira and Lieberman, 1995; Simerly, 2002). The medial preoptic area (MPOA) of the rat is essential for the expression of male sexual behavior. Studies of Gorski *et al.* (1980) reveal a sexually dimorphic region of the MPOA that is significantly larger in male than in female rats. The gender difference in the size of the sexually dimorphic nucleus of the MPOA is androgen sensitive. The nucleus is significantly smaller in males castrated at birth, and equivalent in size in females treated with androgens in late fetal and early postnatal life (Gorski, 1984). The gender difference in the MPOA is apparent in early postnatal life and is associated with an increase in neurogenesis in the males (Jacobson *et al.*, 1985) as well as an increased rate of neuron death in the females (Davis *et al.*, 1996; McCarthy *et al.*, 1997). In mice, ablation of *bax*, a member of the Bcl-2 family of proteins that is required for cell death in developing neurons, eliminates the sex differences in cell number in various hypothalamic and limbic regions (Forger *et al.*, 2004).

An important issue here is the original observation that the absence of exposure to anti-Mullerian hormone or testosterone is associated with the development of the female-typical reproductive system. This idea led many in reproductive biology to assume that the female phenotype in mammals is virtually a "default option." However, studies focusing on processes of sexual differentiation that occur later in development appear to provide support for an active process of feminization (Fitch and Denenberg, 1998; Simerly *et al.*, 1997; Stewart and Cygan, 1980; Toran-Allerand, 1981, 1995). For example, in the mouse, there is an increased number of tyrosine hydroxylase-positive neurons in the anteroventral paraventricular (AVPV) nucleus in the female compared with the male (Simerly *et al.*, 1997). In an estrogen receptor α (ERα) knockout mouse, the number of tyrosine hydroxylase neurons are significantly reduced suggesting an estrogen-dependent process of feminization in a brain region that regulates gonadotropin-releasing hormone (GnRH) activity. These and comparable findings represent a critical conceptual advance that suggests the possibility of variation in the expression of traits that define female reproduction. These findings are also entirely consistent with studies of *within-gender* variation in females in both behavior and endocrine function (see below).

These findings provide a brief outline of the biological basis for sexual differentiation in mammals based on research largely with rodent models. Biological studies have traditionally focused on the essential question of the factors that lead to the development of male- or female-typical patterns of physiology and behavior for any particular species. The cascade of genomically driven events, including organ differentiation and accompanying hormonal secretions, form the basis for the process of sexual differentiation. It is important to note that the processes by which genomic sex is translated into organogenesis and differentiation, and the subsequent downstream hormonal effects on specific target tissues are subject to modification by environmental regulation on, for example, various enzymatic pathways. Such effects could then result in variations in the expression of gender-specific patterns of reproductive function. Studies in ecology, evolutionary biology, and human psychology reveal variation in multiple reproductive functions *within* members of the same sex and species. These variations reflect plasticity in reproductive phenotypes. The basic theme of this chapter is that the process of sexual differentiation in mammals can be influenced by environmental influences, particularly by parental effects. Indeed, these effects appear to derive from a fundamental biological theme whereby environmental conditions prevailing during conception and development can regulate the development of reproductive systems, including behavior.

II. MAIN TEXT

A. Phenotypic plasticity in reproduction

In the hunt for more reliable genotype–phenotype relations, few examples would seem as potentially rewarding as that describing the Y chromosome effect on gonadal differentiation in mammals. However, a preoccupation with sexual differentiation neglects the intriguing issue of individual differences in the expression of traits directly associated with reproduction across members of the same sex (Gross, 1996; Rhen and Crews, 2002). In general, developmental biology has focused on form, rather than function and this likely reflects the close historical link with embryology and anatomy. We might assume that more complex functions, such as behavior, admit more readily to variation between individuals of the same species (although until recently variations in form across members of the same species have also been seriously underestimated; West-Eberhard, 2003). Thus, experiments within developmental biology tend to emphasis the conditions necessary to support the development of form, in this case the sexual differentiation of the reproductive organs and the brain regions that support their function. The focus has thus been on the identification of pathways that culminate in the more readily apparent features of the male and female phenotype. However, the pathway that leads from gonadal differentiation to that of the HPG axis and behavior is subject to considerable regulation, and thus variation in sexual differentiation within members of the same gender should not be surprising.

It is commonly thought that the capacity for phenotypic plasticity evolved to permit diversity in genotype–phenotype relations in response to variations in the level of environmental demands (Agrawal, 2001). Such phenotypic diversity likely reflects complex gene–environment interactions involving protein–DNA interactions at sites (e.g., promoters, enhancers, suppressors) that regulate gene expression. It is interesting to note that across species, increasing complexity is associated more with the size of the noncoding region of the genome than with the number of genes. We presume that this difference reflects the increased complexity of the regulatory regions of the DNA that, in turn, confers enhanced capacity for tissue-specific regulation of gene expression in multiorgan animals. There is also the increased capacity for the generation of various RNA species, all of which adds remarkable potential for diversity in genotype–phenotype relations. In addition, the increased size of the regulatory region of the genome should also correspond to an increased capacity for environmental regulation of gene expression—a process whereby an increasing range of phenotypes might emerge from a common genotype: an increased capacity for phenotypic plasticity. Indeed, Promisolow (2005) examined variation in gene expression in yeast across multiple environments and found that phenotypic plasticity of a gene was indeed positively correlated with the number of transcription factors regulating that gene. Hence, we might assume that with increasing biological complexity, there is a corresponding increase in the capacity for phenotypic diversity. Indeed, evolutionary theory implies variation in at least one critical form of reproductive behavior, that of parental care.

B. Parental investment and sexual selection

Successful reproduction depends on the ability to produce offspring as well as the ability to promote the growth and survival of the offspring to reproductive maturity. The growth and survival of the offspring, by definition, reflects parental investment. Thus, Trivers (1974) defined parental investment as "any investment by the parent in an individual offspring that increases the offspring chances of surviving at the cost of the parent's ability to invest in other offspring." The importance of this issue is underscored by the simple realization that the interests of the mother and any individual offspring are not entirely synonymous (Haig, 1993). While the mother shares 50% of her genome with the offspring, the offspring shares 100% with itself and it is, therefore, in the interest of the offspring to extract maximal resources from the parent. For the parent, the concern of future reproduction is a mitigating consideration. Since resources are limited, investment in any single offspring constrains that available for subsequent progeny.

Lactation is a classic example of parental investment in mammals, since it is commonly associated with a cessation of normal reproductive function in the female (e.g., lactational amenorrhea) and requires considerable energy resources and potential health risks (e.g., insulin resistance). This is an issue of costs and benefits. Increased time and energy invested in lactation might increase the growth and survival of an individual offspring, perhaps enhancing the ability to compete for resources and the opportunity for successful reproduction. But these outcomes are achieved at the cost of limiting the ability to produce subsequent offspring. A critical issue is whether increased parental investment within the environment in which reproduction is occurring will actually enhance reproductive fitness in the offspring.

Parental investment in the young comes at the cost of future opportunities for mating. Importantly, this cost:benefit equation commonly has different implications for males and females (Trivers, 1974). In mammals, increasing the number of mating opportunities and/or the number of mates has a greater benefit for the male than for the female, since the latter is limited by the demands of gestation and lactation. A hypothetical mammalian population of 10 males and 10 females has a substantially greater reproductive capacity than one composed of 18 males and only 2 females. Hence within mammalian societies, males commonly compete for females. Accordingly, evolution has resulted in the emergence of sex-specific strategies that includes neuroendocrine systems that promote behaviors that maximize reproductive fitness for each sex.

Overlaying the evolution of sex-specific patterns of reproduction is the plasticity associated with the fact that for most species there are variations in the quality of the environments within which reproduction occurs. Variations such as the availability of and competition for nutrients or the risk of predation influence the costs associated with reproduction, and can thus determine the level of parental investment. Moreover, as discussed at length below, certain environments involve a high degree of risk for mortality for the offspring (e.g., increased predation) that may render enhanced investment futile. Thus, the degree of parental investment within any species might vary as a function of the quality of the environment. Variations in parental investment influence the development of the offspring including reproductive systems. Herein lies an enormous epigenetic source of phenotypic variation in the offspring. Environmental diversity requires alternative reproductive phenotypes. The merits of such plasticity in reproductive phenotypes are explained by life history theory.

C. Life history and reproduction

Evolutionary theory defines the ultimate challenge of life as that of maximizing reproductive fitness. Reproductive fitness is determined by the ability to survive to reproductive maturity, to reproduce, and to rear the offspring to reproductive age. Reproductive success itself is a function of investments in reproductive processes, as well as in growth and survival. Life history theory (Charnov, 1993; Roff, 1992; Stearns, 1992) attempts to define variations in investment strategies across and within species in the manner in which limited energy resources are allocated to the major challenges of life: (1) growth and development, (2) maintenance, defense, and survival, and (3) reproduction. Resources are limited. Those allocated in the interest of survival limit investment in reproduction, and so on. The challenge is to establish the most effective investment strategies and the efficacy of the solutions vary as a function of the environment. There is no optimal strategy—one size does not fit all.

Selection favors phenotypes that result in life history strategies that maximize reproductive fitness under prevailing environmental conditions. Adaptation to poor-quality environments may require compromises in growth that are not ideal, but which serve the interests of survival and reproduction under conditions of adversity (i.e., "making the best of a bad situation"). Denver (1997) provides a remarkable example of such phenotypic plasticity in the toad. Tadpoles living under conditions of reduced environmental quality (e.g., a rapidly drying pond) respond with early metamorphosis into adult toads. The alternative is to risk the complete exhaustion of resources prior to assuming a viable adult form. As adults, toads undergoing earlier metamorphosis are smaller, at greater risk of predation, and are less competitive in mating. Nevertheless, the animals do survive to adulthood with an opportunity to mate. This is a classic example of a trade-off between growth and survival.

As discussed above, a fundamental trade-off in the area of parental investment is that between investment in current and future offspring. The interests of the offspring are best served by traits that maximize its ability to sequester resources from the parent. In contrast, the reproductive fitness of the parent is best served by a strategy that allocates adequate resources to any single offspring, while limiting constraints on future reproduction. Parenting is therefore a trade-off between the number and quality of offspring. In egg-laying species, there is a common trade-off between the size of an individual egg and the number of eggs (Kaplan, 1998; Mousseau and Fox, 1999). A critical issue is that of the definition of "adequate resources." In the broadest sense, adequate means sufficient to ensure survival to reproductive maturity. But resources that might be sufficient in one environment might be wholly inappropriate in another, more demanding context. Thus, parental investment should be closely linked to environmental demands. Indeed, among insects, offspring that develop from larger eggs are larger at hatching and metamorphosis, and generally have greater reproductive success (Kaplan, 1998). This suggests that more demanding environments, with fewer resources and a greater risk for mortality, should result in greater parental investment in the growth of each offspring. Larger offspring should be better prepared to compete for resources and more resistant to predation or intraspecies aggression. The problem is that more demanding environments are commonly defined by uncertainty in both resource availability and mortality. Thus, Shibata and Rollo (1988) found that the weight of eggs from wild slugs correlates with fitness traits *only in animals reared in a high-quality environment* (see Rossiter, 1998 for several similar examples). Under impoverished conditions, with decreased resources and increased risk of mortality from predation, increased parental investment does not yield any predictable benefit, and a common strategy is that of limiting investment in any individual offspring in favor of producing a greater number of offspring (Chisholm, 1993; Chisholm and Burbank, 2001; Coall and Chisholm, 2003). Ultimately, the degree of parental investment is a life history trait that, as described below, defines the reproductive function of the offspring and forms the basis for the nongenomic inheritance of life history strategies (Boonstra and Boag, 1987; Meaney, 2001; Rossiter, 1998).

D. Phenotypic variation in reproduction

Variations in reproductive tactics may be categorized into those that are conditional as opposed to those that are stable alternative strategies (Moore, 1991). Conditional strategies are directly associated with the prevailing environment. In such cases the reproductive phenotype is highly variable and sexually mature animals can switch between alternative strategies. In a great many species, the most important determinant of such conditional variations in reproductive function is that of social dominance. A classic pattern of alternative mating strategies among males consists of colorful or otherwise attractive, dominant, territorial male phenotypes contrasting with subordinate male phenotypes that appear similar to juveniles or even females. Such males usually obtain matings with females by deception (Rhen and Crews, 2002). Importantly, the phenotype of the subordinate males changes dramatically should they arise in social dominance. Big horn mountain sheep are an excellent example (Geist, 1968) and other detailed examples of conditional phenotypic plasticity in reproductive behavior are described in insects, amphibians, fish, birds, and mammals (see Rhen and Crews, 2002 for a review). Several species of fish exhibit remarkable variability of reproductive tactics with several species presenting extensive phenotype plasticity, ranging from variations related to the breeding season to actual gender role change (Demski, 1986; Gross, 1996; Taborsky, 1994). In fish, gender can actually be a very fluid distinction. For example, sequentially hermaphroditic species can reproduce first as males then switch permanently to a female pattern, or vice versa. In tropical fish, gender is often associated with dominance status, and changes in status can result in a complete switch in gender. In frogs, males may display a calling strategy, using vocalizations to attract a female, or a satellite strategy in which they lurk in proximity attempting to mate with females as they approach calling males (Perrill and Magier, 1988; Wagner, 1992). Males of these species can switch strategies in different social contexts. Males adopt the satellite strategy if adjacent males are larger. The presence of the larger males appears to suppress vocalizations in satellite males, and nonsocial stressors produce a comparable effect. This pattern is common in mammals. There are multiple examples of social suppression of reproduction in both males and females (wolves, big horn sheep, orangutans, and so on). In wolves, active reproduction is the unique domain of the α-female. Other females in the pack, which are often siblings, are subordinate and reproductively suppressed.

A prime example of the relation between social dominance and reproductive phenotype is revealed in studies of naked mole rats (Faulkes and Bennett, 2001), which display a social structure similar to social insects. A single, dominant female or "queen" is the only female that is reproductively active and breeds in a colony of mole rats. The queen suppresses reproduction in subordinate females. These effects are apparent at the level of relevant endocrine systems. Queens are more sensitive to exogenous GnRH than are nonreproductive subordinates, which show suppression of reproductive cycles and ovulation. Nonreproductive females begin cycling when removed from their colony and the suppressive influence of their queen (Margulis *et al.*, 1995).

In many instances, these findings provide evidence for a social suppression of the gender-specific pattern of reproductive activity. Variations in reproductive phenotypes are dependent on concurrent environmental conditions, such as social hierarchies. Alterations in the social order result in a switch in reproductive status in selected individuals. In contrast, the subsequent focus of this chapter is on stable alternative reproductive strategies in which a uniform phenotypic variation is manifested throughout adulthood, thus revealing a life history strategy. While the more stable, or "fixed" (Moore, 1991; Moore *et al.*, 1998; Rhen and Crews, 2002), distinction is justified by the evidence (see below), it is important to note that the expression of such phenotypes can also be modified by concurrent environmental conditions such as social hierarchies. What is apparent in the literature is the degree to which such stable variations in reproductive phenotypes are (1) associated with differences in growth in early development and (2) determined by parental effects. Indeed, the development of such stable phenotypic variations in reproduction represents an example of the parental programming of life history strategies in the offspring.

E. Parental effects on life history strategies

Variations in life history strategies among individuals within a species are often defined by adaptations to environmental conditions prevailing during early development, and reflect the remarkable capacity for phenotypic plasticity. Research in evolutionary biology reveals so-called "parental effects" (Agrawal, 2001; Cameron *et al.*, 2005; Groothuis *et al.*, 2005; Mousseau and Fox, 1998; Qvarnstrom and Price, 2001; Rossiter, 1999). Within evolutionary biology, parental effects are defined as sustained influences on any component of the phenotype of the offspring that is derived from either the mother or the father, apart from nuclear genes (Rossiter, 1998). Such parental effects have been described across a variety of species ranging literally from plants to mammals. Together, these studies suggest that the quality of the prevailing environment, defined primarily as the abundance of nutrients and the risk for mortality, directly influences parent–offspring interactions that, in turn, influence the life history strategies of the offspring (Fig. 7.1). As Hinde (1986) suggests, such effects are likely due to the fact that natural selection shaped offspring to respond to subtle variations in parental behaviors as a forecast of the environmental conditions they will ultimately face following independence from the parent. From a complimentary perspective, Rossiter (1998) suggests that when the pattern of variation in the critical environmental factor is recurrent, then natural selection will favor parental effects that "prepare" offspring for the expected environment *even if that environmental condition is one of uncertainty* (p. 130).

A classic example of a parental effect on defensive tactics emerges from studies with water fleas. In response to chemosignals, or kairomones, from aquatic predators water fleas form impressive, helmet-like growths on their necks and spines along their tails (Tollrian, 1995). These inducible, morphological defenses render the animals less likely to be captured and successfully ingested (Gilbert, 1998; Tollrain and Dodson, 1999). While the mechanism is unknown, the evidence for parental effects is compelling. The F1 and F2 generations of mothers exposed to kairomones until reproduction, and clean water thereafter, exhibited significantly larger helmets than those of mothers consistently maintained in clean water environments (Agrawal *et al.*, 1999). Kairomone exposure of the mother is sufficient to alter the morphology of the completely kairomone-naive offspring. Hence the quality of the maternal environment alters some critical aspect of parent–offspring interaction, and thus the phenotype of the progeny. Moreover, when subsequently exposed to the predator cues, the offspring of the kairomone-exposed mothers exhibit a significantly enhanced morphological defense. It is important to note that the animals in these studies were complete clones. Thus, these findings provide evidence for the preparatory function for epigenetic parental effects.

The most commonly studied phenotypic variations in reproductive tactics are among males and involve the establishment and defense of territories, parental care of the young, and courtship tactics (e.g., vocalizations). The causal mechanisms are known for only a few examples. Variations in reproductive tactics, in some cases, are associated with genetic polymorphisms (Gross, 1996 for a review). For example, in the swordtail (*Xiphiphorus nigrensis*), three alleles at a single Y locus appear to result in small, intermediate-sized, and large males that respectively sneak, sneak and court, and court in order to gain access to females (Ryan *et al.*, 1992). It is not clear whether the genetic polymorphism contributes directly to the alternate mating strategies, or indirectly through, for example, the regulation of growth which is a major determinant of mating tactics among males. Lank *et al.* (1995) have provided another example of alternative mating strategies associated with genomic variants in the lekking male ruff. In these animals the activational effects of testosterone of mating are determined by genotypic variation. Studies of proximate mechanism are required to clearly define the causal pathways by which the genomic variation leads to the alternative reproductive strategy. Nevertheless, these findings reveal evidence for an association between genomic variation and phenotype.

The most definitive research (Bass, 1992) on proximal mechanisms underlying alternative mating tactics has been accomplished with the plainfin midshipman fish (*Porichthys notatus*). Type I midshipman build nests and use courtship vocalizations to attract females. Type I males are parental. Type II males build no nests, but wander as "satellite" males, sneak mating opportunities, and are not parental. These tactics are fixed over the lifespan. Type II males mature earlier, a common strategy under conditions of environmental adversity, are smaller, have a poorly developed vocal circuitry and fewer GnRH neurons, and lower androgen levels. Differences in the pattern of vocalizations are associated with the differential effects of the neuropeptides arginine vasotocin (AVT) and isotocin. In frogs, which show similar variations in vocalizations, such effects are associated with changes in AVT under the control of androgens and estrogens. Indeed, steroidal regulation of neuropeptide systems is a common mechanism underlying the activation of reproductive behaviors.

Importantly, the same degree of phenotypic variation in reproductive tactics can result not from variation in the genome but from that of parental care. In the ground-nesting bee, large larvae develop into males expressing a fighter phenotype that is flightless, has large mandibles, and mates within the nest (Danforth, 1991). Smaller larvae develop into smaller males with functional wings that mate outside the nest. Maternal provisioning determines male larval size and thus adult reproductive tactic. The size and quality of the maternal provision (or propagule) among insects is commonly associated with the quality of the prevailing environment. This suggests that environmental conditions such as nutrient availability can influence the development of reproductive tactics through effects on parental investment. Evidence for such an influence emerges from studies with guppies housed in the presence of a model that simulates a predator. Under such conditions, the male guppy uses a sneaking tactic to access females as opposed to the more conspicuous courtship rituals. Presumably the former is less likely to draw the attention of a predator and could be considered as adaptive.

Perhaps the most dramatic examples of parental effects on reproductive function are observed in reptiles where the position chosen by the mother for her eggs (i.e., oviposition) determines the gender of the offspring (Crews, 2003, 2005). For example, crocodilians show temperature-dependent sex determination where incubation temperature during a critical period of embryonic development determines gonadal sex (Lang and Andrews, 1994). Some turtles and lizards also have temperature-dependent sex determination; others show genotypic sex determination (Rhen and Crews, 2002). One of the best-studied examples is that of the leopard gecko (Crews, 2003, 2005). In the gecko, variation in ambient temperature during embryogenesis determines gonadal sex. Remarkably, both incubation temperature *and* gonadal gender affect reproductive behavior and physiology (Rhen and Crews, 1999, 2000). Female geckos ovariectomized in adulthood and treated with estradiol display female-typical receptive behavior. No such effect occurs in castrated males. Conversely, ovariectomized, androgen-treated females display very little male-typical sexual behavior. These sex differences in hormonal responsiveness in adulthood are determined by sex differences in steroid levels during ontogeny, which reflects a gonadal influence. However, oviposition and egg temperature also produce within-gender variation in reproductive phenotypes. Incubation of leopard gecko eggs at 26 °C produces only females. Incubation at 30 °C produces a female-biased sex ratio (~30% males). Incubation at 32.5 °C produces a male-biased sex ratio (~70% males). And incubation at 34 °C again produces virtually all females. Interestingly, males from the female-biased (i.e., 30 °C) incubation temperature are more sexually active and less aggressive toward females than males from the male-biased (i.e., 32.5 °C) temperature (Flores *et al.*, 1994). Incubation temperature also influences endocrine physiology such that estrogen levels are higher and testosterone levels are lower in adult males from the female-biased versus adult males from the male-biased temperature (Crews *et al.*, 1996; Tousignant and Crews, 1995). Nevertheless, incubation temperature-induced differences in certain behaviors do not depend on differences in circulating hormone levels in adulthood. Males from the male-biased incubation temperature are more territorial than males from the female-biased temperature. However, males from the female-biased incubation temperature mount females more than males from the male-biased incubation temperature. These differences emerge despite common androgen levels and thus suggest that incubation temperature during embryonic development has a permanent effect on the male leopard gecko brain resulting in variable male phenotypes. Indeed, Crews and colleagues provide evidence for differences in neural activity of male and female geckos that accompany intragender differences in phenotype. For example, the size of the preoptic area is significantly larger in males from the male-biased temperature than in males from the female-biased temperature (Coomber *et al.*, 1997).

Maternal effects on selected phenotypic traits are common in avian species, and are likely associated with direct hormonal alterations of the egg (Schwabl, 1993). Thus, for example, Forstmeier *et al.* (2004) found that daughters differed in their "choosiness" in mate-choice experiments depending on whether they originated from eggs produced early or late within the laying sequence of a clutch. Because this effect of laying order occurred independently of hatching order in cross-fostered broods, it must have been caused by consistent within-mother variation in maternal effects transmitted through the egg. There is evidence for altered maternal deposition of androgens in the egg as a function of the quality of the breeding environment, which suggests a potential adaptive value of maternal androgen transfer as a mechanism of nongenetic inheritance, by which the mother can adjust phenotypic development in the offspring in accordance with prevailing environmental conditions (Groothuis *et al.*, 2005). Such effects alter reproductive phenotypes either directly or indirectly through differential growth rates.

A final example emerges from studies of Moore (1995) that describe a remarkable parental effect on phenotypic variation in reproductive behavior in the male rat that is associated with postnatal parental care. The lactating rat licks the anogential region of the pup to stimulate urination, which is subsequently absorbed by the mother as a source of fluid and sodium replenishment. Over the second week of life, but not earlier, male pups typically receive more perineal stimulation than females, a bias that stems from an androgen-dependent sex difference in the signals that elicit maternal licking. Maternal licking of the perineum is promoted by chemosignals in the urine and secreted by preputial glands of pups. Moore and colleagues found that males reared with reduced perineal stimulation show altered copulatory behavior, including a reduced likelihood of intromission during mounts and require more intromissions before ejaculating. Lower levels of pup anogential licking by the mother are associated with reduced innervation of bulbospongiosus muscles that may contribute to the adult pattern of copulatory behavior. The bulbospongiosus muscles surround the bulbus penis, and their reflexive action contributes to erection and ejaculation, to the formation and setting of copulatory plugs, and to the dislodgement of plugs formed by competing males.

These findings provide a brief overview of phenotypic variation in reproductive tactics in a range of species. In certain cases such variations are apparently stable across the life span and are linked to parental effects.

The actual underlying biological processes for such parental effects are generally unknown. The final section of this chapter provides a possible mechanism for what appear to be epigenetic effects on phenotype. A critical issue, of course, is whether there is evidence for such parental effects on phenotypic variations in reproduction in humans.

F. Phenotypic variation in human female reproductive development

The most reliable measure of environmental quality in human research is that of socioeconomic status (SES), which predicts multiple health outcomes. Studies, including those using prospective analyses, reveal significant effects of SES during childhood that are statistically independent of those associated with adult SES, on adult mortality (Davey-Smith *et al.*, 1998; Kaplan and Salonen, 1990; Marmot *et al.*, 2001), metabolic (e.g., body mass index, hip:waist ratio) and cardiovascular (Barker, 1992; Blane *et al.*, 1996; Bosma *et al.*, 1999; Brunner, 1996; Kaplan and Salonen, 1990; Poulton *et al.*, 2002; Power *et al.*, 2005; Rahkonen *et al.*, 1997). Likewise, there are effects of SES in early life on psychological function and mental health. Childhood SES affects alcohol dependence in adulthood and the effects are not reversed with subsequent upward mobility (Poulton *et al.*, 2002). An extensive, prospective study by Gilman *et al.* (2003) found a clear effect of childhood SES on depression (Kessler *et al.*, 1997; Ritsher *et al.*, 2001; Sadowki *et al.*, 1999). Likewise, Bosma *et al.* (1999) found effects of childhood SES on neuroticism and coping style.

Cameron *et al.* (2005) suggest that the essential feature of the process by which SES in childhood influences health is that of parental care (Fig. 7.1) and, as described above, this is apparent in simpler species, suggesting a common strategy of adaptation to conditions of adversity in early life that has been conserved across evolution. This same argument applies to reproductive outcomes (Belsky, 1997b; Coall and Chisholm, 2003; Ellis, 2005). The critical issue is that of parental mediation. In humans, the effects of SES on emotional and cognitive development are mediated by parental factors, to the extent that if such factors are controlled, there is no discernible effect of poverty on child development (Conger *et al.*, 1994; Eisenberg and Earls, 1975; McLloyd, 1998). This is an important consideration for human research, since maternal adversity is common: one in five teens and one in six adults, women experience abuse during pregnancy (Newberger *et al.*, 1992; Parker, 1984).

If parental care is a mediator of the effects of environmental adversity on development, then there should be a predictable relation between the quality of the environment and parental care. There is in fact considerable evidence for such a relation (Fleming, 1999; Repetti *et al.*, 2002). Poor-quality human environments are associated with family dysfunction (Belsky, 1984; Conger *et al.* 1984; Eisenberg and Earls, 1975; McLoyd, 1990; Whiting and Whiting, 1978).

Environmental adversity, including economic hardship and marital strife, compromise the emotional well-being of the parent and thus influence the quality of parent–child relationships (Repetti *et al.*, 2002). High levels of maternal stress are associated with increased parental anxiety, less sensitive childcare (Dix, 1991; Goldstein *et al.*, 1996), and insecure parental attachment (Goldstein *et al.* 1996; Vaughn *et al.*, 1979). Parents in poverty or other environmental stressors experience more negative emotions, irritable, depressed, and anxious moods, which lead to more punitive parenting (Belsky, 1997a; Conger *et al.*, 1984; Fleming, 1999; Grolnick *et al.*, 2002). The greater the number of environmental stressors (e.g., lesser education of parents, low income, many children, being a single parent), the less supportive the mothers are of their children; such mothers are more likely to threaten, push, or grab them, and display more controlling attitudes. Fleming (1988) reported the anxiety of the mother is the best predictor of the mother's attitude toward the child (Field, 1998).

Of course, studies of human families are correlational. Perhaps the most compelling experimental evidence for a direct effect of environmental adversity on parent–infant interactions are the studies of Rosenblum, Coplan, and colleagues with nonhuman primates (Coplan *et al.*, 1996, 1998; Rosenblum and Andrews, 1994). Bonnet macaque mother–infant dyads were maintained under one of three foraging conditions: low foraging demand (LFD), where food was readily available; high foraging demand (HFD) where ample food was available, but required long periods of searching; and variable foraging demand (VFD), a mixture of two conditions on a schedule that did not allow for predictability. At the time these conditions were imposed, there were no differences in the nature of mother–infant interactions. However, following a number of months of these conditions there were highly significant differences in mother–infant interactions. The VFD condition was clearly the most disruptive producing considerable conflict, with significantly more timid and fearful infants. These infants show signs of depression, commonly observed in maternally separated macaque infants, even when the infants are in contact with their mothers.

The critical question is whether adversity-induced alterations in parental care affect reproductive phenotypes of the offspring. There is indeed strong support for such an effect, especially in the pattern of reproductive development in females.

Human populations vary considerably in the age of female sexual maturation, most commonly defined by menarche. Human studies of using adrenarche or pubarche as indicators of sexual maturation reveal the same trend. The age of menarche is about 12 years in modern urban Western societies, but rises to 18–19 years in the more harsh environments of Papua New Guinea or the mountains of Nepal (Worthman, 1999). Resource availability is an obvious explanation for such variations (Ellison, 2001; MacDonald, 1999; Miller *et al.*, 1997). Improved nutritional resources and decreased rates of infection favor an earlier onset of sexual maturation. Secular trends suggest that improved nutrition, hygiene, and health care are associated with an earlier age of menarche over the past 300 years (Eveleth and Tanner, 1990; Tanner, 1962; Worthman, 1999). However, even within human populations within the same region there is considerable variation in the age of sexual maturation that associates with nutritional status only when there is considerable disparity in the availability of energy resources (reviewed in Ellis, 2004). Thus, in studies with human as well as nonhuman populations (Kirkwood and Hughes, 1981), nutrition and age of female sexual maturation are related only under conditions that involve severe dietary restrictions. Likewise, extreme levels of physical activity that place demands on metabolic resources or states of anorexia also delay menarche (Brooks-Gunn and Warren, 1988; Georgopoulos *et al.*, 1999). These are the more extreme cases. Variations in diets within adequately nourished populations have little effect on the age of menarche (Ellis, 2004).

G. Environmental adversity and reproductive development

Sexual maturation and growth occur with a social and economic context that influence developmental trajectories. Developmental data suggest a strong relation between patterns of growth and development on the one hand, and reproduction on the other. This relation is entirely consistent with life history theory (see above), which suggests that assuming adequate current nutritional support conditions that imperil survival will accelerate the timing of sexual maturation. The fatal error of life history is that of the complete failure to reproduce resulting in the loss of the lineage. Thus evolutionary theory suggests that selection will favor mechanisms for reducing the risk for lineage extinction in environments that offer a high risk for mortality and uncertain resource availability (Chisholm and Burbank, 2001; Maynard Smith, 1982): "When the future is risky and uncertain the optimal reproductive strategy is to take whatever resources are available and quickly convert them into offspring" (Chisholm and Burbank, 2001, p. 207). Strong support for this prediction is found in the literature on the relation between birth weight (a proxy measure for fetal adversity, Matthews and Meaney, 2005) and female puberty in humans.

The complexity of the relation between nutrition and female sexual maturation is reflected in the fact that while there is a secular trend for earlier menarche, there is no such change in birth weight (Ellis, 2005). Malnutrition certainly delays sexual maturation and reproductive activity. Severe energy demands or malnutrition preclude reproduction. Nutrient deprivation in early life leads to more complex scenarios. Early periods of growth retardation followed by exposure to adequate nutritional resources are a formula for the acceleration of sexual maturation in females and such effects are apparent in studies on birth weight and age of menarche. In developed countries where energy resources are not rate limiting, low birth weight is reliably associated with an early age of menarche (Chisholm, 1993; Coall and Chisholm, 2003; Ellis, 2005). Importantly, as with other outcome measures associated with birth weight, this relation is continuous across birth weights and is not restricted to a unique sample of very low birth weight babies (Phillips, 1998). Thus, the more successful the growth of the fetus, the later the onset of puberty.

Low birth weight (birth weight that is less than expected for gestational age) followed by a postnatal period of adequate nutritional resources is commonly associated with a period of catch-up growth over the first 3 years of life. As predicted by Chisholm and Burbank (2001), it is precisely this population of children that show an advanced age of sexual maturation. This same population shows an increased risk for obesity and metabolic diseases (Barker *et al.*, 1989, 2002; Gluckman and Hanson, 2004a; Hales and Barker, 2001; Phillips, 1998). Compared with controls (i.e., children born at weights that are average for gestational age) children showing low birth weight and postnatal catch-up growth reveal evidence for hyperinsulinemia, dyslipidemia, and increased body fat with reduced lean body mass. This condition is also associated with increased leptin levels (Ong *et al.*, 1999; Pulzer *et al.*, 2001). Increased insulin activity is thought to advance adrenarche by stimulating the activity of P450c17 that promotes adrenal androgen [dihydroepiandrosterone (DHEA)] synthesis and release. (Ibanez *et al.*, 2000). The relation between low birth weight and early sexual maturation is also evident in prospective studies conducted with children well before the onset on even the earlier stages of pubertal development (Veening *et al.*, 2004). Adrenarche reliably predicts the onset of endocrine events that culminate in pubarche (eruption of pubic hair) and ultimately sexual maturation in females. Thus, the age of adrenarche is significantly advanced in low birth weight girls that show postnatal catch-up growth (Ong *et al.*, 2004). Interestingly, this condition likely derives from an effect of endocrine signals associated with nutrition and growth (leptin, insulin) on reproductive systems.

The relation between interuterine growth retardation, resulting in diminished birth weight, and an early onset of sexual maturation may seem somewhat counterintuitive. Note, however, that this relation is only apparent under conditions of adequate postnatal nutrition, and indeed is best reflected in those children that reveal postnatal catch-up growth. Interuterine growth retardation is a reliable consequence of materno-fetal function under conditions of environmental adversity. Indeed, increased activity of the maternal and/or fetal hypothalamic-pituitary-adrenal (HPA) axis is considered the proximal cause of interuterine growth retardation (Goland *et al.*, 1993, 1995; Matthews and Meaney, 2005; Meaney *et al.*, in press; Seckl, 1998). Thus, poverty is associated with a significantly increased risk for inter-uterine growth retardation (IUGR) and the major predictors of low birth weight, maternal protein deprivation, tobacco and alcohol consumption, and maternal stress/anxiety are related to

SES and each activates the HPA axis. Williams *et al.* (1997) found that lower SES is associated with a higher ratio of placental weight/fetal weight. Placental hypertrophy is considered a fetal adaptation to restricted maternal provisioning (bear in mind the placenta is largely of fetal origin). Predictably, there are increased chord levels of both corticotrophin-releasing factor (CRF) and cortisol in small babies compared with controls (Goland *et al.*, 1993, 1995). Moreover, studies with both rodent and primate models reveal that the risk for IUGR associated with factors such as protein deprivation is directly mediated by maternal HPA activity (Langley-Evans, 1997; Seckl, 1998).

The relation between materno-fetal environmental adversity and IUGR is critical for the understanding of the adaptive nature of early menarche in low birth weight children. Life history theory suggests that when environmental conditions predict an increased risk for mortality, it is advantageous to utilize existing energy resources to accelerate reproduction (Worthman, 1999). The alternative associated with delayed sexual maturation is that death will occur prior to reproduction—lineage extinction. Evidence for the trade-off in life history strategies is apparent in the finding that girls that show early puberty are significantly shorter than their peers, and that this relation is evident even before the onset of puberty.

In a comparative analysis across 48 mammalian species, Promislow and Harvey (1990) found that high mortality rates predicted an early age of sexual maturity. Perhaps the ideal human study was that showing that children adopted from developing countries into affluent western families experienced earlier puberty than do children from either the country of origin or the host country (Mul *et al.* 2002). Thus, if energy resources permit, the ideal strategy under conditions of adversity in early development is to enhance the age of reproduction, even if, as in human populations, early reproduction is associated with increased health risks for the offspring (Coall and Chisholm, 2003). Worthman (1999) suggested that girls experiencing deprivation would react to an increase in resource availability by hastening reproduction, thus exploiting what might be a unique window of reproductive opportunity. Indeed, the relation between environmental adversity and early sexual maturation in human females is also revealed in studies of the influences of postnatal adversity.

H. Environmental adversity and sexual maturation

Studies of the relation between birth weight and puberty reflect the influence of social and economic context on female reproductive function. And as with other developmental outcomes, there is evidence that socioeconomic conditions are mediated by effects on parent–offspring interactions occurring during both the pre- and postnatal periods. Indeed, the ultimate consequences for reproduction depend on environmental adversity prior to *and* following birth. Belsky *et al.* (1991)

suggested that environmental adversity is associated with a decreased quality of parental care that directly leads to early menarche and earlier onset of sexual activity in human females. There is strong evidence that insecure attachment is more prevalent in populations living under conditions of risk and uncertainty (i.e., poverty and inequality: Belsky, 1997a; McLoyd, 1990; Repetti, *et al.*, 2002; and see above). The age of menarche is influenced by the quality of family function (Ellis *et al.*, 1999), number of major life events (Coall and Chisholm, 2003; Surbey, 1990); family conflict (Graber *et al.*, 1995; Moffitt *et al.*, 1992), marital conflict (Kim and Smith, 1998a,b; Wierson *et al.*, 1993), and negative family relationships (Ellis and Garber, 2000; Kim *et al.*, 1997). Despite some exceptions (Campbell and Udry, 1995; Graber *et al.*, 1995), the majority of studies show that childhood psychosocial stress predicts earlier menarche. There is strong evidence for familial influences on sexual maturation in human females (Belsky *et al.*, 1991; Chisholm, 1999; Draper and Harpending, 1982; Ellis, 2004; Ellis *et al.*, 1999; Graber *et al.*, 1995; Hulanicka, 1999; Steinberg, 1988; Wilson and Daly, 1997). Females reared in poor-quality familial environments (father absent, conflict, abuse, parental psychopathology, and so on) show an earlier age of menarche. In contrast, familial warmth and cohesion are associated with a later onset of sexual development. Parental separation is associated with an earlier age of menarche in human females, and interestingly, the younger the age at which separation was experienced, the greater the effect on the age of menarche (Quinlan, 2003). Childhood abuse also predicts an earlier onset of puberty in girls (Herman-Giddens *et al.*, 1988; Jorm *et al.*, 2004; Romans *et al.*, 2003; Turner *et al.*, 1999; Zabin *et al.*, 2005). In a prospective, longitudinal study of girls, familial warmth at 5 years of age was negatively correlated with the age of puberty (Ellis *et al.*, 1999; also see Graber *et al.*, 1995). Steinberg (1998) reported that mother–daughter relations predicted pubertal development. Likewise, there is evidence that increased paternal investment increases the age of pubertal development in the female offspring (Marlowe, 2003). Increased time spent in child care by the father is associated with later age of sexual maturation in daughters (Ellis *et al.*, 1999). These findings are consistent with a substantial literature in which data on the quality of parent–child interactions was assessed either at the time of puberty or retrospectively (reviewed in Ellis, 2005).

Belsky *et al.* (1991) suggest that the degree of early psychosocial stress is critical for the development of alternative reproductive strategies. They proposed that inconsistent or insensitive parenting increases risk for insecure attachment and that such forms of parenting are more common under conditions of environmental risk and uncertainty. Thus psychosocial stress associates with insecure attachment and accelerates sexual maturity. Chisholm, 1999 argued that ultimately the most consistent cause of insecure attachment would be the risky and uncertain environments that cause high mortality rates.

Because children do not directly perceive mortality rates, Chisholm proposed that the attachment process functions as a phenotypic mechanism enabling the child to gauge environmental risk and uncertainty indirectly.

Importantly, early menarche predicts an earlier age of onset of sexual behavior and reproduction. Thus, the timing of puberty is associated with initial dating, genital contact, and sexual intercourse. Most importantly, early menarche is associated with the age at first pregnancy and birth. Women with early menarche tend to be younger at first intercourse (Andersson-Ellstrom *et al.*, 1996; Bingham *et al.*, 1990; Phinney *et al.*, 1990; Udry, 1979), first pregnancy, and birth of first child (Roosa *et al.*, 1997; Ryder and Westoff, 1971; Udry and Cliquet, 1982). Likewise, an impoverished family environment predicts an increased onset of sexual activity. Single-parent households, for example, predict an early onset of sexual activity (Newcomer and Udry, 1997). Thus, while most studies focus on pubertal development, there is evidence for the idea that accelerated sexual maturation is associated with early onset of sexual activity.

These findings reveal that the timing of reproductive development in the human female shows considerable phenotypic plasticity determined, in part, by social and economic (e.g., resources availability) conditions. Environmental adversity appears to significantly advance the age of sexual maturation. This appears to be adaptive (Chisholm, 1996; Coall and Chisholm, 2003; Ellis, 2005). Environmental adversity is associated with an increased risk of mortality, and should thus hasten the onset of reproductive activity. Environmental adversity is also associated with alterations in parent–child interactions (see above). Importantly, there is evidence for the idea that the effects of environmental adversity on child development are mediated by effects on parent–offspring interactions. These findings also fit well with predictions from life history theory. Under harsh environmental conditions that predict an increased risk for early mortality in the offspring and when increased parental investment does not significantly offset the risk for mortality, the optimal strategy is to maximize fertility at the cost of investment in individual offspring. These patterns of reproductive behavior are reminiscent of *r* and *K* reproductive strategies (McCarthy, 1965). Life requires a series of energy investments in three domains: growth, mating, and parental care. The *r* and *K* reproductive strategies represent differential investment in mating versus parental care. The *r* strategy emphasizes increased investment in mating and maximizes the *quantity* of offspring. The *K* strategy emphasizes parental care and the *quality* of offspring. While originally conceived as descriptions of species differences, more recent studies focusing on phenotypic variation in mating tactics reveal that both strategies are often represented within the same species, depending on prevailing environment conditions (Chisholm, 1996; Coall and Chisholm, 2003). It appears that *r* and *K* approaches lie along a continuum and that the point along this continuum where the individual lies is defined, in part, by the quality of parental investment in early life (Belsky *et al.*, 1991; Cameron *et al.*, 2005; Coall and Chisholm, 2003). The rationale for such "phenotypic plasticity" is that in adverse environmental conditions with high risk and uncertainty, when the probability of extended periods of growth and survival is low, the optimal strategy is to maximize offspring quantity through enhanced mating. Maximizing offspring quantity enhances the chances that at least some offspring will survive to reproductive maturity. Moreover, since such adverse environments are characterized by high, unavoidable risks, parental investment in offspring quality is seen as futile (Coall and Chisholm, 2003). Increased risk of mortality thus favors a shift in parental investment away from offspring quality to quantity (Coall and Chisholm, 2003; Gangestad and Simpson, 2000). In contrast, more propitious environmental conditions favor greater investment in individual offspring at the cost of mating. In more favorable environments, competition for resources tends to determine success, and offspring quality is a relevant consideration in defining the reproductive fitness of the progeny. Thus the quality of the prevailing environment defines the level of parental care (or investment), which is, in turn, reflected in the mating and parental behavior of the offspring. Indeed, these findings provide evidence for an inherited environmental effect on reproductive development. Rossiter (1999) noted that such effects commonly derive from more stable components of the parental environment (population density, availability of nutrients, risk of predation, and so on). Research with plants and insect models suggest that inherited environmental effects on defensive responses are indeed adaptive (Agrawal *et al.*, 1999; Mousseau and Fox, 1999).

These considerations suggest that the same conditions that favor early sexual development in human females should also compromise the development of parental care. Thus, decreased parental investment in female offspring should results in comparable pattern of parenting in the daughters. In humans and other species there is evidence that individual differences in parental care beget individual differences in parental care (Fleming, 1999; Meaney, 2001). Conditions that commonly characterize abusive and neglectful homes involve economic hardship, martial strife, and a lack of social and emotional support breed neglectful parents. Perhaps the best predictor of child abuse and neglect is the parents own history of childhood trauma. More subtle variations in parental care also show continuity across generations. Scores on the Parental Bonding Index, a measure of parent–child attachment, are highly correlated across generations of mothers and daughters (Miller *et al.*, 1997). In nonhuman primates there is also strong evidence for the transmission of stable individual differences in maternal behavior (Berman, 1990; Fairbanks, 1996; Maestripieri, 1999). Thus, there are familial influences on the parental care of the offspring that form the basis for the possible transmission of individual differences in parenting across generations. To the extent that parenting influences pubertal development, such factors may also contribute to the significant correlation between the age of menarche in the mother and her daughter. Indeed, the relation exists for birth weight: low birth weight mothers tend to have low birth weight babies.

I. Parental care and reproductive development

These findings suggest a relation between parental care and reproductive development in the offspring. An important question concerns the identity of critical mechanisms for such effects. Phenotypic plasticity in reproductive tactics have been studied from the point of ultimate causation (Tinbergen, 1972), which refers to the impact of such variations on reproductive fitness. In contrast, proximal causation, which refers to more immediate cause–effect relations, such as genomic polymorphisms or hormonal influences is less well understood, and this is particularly true for mammals. While the evidence is consistent with a parental effect on the reproductive strategies of the female offspring, there is no direct evidence in any mammalian species of an effect of variations in parental care on reproductive development. Nevertheless, some of the fundamental relations apparent in human reproductive development are observed in nonhuman species that are amendable to studies of proximate causation. Recent studies on the effects of variations in maternal care in the rat provide both evidence for a direct effect of maternal care as well as the elucidation of a possible molecular mechanism for such effects. In the rat, for example, there are maternal effects on the parental care of the offspring (Fleming *et al.*, 1999; Francis *et al.*, 1999).

Maternal care in the rat involves the maintenance of the nest site and frequent nursing bouts over the course of the day. Milk delivery is the hallmark of the nursing bout. However, another important component of the nursing bout is the maternal licking/grooming (LG) of the pups. Pup LG may be considered as a rudimentary form of parental "nurturance," since this maternal behavior serves to enhance somatic growth and brain development through effects on multiple endocrine systems. Thus, the tactile stimulation associated with pup LG leads to an increased release of anabolic hormones, such as growth hormones and thyroid hormones, and decreased release of the catabolic glucocorticoids (Levine, 1994; Schanberg *et al.*, 1984).

There is considerable variation among lactating Long-Evans rats in the frequency of pup LG over the first week of life (Caldji *et al.*, 1998; Champagne *et al.*, 2003a; Liu *et al.*, 1997; Myers *et al.*, 1989). Variations in pup LG are scored during daily observations and the results summed over the first week of postnatal life. Lactating females for which the LG score is 1 SD greater than the mean of the breeding cohort are deemed to be High LG mothers. Conversely, those for which the LG scores are 1 SD below the mean are considered as Low LG mothers. High LG mothers commonly engage in pup LG at about twice the frequency as Low LG mothers, and this difference persists only for the first week of life (Caldji *et al.*, 1998; Champagne *et al.*, 2003a).

The differences in maternal behavior in the high and Low LG mothers are not unique to the first litter (Champagne *et al.*, 2003). Across mothers, the frequency of pup LG is highly correlated between the first and second litters ($r = +0.84$) and even between the first and third litters correlation of +0.72. Thus, the variations in maternal behavior represent stable individual differences in a specific form of reproductive behavior. In context of the discussion provided above, this trait may be considered a reproductive tactic. These findings are comparable to those of primate studies in which individual differences in maternal behavior remained consistent across infants.

In the rat and monkey, individual differences in maternal behavior are transmitted from mother to daughter. Human clinical research suggests that the social, emotional, and economic context are overriding determinants of the quality of the relationship between parent and child (Eisenberg and Earls, 1975). Human parental care is disturbed under conditions of chronic stress. Conditions which most commonly characterize abusive and neglectful homes involve economic hardship, martial strife, and a lack of social and emotional support (Eisenberg, 1990). Such homes, in turn, breed neglectful parents. Perhaps the best predictor of child abuse and neglect is the parents own history of childhood trauma. In a remarkable study, Maestripieri (2005) documented the prevalence of infant abuse in the rhesus monkey and used a cross-fostering model to show that it was clearly associated with the maternal behavior of the rearing mother.

More subtle variations in parental care also show continuity across generations. Scores on the Parental Bonding Index, a measure of the quality of parent–child interactions, are highly correlated across generations of mothers and daughters (Miller *et al.*, 1997). In nonhuman primates there is also strong evidence for the transmission of stable individual differences in maternal behavior (Berman, 1990; Fairbanks, 1996; Maestripieri, 1999). Measures such as time spent with the infant and infant rejection are significantly correlated across mothers and daughters. In humans, parenting style begets parenting style (Fleming *et al.*, 2002). Mothers who received inconsistent care engaged in more instrumental and less affectionate behavior (Krpan *et al.*, 2005).

In the case of the rat, the lactating female offspring of High LG mothers exhibit significantly higher levels of pup LG than do those of Low LG mothers (Francis *et al.*, 1999). The effect of maternal care on maternal behavior in the offspring is also apparent in an experimental model referred to as the pup sensitization paradigm (Bridges, 1996). Adult female rats are not spontaneously maternal. Instead, the hormonal conditions prevailing during pregnancy produce a maternal state in which the mother is immediately responsive to her pups following delivery. This maternal state can be environmentally induced. Adult female rats exposed continuously to pups become fully maternal over a period ranging from 4 to 12 days. The adult virgin female offspring of High LG mothers become fully maternal in response to the presence of pups in a significantly shorter period than do those of Low LG mothers (Champagne *et al.*, 2001).

Cross-fostering studies in the rat provide evidence for a direct effect of maternal care (Francis *et al.*, 1999). As adults, the female offspring of Low LG dams reared by High LG mothers did not differ from normal, High LG offspring in the frequency of pup LG. The frequency of pup LG in animals reared by High LG mothers was significantly higher than in any of the Low LG groups, and again this included female pups originally born to High LG mothers, but reared by Low LG dams. Individual differences in maternal behavior mapped onto those of the rearing mother, rather than the biological mother. These findings suggest that maternal–infant interactions may program specific forms of reproductive behavior. Fleming and colleagues provided direct experimental support for this conclusion. As lactating mothers, female rats artificially reared in isolation with no maternal care following the first day of life show significantly reduced responsiveness to pups and reduced pup LG (Lovic *et al.*, 2001; Melo *et al.*, 2006). The effects of artificial rearing on maternal behavior are greatly reduced by providing the female pups with stroking, which mimics the tactile stimulation associated with maternal LG, as well as with social contact with peers. These findings suggest a direct relation between the quality of maternal care received in early life and that expressed as an adult (Fleming *et al.*, 2002). Moreover, the inheritance of individual differences in maternal care in the rat is potentially nongenomic and is associated with tactile stimulation derived from pup LG.

These findings certainly do not imply that genomic variations are unrelated to variations in maternal behavior in this or other species. Nevertheless, within this specific model such factors appear to be of limited influence as revealed by the results of cross-fostering studies. Moreover, concerted efforts by our laboratory at selective breeding have failed to maintain the phenotypic variation in pup LG over more than 2–3 generations.

The transgenerational effects on maternal behavior are interesting to consider in light of what is known about the neuroendocrine mechanisms underlying the differences in maternal behavior in the rat. The elaborate pattern of maternal care that is evident shortly after parturition in the rat is orchestrated through a complex pattern of endocrine events during gestation. Throughout most of pregnancy, progesterone levels are high and accompanied by moderate levels of estrogen. Then, days prior to parturition, progesterone levels fall and there occurs a surge in estrogen levels (Bridges, 1996). Both events are obligatory for the onset of maternal behavior and of particular importance are the effects of estrogen at the level of the MPOA. This brain region is critical for the expression of maternal behaviors in the rat (Numan and Sheehan, 1997), including pup LG (Fleming, 1999).

The influence of ovarian hormones on the onset of maternal behavior in the rat is mediated in part by effects on central oxytocinergic systems (Pedersen, 1995). Estrogen increases oxytocin synthesis in the parvocelluar neurons of the PVNh that project to the MPOA as well as other brain regions that regulate maternal behavior (Shahrohk and Meaney, 2006). Estrogen also increases oxytocin receptor gene expression and receptor binding in the MPOA (Bale *et al.*, 1995; Champagne *et al.*, 2001, 2003b; Fahrbach *et al.*, 1984) and this effect appears to be mediated through the ERα (Young *et al.*, 1997). Intracereboventricular (ICV) administration of oxytocin rapidly stimulates maternal behavior in virgin rats (Fahrbach *et al.*, 1984; Pedersen and Prange, 1979) and the MPOA appears to be a critical site. This effect is estrogen dependent. The effect of oxytocin is abolished by ovariectomy and reinstated with estrogen treatment. Moreover, treatment with oxytocin-antisera or receptor antagonists blocks the effects of ovarian steroid treatments on maternal behavior (Fahrbach *et al.*, 1985; Pedersen *et al.*, 1985). Among lactating females, there are significantly higher levels of oxytocin receptors in the MPOA, the bed nucleus of the stria terminalis, and the lateral septum of all animals (Pedersen, 1995 for a review); however, the lactation-induced increase in receptors levels is substantially greater in the High LG mothers (Francis *et al.*, 2000). Each of these brain regions is implicated in the expression of maternal behavior in the rat (Fleming, 1999; Numan and Sheehan, 1997; Pedersen, 1995). Not surprisingly, the oxytocin receptor binding levels are highly correlated with the frequency of pup LG (Champagne and Meaney, 2006). Importantly, central infusion of an oxytocin receptor antagonist on day 3 of lactation completely eliminates the differences in maternal LG between High and Low LG mothers (Champagne *et al.*, 2001).

The ascending mesolimbic dopamine appears to be a relevant target for the effects of oxytocin on maternal LG. Oxytocin neurons in the MPOA project directly to the ventral tegmental area (Numan and Sheehan, 1997; Shahrohk and Meaney, 2006), the origin of the mesocorticolimbic dopamine system. Preliminary findings suggest that oxytocin facilitates dopamine release through effects on oxytocin receptors in VTA neurons. Oxytocin is known to enhance dopamine-mediated behaviors. Dopamine levels in the nucleus accumbens are directly related to the frequency of pup LG (Champagne *et al.*, 2004). Not surprisingly, the magnitude of the dopamine signal in the nucleus accumbens accompanying pup LG is significantly greater in High LG mothers. Moreover, central infusion of a dopamine transporter blocker in lactating high and Low LG females completely eliminated the differences in both dopamine levels in the nucleus accumbens and pup LG (Champagne *et al.*, 2004). These findings are consistent with earlier reports that lesions of the nucleus accumbens significantly reduce pup LG (Fleming, 1999).

Differences in estrogen sensitivity mediate the differential effects of lactation on the induction of oxytocin receptors in High and Low LG females.

Among ovariectomized females provided estrogen replacement, there is a significantly greater estrogen effect on oxytocin receptor levels in the MPOA in high compared with Low LG animals (Champagne *et al.*, 2001). The effect is apparent across a wide range of doses, and indeed there is no significant effect of estrogen on oxytocin receptor levels in the MPOA of Low LG females. The fact that such differences occurred even in the nonlactating, ovariectomized state suggests the existence of stable differences in estrogen sensitivity. Indeed, either among lactating High LG mothers or in the virgin female offspring of High LG dams, the expression of ERα, but not ERβ, is significantly increased in the MPOA [208]. The effect is apparent at the level of mRNA and protein. More recent studies (McAllister, Diorio and Meaney, submitted for publication) reveal that the differences in ERα expression in the medial preoptic region are confined to the caudal regions of the medial and central subdivisions of the medial preoptic nucleus. In these regions ERα mRNA expression is about twofold greater in the virgin offspring of high compared with Low LG mothers.

The working hypothesis suggests that the increased pup LG of High LG mothers is associated with increased ERα expression in the MPOA. This effect is apparent in virgin female offspring of High LG mothers. During gestation, the increase in circulating estrogen appears to enhance oxytocin receptor binding to a greater extent in the High LG mothers, leading to increased oxytocin activation of the ascending mesolimbic dopamine system and increased dopamine release in the nucleus accumbens. According to this model, the critical feature for the transmission of the individual differences in maternal behavior from the mother to *her* female offspring is the differences in ERα expression in the MPOA.

J. Maternal effects on HPG function and mating behavior

In an effort to establish the regional specificity we examined ERα expression in multiple brain regions. To our surprise, we found that ERα mRNA expression is significantly increased in the ventromedial nucleus of the hypothalamus (VMNh) in the offspring of Low LG mothers. Further studies revealed a significant increase in the AVPV region of the female offspring of Low LG mothers. Estrogen acts through ERs in the AVPV to regulate GnRH with downstream effects on pulsitile LH and ovarian hormone production (Simerly, 2002). The differences in ERα expression in the AVPV appear to be functionally relevant. In intact cycling adult females, circulating levels of LH at proestrus are significantly higher in offspring of low compared with High LG mothers. Likewise, there are increased proestrus levels of both estrogen and progesterone in the Low LG offspring. Endocrine studies suggest increased sensitivity to the positive feedback effects of estrogen on neural systems that regulate LH release in the offspring of Low LG mothers. Thus, among ovariectomized animals a single bolus injection of estrogen produces a significantly greater increase in plasma LH levels in the offspring of low compared with High LG dams.

Since estrogen and progesterone regulate mating over the estrus cycle in the rat, these findings suggest possible effects of maternal care over the first week of life on sexual behavior in the female offspring. Females were tested in proestrus with males in the confinements of a smaller, traditional testing arena. Under these circumstances, the female offspring of Low LG mothers exhibited an increased lordosis response to male mounts. The female offspring of Low LG dams also exhibited increased rates of proceptive behaviors (ear wiggling, hopping; Erskine, 1989) that serve to attract the male and enhance male copulation. In contrast, the female offspring of High LG dams exhibited increased levels of agonistic behavior toward the males.

Solicitation behaviors in the female rat are highly dependent on context. In smaller confines, the most common pattern of paracopulatory (or proceptive) behavior is that of hopping, darting, and ear wiggling. However, when a receptive female is tested in a larger area that affords the opportunity to retreat from the male, the approach-withdrawal pattern prevails and reveals the females ability to pace the mating with the male (Erskine, 1989). Female rats pace the rate of male intromissions and thus ejaculation by withdrawal from the male following each intromission. The latency to return to the male is longer after ejaculation than after an intromission, which in turn is longer than after a mount with an intromission (Erskine *et al.*, 2004; Yang and Clemens, 1996). As testing proceeds over the courses of multiple ejaculatory sequences, the interintromission interval increases (Coopersmith *et al.*, 1996). Testing in the pacing chamber revealed considerable differences in sexual behavior as a function of maternal care. We examined the adult offspring of low, mid, and High LG mothers in order to examine the effect of postnatal maternal care over a wider range of the population. As in the previous test, the critical measure of receptivity (lordosis rating) suggested *decreased sexual receptivity* in the adult female offspring of High LG mothers by comparison to those reared by either low or mid LG dams. The most impressive differences in pacing were in the intervals between intromissions. Over the entire session, the average interintromission interval was substantially longer in the female offspring of High LG mothers, with no difference between females reared by low or mid LG dams. A similar effect was observed in the second half of the test session, when variation in the intervals between intromissions is generally greater. Importantly, there were also significant group differences in the rate of pregnancy following mating in the pacing chamber. While 50% of the female offspring of High LG mothers became pregnant, over 80% of those of Low LG mothers were pregnant. These findings suggest that the differences in sexual behavior between the offspring of high and Low LG mothers are indeed functionally relevant for reproductive success.

The importance of the differences in HPG function was further assessed through measures of the estrous cycles of over 150 high and Low LG offspring. The Low LG showed a trend for a more regular estrous cycle than did High LG offspring. To further verify the maternal influence on reproductive capacity, proestrus offspring of High and Low LG mothers received 5–7 intromissions from a male at 4- to 5-min intervals. This intromission interval is optimal for pregnancy in the female rat. Under these conditions 75% of female rats usually reach pseudopregnancy. *Pseudopregnancy* (the continuous presence of diestrus for 8 consecutive days) was seen in 67% of Low LG offspring compared to only 27% in High LG offspring. This result suggests a maternal effect on fecundity that is independent of the timing of intromissions during mating.

These findings reveal evidence for the maternal programming of sexual behavior in the female rat. Moreover, maternal care is associated not only with alterations in sexual behavior in the adult rat but also in the timing of the onset of sexual behavior. The female offspring of Low LG mothers show vaginal opening (an unambiguous indication of pubertal development in the rat) significantly earlier in life than do the offspring of High LG dams. These findings provide a stunning parallel to the human literature (see below) in which the onset of reproductive function as well as sexual activity were influenced by parental care in early life.

K. Molecular mechanisms

To summarize, the offspring of Low LG mothers show evidence for increased sexual receptivity and decreased maternal LG. The offspring of High LG mothers show precisely the opposite profile. And these differences in reproductive behaviors map onto those in ERα expression. These findings suggest that maternal effects on reproductive behavior are mediated by tissue-specific differences in ERα expression in brain regions that regulate maternal and sexual behaviors. The remarkable feature of this effect is that the same stimulus input (maternal care) appears to regulate the expression of the same gene, the ERα gene, in exactly the opposite manner depending on brain region (MPOA vs VMNh/AVPV).

A critical issue concerns the mechanism for the apparent maternal effect on ERα expression. Studies in the rat suggest a maternal effect on glucocorticoid receptor (GR) expression in the hippocampus and HPA function (Liu *et al.*, 1997; Weaver *et al.*, 2004, 2005). Glucocorticoids act on GRs in the hippocampus to inhibit the synthesis of hypothalamic corticotropin-releasing factor (CRF), and thus diminish the magnitude of pituitary-adrenal responses to stress. In the rat, the adult offspring of High LG mothers increased hippocampal GR expression, enhanced glucocorticoid feedback sensitivity, decreased hypothalamic corticotrophin-releasing factor (CRF) expression, and more modest

HPA stress responses compared to animals reared by Low LG mothers. Direct infusion of a GR antagonist into the hippocampus eliminates the maternal effect on pituitary-adrenal responses to stress.

Increased maternal LG is associated with demethylation of the 5′ CpG dinucleotide within a nerve growth factor-inducible protein A (NGFI-A) transcription factor response element located within the exon 17 GR promoter (Weaver *et al.*, 2004, 2005, 2007). The difference in the methylation status of this CpG site between the offspring of high and Low LG mothers emerges over the first week of life, is reversed with cross-fostering, persists into adulthood, and is associated with altered histone acetylation and NGFI-A binding to the GR promoter (Weaver *et al.*, 2004). This epigenomic programming of the exon 17 GR promoter by maternal care might serve as a model for a novel mechanism through which the social environment programs adaptation at the level of the genome (Szyf *et al.*, 2005). DNA methylation is commonly associated with the silencing of gene transcription (Bird, 2001; Bird and Wolffe, 1999; Jones *et al.*, 1998; Razin, 1998; Reik *et al.*, 2000; Strobl, 1990; Szyf, 2001). In part, such effects are mediated through the binding of methylated DNA-binding proteins to methylated regions of the DNA. Methylated DNA-binding proteins attract a repressor complex that includes histone deacetylases. The histone deacetylases prevent the acetylation of histone tails that initiates the chromatin remodeling necessary for transcription factor binding and the initiation of gene expression. This model suggests that the inhibition of histone deacetylase activity might reverse the effects of DNA methylation. Indeed, in adult offspring of Low LG mothers the infusion of the histone deacetylase inhibitor, trichostatin A (TSA), reverses the effects of maternal care on both GR expression and HPA function (Weaver *et al.*, 2004). Thus, central infusion of TSA over a 4-day period increases histone acetylation of and NGFI-A binding to the exon 17 GR promoter. These effects are associated with increased hippocampal GR expression and more modest HPA responses to acute stress. In each measure, the TSA infusion reversed the difference between the offspring of the High and Low LG mothers. Conversely, central infusion of L-methionine, which increases the endogenous levels of the methyl donor SAM, increases the methylation of the 5′ CpG of the NGFI-A response element in the exon 17 GR promoter, decreases histone acetylation and NGFI-A binding, reduces GR expression, and increases corticosteroid responses to stress (Weaver *et al.*, 2005).

These findings suggest a direct relation between the CpG methylation of the exon 17 promoter, NGFI-A binding, and GR expression. Preliminary findings suggest a comparable effect of maternal care on the methylation status of the exon 1B promoter of the ERα gene in the MPOA of the female rat (Champagne *et al.*, 2006). Levels of cytosine methylation were determined using sodium bisulfite mapping (Frommer *et al.*, 1992) and revealed increased levels of methylation at multiple sites, including a consensus sequence for the stat5b transcription factor of the exon 1B promoter in the adult female offspring of the Low LG mothers. Stat5b is a potent mediator of the regulatory effects of multiple extracellular signals, especially cytokines, on ERα expression in myometrium (Frasor and Gibori, 2003). Across several tissues, the level of cytosine methylation predicts ERα expression. Importantly, the results of chromatin immunoprecipitation assays revealed increased stat5b binding to the exon 1B ERα promoter in the adult female offspring of High LG mothers.

III. CONCLUSION

Reproduction represents a series of energy investments in mating and parental care. The *r* and *K* reproductive strategies in the female represent differential investment in mating versus parental care. The *r* strategy emphasizes a strategy that maximizes the *quantity* of offspring. In contrast, the *K* strategy emphasizes the *quality* of offspring. The former involves increased investment in mating, the later in parental care. While originally conceived as descriptions of species differences, more recent studies focusing on phenotypic variation with species reveal that both strategies can be represented within the same species, depending on the prevailing environment conditions. Indeed it is reasonable to consider the *r* and *K* approaches as lying along a continuum and that the point along this continuum is, in certain species, defined by the quality of the environment and parental investment. Both strategies may be successful depending on the prevailing environmental conditions. In some species, and this may include humans, there is evidence for a stable influence of environmental factors on reproductive phenotypes. Hence, the same environmental conditions that compromise the early growth of the offspring and that "program" enhanced defensive responses may also alter the development of reproductive strategies (Fig. 7.1). The findings with the rat suggest that maternal behavior may program variations in reproductive strategies through tissue-specific effects on gene expression.

Belsky *et al.* (1991) and others argued for a comparable relation between the quality of the environment and reproduction in humans and further cited the importance of parent–child attachment as the mediating variable. In brief, their argument is that the quality of the environment influences parenting and thus the development of individual differences in attachment. This view does not necessarily contrast with explanations at the level of gene expression. Alterations in parent–child attachment are likely to be reflected in neural development, perhaps involving differences in gene expression. Moreover, in the rat maternal care during early development alters the expression of genes in brain regions that regulate maternal care, providing a mechanism for the "inheritance" of individual differences in maternal care. It is possible that those patterns of maternal care in humans that alter reproductive behavior in the female offspring reflect underlying differences in ERα expression and then give rise to variations in attachment that, in turn, predict variations in specific phenotypic variations in reproductive strategies.

Acknowledgments

Research on this chapter in the author's laboratory is supported by a grant from the Canadian Institutes for Health Research.

References

Agrawal, A. A. (2001). Phenotypic plasticity in the interactions and evolution of species. *Science* **294,** 321–326.

Agrawal, A. A., Laforsch, C., and Tollrian, R. (1999). Transgenerational induction of defenses in animals and plants. *Nature* **401,** 60–63.

Andersson-Ellstrom, A., Forssman, L., and Milsom, I. (1996). Age of sexual debut related to life-style and reproductive health factors in a group of Swedish teenage girls. *Acta. Obstet. Gynecol. Scand.* **75,** 484–489.

Arnold, A. P. (2004). Sex chromosomes and brain gender. *Nat. Neurosci. Rev.* **5,** 1–8.

Bale, T. L., Pedersen, C. A., and Dorsa, D. M. (1995). CNS oxytocin receptor mRNA expression and regulation by gonadal steroids. *Adv. Exp. Med. Biol.* **395,** 269–280.

Balthazart, J., and Ball, G. F. (1995). Sexual differentiation of brain and behavior in birds. *Trends Endocrinol. Metab.* **6,** 21–29.

Barker, D. J. P. (1992). "Fetal and Infant Origins of Adult Disease." BMJ Books, London.

Barker, D. J. P., Winter, P. D., Osmond, C., Margetts, B., and Simmonds, S. J. (1989). Weight in infancy and death from ischaemic heart disease. *Lancet* **2,** 577–580.

Barker, D. J., Forsen, T., Eriksson, J. G., and Osmond, C. (2002). Growth and living conditions in childhood and hypertension in adult life: A longitudinal study. *J. Hypertension* **20,** 1951–1956.

Bass, A. (1992). Dimorphic male brains and alternative reproductive tactics in a vocalizing fish. *Trends Neurosci.* **15,** 139–145.

Belsky, J. (1984). The determinants of parenting: A process model. *Child Dev.* **55,** 83–96.

Belsky, J. (1997a). Attachment, mating, and parenting: An evolutionary interpretation. *Hum. Nat.* **8,** 361–381.

Belsky, J. (1997b). Theory testing, effect-size evaluation, and differential susceptibility to rearing influence: The case of mothering and attachment. *Child Dev.* **64,** 598–600.

Belsky, J., Steinberg, L., and Draper, P. (1991). Childhood experience, interpersonal development, and reproductive strategy: An evolutionary theory of socialization. *Child Dev.* **62,** 647–670.

Berman, C. M. (1990). Intergenerational transmission of maternal rejection rates among free-ranging rhesus monkeys on Cayo Santiago. *Anim. Behav.* **44,** 247–258.

Bingham, C. R., Miller, B. C., and Adams, G. R. (1990). Correlates of age at first sexual intercourse in a national sample of young women. *J. Adol. Res.* **5,** 18–33.

Bird, A. (2001). Methylation talk between histones and DNA. *Science* **294,** 2113–2115.

Bird, A., and Wolffe, A. P. (1999). Methylation-induced repression—belts, braces, and chromatin. *Cell* **99,** 451–454.

Blane, D., Hart, C. L., Davey-Smith, G., Gillis, C. R., Hole, D. J., and Hawthorne, V. M. (1996). Association of cardiovascular disease risk factors with socioeconomic position during childhood and during adulthood. *Br. Med. J.* **313,** 1434–1438.

Boonstra, R., and Boag, P. T. (1987). A test of the Chitty hypothesis: Inheritance of life history traits in meadow voles Microtus pennsylvanicus. *Evolution* **41,** 929–947.
Bosma, H., van De Mheen, H. D., and Mackenback, J. P. (1999). Social class in childhood and general health in adulthood: Quationnaire study of contribution of psychological attributes. *Br. Med. J.* **318,** 18–22.
Breedlove, S. M., and Hampson, E. (2002). Sexual differentiation of the brain and behavior. *In* "Behavioral Endocrinology" (J. B. Becker, S. M. Breedlove, D. Crews, and M. M. McCarthy, eds.). MIT Press, Cambridge, Massachusetts.
Bridges, R. S. (1996). Biochemical basis of parental behavior in the rat. *Adv. Study Behav.* **25,** 215–242.
Brooks-Gunn, J., and Warren, M. P. (1988). Mother–daughter differences in menarcheal age in adolescent girls attending national dance company schools and non-dancers. *Ann. Hum. Biol.* **15,** 35–43.
Brunner, E. (1996). Stress and the biology of inequality. *Br. Med. J.* **314,** 1472–1476.
Caldji, C., Tannenbaum, B., Sharma, S., Francis, D., Plotsky, P. M., and Meaney, M. J. (1998). Maternal care during infancy regulates the development of neural systems mediating the expression of behavioral fearfulness in adulthood in the rat. *Proc. Natl. Acad. Sci. USA* **95,** 5335–5340.
Cameron, N., Parent, C., Champagne, F. A., Fish, E. W., Ozaki-Kuroda, K., and Meaney, M. J. (2005). The programming of individual differences in defensive responses and reproductive strategies in the rat through variations in maternal care. *Neurosci. Biobehav. Rev.* **29,** 843–865.
Campbell, B. C., and Udry, J. R. (1995). Stress and age at menarche of mothers and daughters. *J. Biosocial. Sci.* **27,** 127–134.
Charnov, E. L. (1993). "Life History Invariants: Some Explorations of Symmetry in Evolutionary Ecology." Oxford University Press, Oxford.
Champagne, F. A., Diorio, J., Sharma, S., and Meaney, M. J. (2001). Variations in maternal care in the rat are associated with differences in estrogen-related changes in oxytocin receptor levels. *Proc. Nat. Acad. Sci. USA* **98,** 12736–12741.
Champagne, F. A., Francis, D. D., Mar, A., and Meaney, M. J. (2003a). Naturally-occurring variations in maternal care in the rat as a mediating influence for the effects of environment on the development of individual differences in stress reactivity. *Physiol. Behav.* **79,** 359–371.
Champagne, F. A., Weaver, I. C. G., Diorio, J., Sharma, S., and Meaney, M. J. (2003b). Natural variations in maternal care are associated with estrogen receptor alpha expression and estrogen sensitivity in the MPOA. *Endocrinology* **144,** 4720–4724.
Champagne, F. A., Stevenson, C., Gratton, A., and Meaney, M. J. (2004). Individual differences in maternal behavior are mediated by dopamine release in the nucleus accumbens. *J. Neurosci.* **24,** 4113–4123.
Champagne, F. A., Weaver, I. C. G., Diorio, J., Dymov, S., Szyf, M., and Meaney, M. J. (2006). Maternal care regulates methylation of the estrogen receptor alpha 1b promoter and estrogen receptor alpha expression in the medial preoptic area of female offspring. *Endocrinology* **147,** 2909–2915.
Chisholm, J. S. (1993). Death, hope, and sex: Life-history theory and the development of reproductive strategies. *Curr. Anthropol.* **34,** 1–24.
Chisholm, J. S. (1996). The evolutionary ecology of attachment organization. *Hum. Nat.* **7,** 1–38.
Chisholm, J. S. (1999). "Death, Hope, and Sex: Steps to an Evolutionary Ecology of Mind and Morality." Cambridge University Press, New York.
Coall, D. A., and Chisholm, J. S. (2003). Evolutionary perspectives on pregnancy: Maternal age at menarche and infant birth weight. *Soc. Sci. Med.* **57,** 1771–1781.
Chisholm, J. S., and Burbank, V. K. (2001). Evolution and inequality. *Int. J. Epidemiol.* **30,** 206–211.
Conger, R. D., McCarty, J. A., Yang, R. K., Lahey, B. B., and Kropp, J. P. (1984). Perception of child, childrearing values, and emotional distress as mediating links between environmental stressors and observed maternal behavior. *Child Dev.* **55,** 2234–2247.

Conger, R., Ge, X., Elder, G., Lorenz, F., and Simons, R. (1994). Economic stress, coercive family process and developmental problems of adolescents. *Child Dev.* **65,** 541–561.

Coomber, P., Crews, D., and Gonzalez-Lima, F. (1997). Independent effects of incubation temperature and gonadal sex on the volume and metabolic capacity of brain nuclei in the leopard gecko (Eublepharis macularius), a lizard with temperature-dependent sex determination. *J. Comp. Neurol.* **380,** 409–421.

Coopersmith, C., Candurra, C., and Erskine, M. (1996). Effects of paced mating and intromissive stimulation on feminine sexual behavior and estrus termination in the cycling rat. *J. Comp. Psychol.* **110,** 176–186.

Coplan, J. D., Andrews, M. W., Rosenblum, L. A., Owens, M. J., Friedman, S., Gorman, J. M., and Nemeroff, C. B. (1996). Persistent elevations of cerebrospinal fluid concentrations of corticotropin-releasing factor in adult nonhuman primates exposed to early-life stressors: Implications for the pathophysiology of mood and anxiety disorders. *Proc. Nat. Acad. Sci. USA* **93,** 1619–1623.

Coplan, J. D., Trost, R. C., Owens, M. J., Cooper, T. B., Gorman, J. M., Nemeroff, C. B., and Rosenblum, L. A. (1998). Cerebrospinal fluid concentrations of somatostatin and biogenic amines in grown primates reared by mothers exposed to manipulated foraging conditions. *Arch. Gen. Psychiat.* **55,** 473–477.

Crews, D. (2003). The development of phenotypic plasticity: Where biology and psychology meet. *Dev. Psychobiol.* **43,** 1–10.

Crews, D. (2005). Evolution of neuroendocrine mechanisms that regulate sexual behavior. *Trends Endocrinol. Metab.* **16,** 354–361.

Crews, D., Coomber, P., Baldwin, R., Azad, N., and Gonzalez-Lima, F. (1996). Effects of gonadectomy and hormone treatment on the morphology and metabolic capacity of brain nuclei in the leopard gecko (Eublepharis macularius) a lizard with temperature-dependent sex determination. *Horm. Behav.* **30,** 474–486.

Danforth, B. N. (1991). The morphology and behavior of dimorphic males in *Perdita portalis* (Hymenoptera; Andrenidae). *Behav. Ecol. Sociobiol.* **29,** 235–247.

Davey-Smith, G., Hart, C., Blane, D., and Hole, D. (1998). Adverse socioeconomic conditions in childhood cause specific adult mortality: Prospective observation study. *Br. Med. J.* **316,** 1631–1635.

Davis, E. C., Popper, P., and Gorski, R. A. (1996). The role of apoptosis in sexual differentiation of the rat sexually-dimorphic nucleus of the preoptic area. *Brain Res.* **734,** 10–18.

Demski, L. S. (1986). Reproductive patterns in teleost fishes. *In* "Psychobiology of Reproductive Behavior: An Evolutionary Perspective" (D. Crews, ed.), pp. 1–27. Prentice Hall, New Jersey.

Denver, R. J. (1997). Proximate mechanisms of phenotypic plasticity in amphibian metamorphosis. *Am. Zool.* **37,** 172–184.

Dix, T. (1991). The affective organization of parenting adaptive and maladaptive processes. *Psych. Bull.* **110,** 3–25.

Draper, P., and Harpending, H. (1982). Father absence and reproductive strategy: An evolutionary perspective. *J. Anthropol. Res.* **38,** 255–273.

Eisenberg, L., and Earls, F. J. (1975). Poverty, social depreciation and child development. *In* "American Handbook of Psychiatry" (D. A. Hamburg, ed.), Vol. 6, pp. 275–291. Basic Books, New York.

Ellis, B. J. (2004). Timing of pubertal maturation in girls: An integrated life history approach. *Psych. Bull.* **130,** 920–958.

Ellis, B. J. (2005). Timing of pubertal maturation in girls: An integrated life history approach. *Psychol. Bull* **130,** 920–958.

Ellis, B. J., and Garber, J. (2000). Psychosocial antecedents of variation in girls' pubertal timing: Maternal depression, stepfather presence, and marital and family stress. *Child Dev.* **71,** 485–501.

Ellis, B. J., McFadyen-Ketchum, S., Dodge, K. A., Pettit, G. S., and Bates, K. E. (1999). Quality of early family relationships and individual differences in the timing of pubertal maturation in girls: A longitudinal test of an evolutionary model. *J. Pers. Soc. Psychol.* **77,** 387–401.

Ellison, P. T. (2001). "On Fertile Ground: A Natural History of Human Reproduction." Harvard University Press, Cambridge, Massachusetts.

Erskine, M. S. (1989). Solicitation behavior in the estrous female rat: A review. *Horm. Behav.* **23,** 473–502.

Erskine, M. S., Lehmann, M. L., Cameron, N., and Polston, E. (2004). Coregulation of female sexual behavior and pregnancy induction: An exploratory synthesis. *Brain Res. Rev.* **153,** 295–315.

Eveleth, P. B., and Tanner, J. M. (1990). "World-wide Variation in Human Growth," 2nd Edn., Cambridge University Press, Cambridge, England.

Fahrbach, S. E., Morrell, J. I., and Pfaff, D. W. (1984). Oxytocin induction of short-latency maternal behavior in nulliparous, estrogen-primed female rats. *Horm. Behav.* **18,** 267–286.

Fahrbach, S. E., Morrell, J. I., and Pfaff, D. W. (1985). Possible role for endogenous oxytocin in estrogen-facilitated maternal behavior in rats. *Neuroendocrinology* **40,** 526–532.

Fairbanks, L. M. (1996). Individual differences in maternal style. *Adv. Study Behav.* **25,** 579–611.

Faulkes, C. G., and Bennett, N. C. (2001). Family values: Group dynamics and social control of reproduction in African mole-rats. *Trends Ecol. Evol.* **16,** 184–190.

Faulkes, C. G., Abbott, D. H., and Jarvis, J. U. (1990a). Social suppression of ovarian cyclicity in captive and wild colonies of naked mole-rats, Heterocephalus glaber. *J. Reprod. Fertil.* **88,** 559–568.

Faulkes, C. G., Abbott, D. H., Jarvis, J. U., and Sherriff, F. E. (1990b). LH responses of female naked mole-rats, Heterocephalus glaber, to single and multiple doses of exogenous GnRH. *J. Reprod. Fertil.* **89,** 317–323.

Field, T. (1998). Maternal depression effects on infants and early interventions. *Prev. Med.* **27,** 200–203.

Fitch, R. H., and Denenberg, A. (1998). Role for ovarian hormones in sexual differentiation of the brain. *Behav. Brain Sci.* **21,** 311–327.

Fleming, A. S. (1988). Factors influencing maternal responsiveness in humans: Usefulness of an animal model. *Psychoneuroendocrinology* **13,** 189–212.

Fleming, A. S. (1999). The neurobiology of mother–infant interactions: Experience and central nervous system plasticity across development and generations. *Neurosci. Biobehav. Rev.* **23,** 673–685.

Fleming, A. S., Kraemer, G. W., Gonzalez, A., Lovic, V., Rees, S., and Melo, A. (2002). Mothering begets mothering: The transmission of behavior and its neurobiology across generations. [Review] [90 refs] [Journal Article. Review] *Pharmacol. Biochem. Behav.* **73**(1), 61–75.

Fleming, A. S., O'Day, D. H., and Kraemer, G. W. (1999). Neurobiology of mother-infant interactions: Experience and central nervous system plasticity across development and generations. *Neurosci. Biobehav. Rev.* **23,** 673–685.

Flores, D., Tousignant, A., and Crews, D. (1994). Incubation temperature affects the behavior of adult leopard geckos (*Eublepharis macularius*). *Physiol. Behav.* **55,** 1067–1072.

Forger, N. G., Rosen, G. J., Waters, E. M., Jacob, D., Simerly, R. B., and de Vries, G. J. (2004). Deletion of Bax eliminates sex differences in the mouse forebrain. *Proc. Natl. Acad. Sci. USA* **101,** 13666–13671.

Forstmeier, W., Coltman, D. W., and Birkhead, T. R. (2004). Maternal effects influence the sexual behavior of sons and daughters in the zebra finch. *Evolution* **58,** 2574–2583.

Francis, D. D., Diorio, J., Liu, D., and Meaney, M. J. (1999). Nongenomic transmission across generations in maternal behavior and stress responses in the rat. *Science* **286,** 1155–1158.

Francis, D. D., Champagne, F., and Meaney, M. J. (2000). Variations in maternal behaviour are associated with differences in oxytocin receptor levels in the rat. *J. Neuroendocrinol.* **12,** 1145–1148.

Frasor, J., and Gibori, G. (2003). Prolactin regulation of estrogen receptor expression. *Trends Endo. Metabol.* **14,** 118–123.

Frommer, M., McDonald, L. E., Millar, D. S., Collis, C. M., Watt, F., Grigg, G. W., Molloy, P. L., and Paul, C. L. (1992). A genomic sequencing protocol that yields a positive display of 5-methylcytosine residues in individual DNA strands. *Proc. Natl. Acad. Sci. USA* **89**(5), 1827–1831.

Gangestad, S. W., and Simpson, J. A. (2000). The evolution of human mating. Trade-offs and strategic pluralism. *Brain Behav. Sci.* **23,** 573–644.

Geist, V. (1968). Horn-like structures as rank symbols, guards and weapons. *Nature* **220,** 813–814.

Georgopoulos, N., Markou, K., Theodoropoulou, A., Paraskevopoulou, P., Varaki, L., Kazantzi, Z., Leglise, M., and Vagenakis, A. G. (1999). Growth and pubertal development in elite female rhythmic gymnasts. *J. Clin. Endocrinol. Metab.* **84,** 4525–4530.

Gilbert, J. J. (1998). Karimone-induced morphological defenses in rotifers. *In* "The Ecology and Evolution of Inducible Defenses" (R. Tollrian and C. D. Harvell, eds.). Princeton University Press, Princeton, New Jersey.

Gilman, S. E., Kawachi, I., Fitzmaurice, G. M., and Buka, S. L. (2003). Family disruption in childhood and risk of adult depression. *Am. J. Psychiatry* **160,** 939–946.

Gluckman, P. D., and Hanson, M. A. (2004a). Living with the past: Evolution, development, and patterns of disease. *Science* **305,** 1733–1736.

Gluckman, P. D., and Hanson, M. A. (2004b). The developmental origins of the metabolic syndrome. *Trends Endocrinol. Metab.* **15,** 183–187.

Godwin, J., and Crews, D. (1997). Sex differences in the nervous system of reptiles. *Cell. Mol. Neurobiol.* **17,** 1649–1669.

Goland, R. S., Jozak, S., Warren, W. B., Conwell, I. M., Stark, R. I., and Tropper, P. J. (1993). Elevated levels of umbilical cord plasma corticotropin-releasing hormone in growth-retarded fetuses. *J. Clin. Endocrinol. Metab.* **77,** 1174–1179.

Goland, R. S., Tropper, P. J., Warren, W. B., Stark, R. I., Jozak, S. M., and Conwell, I. M. (1995). Concentrations of corticotropin-releasing hormone in the umbilical-cord blood of pregnancies complicated by preeclampsia. *Reprod. Fertil. Dev.* **7,** 1227–1230.

Goldstein, L. H., Diener, M. L., and Mangelsdorf, S. C. (1996). Maternal characteristics and social support across the transition to motherhood: Associations with maternal behavior. *J. Fam. Psychol.* **10,** 60–71.

Gorski, R. A. (1984). Critical role for the medial preoptic area in the sexual differentiation of the brain. *Prog. Brain Res.* **61,** 129–146.

Gorski, R. A., Harlan, R. E., Jacobson, C. D., Shryne, J. E., and Southam, A. M. (1980). Evidence for the existence of a sexually dimorphic nucleus in the preoptic area of the rat. *J. Comp. Neurol.* **193,** 529–539.

Goy, R. W., and McEwen, B. S. (1980). "Sexual Differentiation of the Brain." MIT Press, Cambridge.

Graber, J. A., Brooks-Gunn, J., and Warren, M. P. (1995). The antecedents of menarcheal age: Heredity, family environment, and stressful life events. *Child Dev.* **66,** 346–359.

Gross, M. R. (1996). Alternative reproductive strategies and tactics: Diversity within sexes. *Trends Ecol. Evol.* **11,** 92–98.

Groothuis, T. G. G., Muller, W., von. Engelhardt, N., Carere, C., and Eising, C. (2005). Maternal hormones as a tool to adjust offspring phenotype in avian species. *Neurosci. Biobehav. Rev.* **29,** 329–352.

Grolnick, W. S., Gurland, S. T., DeCourcey, W., and Jacob, K. (2002). Antecedents and consequences of mothers' autonomy support: An experimental investigation. *Dev. Psychol.* **38,** 143–155.

Haig, D. (1993). Genetic conflicts in human pregnancy. *Q. Rev. Biol.* **68,** 495–532.

Hales, C. N., and Barker, D. J. P. (2001). The thrifty phenotype hypothesis. *Br. Med. Bull.* **60,** 5–20.

Herman-Giddens, M. E., Sandler, A. D., and Friedman, N. E. (1988). Sexual precocity in girls. An association with sexual abuse? *Am. J. Dis. Child* **142,** 431–433.

Hinde, R. A. (1986). Some implications of evolutionary theory and comparative data for the study of human prosocial and aggressive behaviour. *In* "Development of Anti-Social and Prosocial Behaviour" (D. Olweus, J. Block, and M. Radke-Yarrow, eds.), pp. 13–32. Academic Press, Orlando.

Hulanicka, B. (1999). Acceleration of menarcheal age of girls from dysfunctional families. *J. Reprod. Infant. Psychol.* **17,** 119–132.

Ibanez, L., Dimartino-Nardi, J., Potau, N., and Saenger, P. (2000). Premature adrenarche—normal variant or forerunner of adult disease? *Endocr. Rev.* **21,** 671–696.

Imperato-McGinley, J. I. (1979). Androgens and the evolution of male gender identity among male pseudohermaphrodites with 5a-reductase deficiency. *N. Engl. J. Med.* **300,** 1233–1237.

Jacobson, C. D., Davis, F. C., and Gorski, R. A. (1985). Formation of the sexually dimorphic nucleus of the preoptic area: Neuronal growth, migration and changes in cell number. *Dev. Brain Res.* **21,** 7–18.

Jones, P. L., Veenstra, G. J., Wade, P. A., Vermaak, D., Kass, S. U., Landsberger, N., *et al.* (1998). Methylated DNA and MeCP2 recruit histone deacetylase to repress transcription. *Nat. Genet.* **19,** 187–191.

Jorm, A. F., Christensen, H., Rodgers, B., Jacomb, P. A., and Easteal, S. (2004). Association of adverse childhood experiences, age of menarche and adult reproductive behavior: Does the androgen receptor gene play a role? *Am. J. Med. Genet. B Neuropsych. Genet.* **125,** 105–111.

Kaplan, R. (1998). Maternal effects, developmental plasticity, and life history evolution: An amphibian model. *In* "Maternal Effects as Adaptations" (T. A. Mousseau and C. W. Fox, eds.), pp. 244–260. Oxford University Press, Oxford.

Kaplan, G. A., and Salonen, J. T. (1990). Socioeconomic conditions in childhood and ischaemic heart disease in middle age. *Br. Med. J.* **301,** 1121–1131.

Kelley, D. B. (1997). Generating sexually differentiated songs. *Curr. Opin. Neurobiol.* **7,** 839–843.

Kessler, R. C., Davis, C. G., and Kendler, K. S. (1997). Childhood adversity and adult psychiatric disorder in the US national comorbidity study. *Psychol. Med.* **27,** 1101–1119.

Kim, K., and Smith, P. K. (1998a). Childhood stress, behavioural symptoms and mother–daughter pubertal development. *J. Adolesc.* **21,** 231–240.

Kim, K., and Smith, P. K. (1998b). Retrospective survey of parental marital relations and child reproductive development. *Int. J. Behav. Dev.* **22,** 729–751.

Kim, K., Smith, P. K., and Palermiti, A. L. (1997). Conflict in childhood and reproductive development. *Evol. Hum. Behav.* **18,** 109–142.

Kirkwood, R. N., and Hughes, P. E. (1981). A note on the influence of boar age on its ability to advance puberty in the gilt. *Anim. Prod.* **32,** 211–213.

Kow, L. M., and Pfaff, D. W. (1998). Mapping of neural and signal transduction pathways for lordosis in the search for estrogen actions on the central nervous system. *Behav. Brain Res.* **92,** 169–180.

Krpan, K. M., Coombs, R., Zinga, D., Steiner, M., and Fleming, A. S. (2005). Experiential and hormonal correlates of maternal behavior in teen and adult mothers. [Clinical Trial. Comparative Study. Journal Article. Research Support, Non-U.S. Gov't] *Horm. Behav.* **47**(1), 112–122.

Lang, J. W., and Andrews, H. V. (1994). Temperature-dependent sex determination in crocodilians. *J. Exp. Zool.* **270,** 28–44.

Langley-Evans, S. C. (1997). Hypertension induced by foetal exposure to a maternal low-protein diet, in the rat, is prevented by pharmacological blockade of maternal glucocorticoid synthesis. *J. Hypertens.* **15,** 537–544.

Lank, D. B., Smith, C. M., Hanotte, O., Burke, T., and Cooke, F. (1995). Genetic polymorphism for alternative mating behavior in a lekking male ruff *Philomachus pugnax*. *Nature* **378,** 59–62.

Levine, S. (1994). The ontogeny of the hypothalamic-pituitary-adrenal axis. The influence of maternal factors. *Ann. N. Y. Acad. Sci.* **746,** 275–288.

Levine, J. E. (1997). New concepts of the neuroendocrine regulation of gonadotropin surges in rats. *Biol. Reprod.* **56,** 293–302.

Liu, D., Diorio, J., Tannenbaum, B., Caldji, C., Francis, D., Freedman, A., Sharma, S., Pearson, D., Plotsky, P. M., and Meaney, M. J. (1997). Maternal care, hippocampal glucocorticoid receptors, and hypothalamic-pituitary-adrenal responses to stress. *Science* **277,** 1659–1662.

Lovic, V., Gonzalez, A., and Fleming, A. S. (2001). Maternally separated rats show deficits in maternal care in adulthood. [Journal Article. Research Support, Non-U.S. Gov't] *Dev. Psychobiol.* **39**(1), 19–33.

MacDonald, K. (1999). An evolutionary perspective on human fertility. Population and Environment: A Journal of Interdisciplinary Studies, **21,** 223–246.

Madeira, M. D., and Lieberman, A. R. (1995). Sexual dimorphism in the mammalian limbic system. *Prog. Neurobiol.* **45,** 275–333.

Maestripieri, D. (1999). The biology of human parenting: Insights from nonhuman primates. *Neurosci. Biobehav. Rev.* **23,** 411–422.

Maestripieri, D. (2005). Early experience affects the intergenerational transmission of infant abuse in rhesus monkeys. *Proc. Natl. Acad. Sci. USA* **102,** 9726–9279.

Margulis, S. W., Salzman, W., and Abbott, D. H. (1995). Behavioral and hormonal changes in female naked mole-rats (Heterocephalus glaber) following removal of the breeding female from a colony. *Horm. Behav.* **29,** 227–247.

Marlowe, F. W. (2003). The mating system of foragers in the standard cross-cultural sample. *J. Comp. Soc. Sci.* **37,** 282–306.

Marmot, M., Shipley, M., Brunner, E., and Hemingway, H. (2001). Relative contribution of early life and adult socioeconomic factors to adult morbidity in the Whitehall II study. *J. Epidemiol. Community Health* **55,** 301–307.

Matthews, S. G., and Meaney, M. J. (2005). Maternal adversity, vulnerability and disease. *In* "Perinatal Stress, Mood and Anxiety Disorders" (A. Riecher Rossler and M. Steiner, eds.), pp. 28–49. Karger.

Maynard Smith, J. (1982). "Evolution and the Theory of Games." Cambridge University Press.

McCarthy, J. C. (1965). Effects of concurrent lactation of litter size and prenatal mortality in an inbred stain of mice. *J. Reprod. Fertil.* **147,** 29–39.

McCarthy, M. M. (1994). Molecular aspects of sexual differentiation of the rodent brain. *Psychoneuroendocinology* **19,** 415–427.

McCarthy, M. M., Davis, A. M., and Mong, J. A. (1997). Excitatory neurotransmission and sexual differentiation of the brain. [Review] [84 refs] [Journal Article. Review] *Brain Res. Bull.* **44**(4), 487–495.

McLoyd, V. C. (1990). The impact of economic hardship on black families and children: Psychological distress, parenting, and socio-emotional development. *Child Dev.* **61,** 311–346.

McLloyd, V. C. (1998). Socioeconomic disadvantage and child development. *Am. Psychol.* **53,** 185–204.

Meaney, M. J. (2001). The development of individual differences in behavioral and endocrine responses to stress. *Ann. Rev. Neurosci.* **24,** 1161–1192.

Meaney, M. J., Szyf, M., and Seckl, J. R. (In press). Epigenetic mechanisms of perinatal programming of hypothalamic-pituitary-adrenal function and health. *Trends Mol. Med.*

Melo, A. I., Lovic, V., Gonzalez, A., Madden, M., Sinopoli, K., and Fleming, A. S. (2006). Maternal and littermate deprivation disrupts maternal behavior and social-learning of food preference in adulthood: Tactile stimulation, nest odor, and social rearing prevent these effects. [Journal Article. Research Support, Non-U.S. Gov't] *Dev. Psychobiol.* **48**(3), 209–219.

Miller, L., Kramer, R., Warner, V., Wickramaratne, P., and Weissman, M. (1997). Integenerational transmission of parental bonding among women. *J. Am. Acad. Child Adolesc. Psychiatry* **36,** 1134–1139.

Moffitt, T. E., Caspi, A., Belsky, J., and Silva, P. A. (1992). Childhood experience and the onset of menarche: A test of a sociobiological model. *Child Dev.* **63,** 47–58.

Moore, M. C. (1991). Application of organizational-activational theory to alternating reproduction strategy: A review. *Horm. Behav.* **25,** 154–79.

Moore, C. L. (1995). Maternal contributions to mammalian reproductive development and the divergence of males and females. *Adv. Study Behav.* **24,** 47–118.

Moore, M. C., Hews, D. K., and Knapp, R. (1998). Hormonal control and evolution of alternative male phenotypes: Generalizations of models for sexual differentiation. *Am. Zool.* **38,** 133–151.

Mousseau, T. A., and Fox, C. W. (1998). The adaptive significance of maternal effects. *Trends Evol. Ecol.* **13,** 403–407.

Mousseau, T. A., and Fox, C. W. (1999). "Maternal Effects as Adaptations." Oxford University Press, London.

Mul, D., Oostdijk, W., and Drop, S. L. S. (2002). Early puberty in adopted children. *Horm. Res.* **57,** 1–9.

Myers, M. M., Brunelli, S. A., Squire, J. M., Shindeldecker, R. D., and Hofer, M. A. (1989). Maternal behavior of SHR rats and its relationship to offspring blood pressures. *Dev. Psychobiol.* **22,** 29–53.

Newberger, E. H., Barkan, S. E., and Lieberman, E. S. (1992). Abuse of pregnant women and adverse birth outcomes: Current knowledge and implications for practice. *JAMA* **267,** 2370–2372.

Newcomer and Udry, (1987).

Numan, M., and Sheehan, T. P. (1997). Neuroanatomical circuitry for mammalian maternal behavior. *Ann. N. Y. Acad. Sci.* **807,** 101–125.

Ong, K. K., Ahmed, M. L., Sherriff, A., *et al.* (1999). Cord blood leptin is associated with size at birth and predicts infancy weight gain in humans. ALSPAC Study Team. Avon Longitudinal Study of Pregnancy and Childhood. *J. Clin. Endocrinol. Metab.* **84,** 1145–1148.

Ong, K. K., Potau, N., Petry, C. J., Jones, R., Ness, A. R., Honour, J. W., de Zegher, F., Ibanez, L., and Dunger, D. B.; The AVON Longitudinal Study of Parents and Children (ALSPAC) Study Team (2004). Opposing influences of prenatal and postnatal weight gain on adrenarche in normal boys and girls. *J. Clin. Endo. Metabol.* **89,** 2647–2651.

Parker, G. A. (1984). Evolutionarily stable strengths. *In* "Behavioral Ecology: An Evolutionary Approach" (J. R. Krebs and N. B. Davies, eds.), pp. 30–61. Sinauer, Sunderland, MA.

Pedersen, C. A. (1995). Oxytocin control of maternal behavior: Regulation by sex steroids and offspring stimuli. *Ann. N. Y. Acad. Sci.* 126–145.

Pedersen, C. A., and Prange, A. J., Jr. (1979). Induction of maternal behavior in virgin rats after intracerebroventricular administration of oxytocin. *Proc. Natl. Acad. Sci. USA* **76,** 6661–6665.

Pedersen, C. A., Caldwell, J. D., Johnson, M. F., Fort, S. A., and Prange, A. J., Jr. (1985). Oxytocin antiserum delays onset of ovarian steroid-induced maternal behavior. *Neuropeptides* **6,** 175–182.

Perrill, S. A., and Magier, M. (1988). Male mating behavior in Acris crepitans. *Copeia* **1,** 245–248.

Pheonix, C. H., Goy, R. W., Gerall, A. A., and Young, W. C. (1959). Organizing action of prenatally administered testoerorone propionate on the tissues mediating behavior in the female guinea pig. *Endocrinology* **65,** 369–382.

Phillips, D. J. W. (1998). Birth weight and the future development of diabetes. *Diabet. Care* **21,** B150–B155.

Phinney, V. G., Jensen, L. C., Olsen, J. A., and Cundick, B. (1990). The relationship between early development and psychosexual behaviors in adolescent females. *Adolescence* **25,** 321–332.

Poulton, R., Caspi, A., Milne, B. J., *et al.* (2002). Association between children's experience of socioeconomic disadvantage and adult health: A life-course study. *Lancet* **360,** 1640–1645.

Power, C., Graham, H., Due, P., *et al.* (2005). The contribution of childhood and adult socioeconomic position to adult obesity and smoking behaviour: An international comparison. *Int. J. Epidemiol.* **34,** 335–344.

Promislow, D. E. L., and Harvey, P. H. (1990). Living fast and dying young: A comparative analysis of life-history variation among mammals. *J. Zool.* London, **220,** 417–437.

Promisolow, D. (2005). A regulatory network of yeast phenotypic plasticity. *Am. Nat.* **165,** 515–523.

Pulzer, F., Haase, U., Knupfer, M., Kratzsch, J., Richter, V., Rassoul, F., Kiess, W., and Keller, E. (2001). Serum leptin in formerly small-for-gestational-age children during adolescence: Relationship to gender, puberty, body composition, insulin sensitivity, creatinine, and serum uric acid. *Metabol. Clin. Exptl.* **50,** 1141–1146.

Quinlan, R. J. (2003). Father absence, parental care, and female reproductive development. *Evol. Human Behav.* **24,** 376–390.

Qvarnstrom, A., and Price, T. D. (2001). Maternal effects, paternal effects and sexual selection. *Trends Ecol. Evol.* **16,** 95–100.

Rahkonen, O., Lahelma, E., and Huuhka, M. (1997). Past or present? Childhood living conditions and adult socioeconomic status as determinants of adult health. *Soc. Sci. Med.* **44,** 327–336.

Razin, A. (1998). CpG methylation, chromatin structure and gene silencing—a three-way connection. *EMBO J.* **17,** 4905–4908.

Reik, W., Dean, W., and Walter, J. (2000). Epigenetic reprogramming in mammalian development. *Science* **293,** 1089–1093.

Repetti, R. L., Taylor, S. E., and Seeman, T. E. (2002). Risky families: Social environments and the mental and physical health of offspring. *Psychol. Bull.* **128,** 330–366.

Rhen, T., and Crews, D. (1999). Embryonic temperature and gonadal sex organize male-typical sexual and aggressive behavior in a lizard with temperature-dependent sex determination. *Endocrinology* **140,** 4501–4508.

Rhen, T., and Crews, D. (2000). Organization and activation of sexual and agonistic behavior in the leopard gecko, Eublepharis macularius. *Neuroendocrinology* **71,** 252–261.

Rhen, T., and Crews, D. (2002). Variation in reproduction behaviour within a sex: Neural systems and endocrine activation. *J. Neuroendocrinol.* **14,** 507–513.

Ritsher, J. E. B., Warner, V., Johnson, J. G., and Dohrenwend, B. P. (2001). Intergenerational longitudinal study of social class and depression: A test of social causation and social selection models. *Br. J. Psychiatry* **178,** 584–590.

Roff, D. A. (1992). "The Evolution of Life Histories: Theory and Analysis." Chapman and Hall, New York.

Romans, S. E., Martin, M., Gendall, K., and Herbison, G. P. (2003). Age of menarche: The role of some psychosocial factors. *Psychol. Med.* **33,** 933–939.

Roosa, M. W., Tein, J., Reinholtz, C., and Angelini, P. J. (1997). The relationship of childhood sexual abuse to teenage pregnancy. *J. Marr. Fam.* **59,** 119–130.

Rosenblum, L. A., and Andrews, M. W. (1994). Influences of environmental demand on maternal behavior and infant development. *Acta Paediatr. Suppl.* **397,** 57–63.

Rossiter, M. C. (1998). The role of environmental variation in parental effects expression. *In* "Maternal Effects as Adaptations." (T. A. Mousseau and C. W. Fox, eds.), pp. 112–136. Oxford University Press, Oxford, New York.

Rossiter, M. C. (1999). *In* "Maternal Effects as Adaptations" (T. A. Mousseau and C. W. Fox, eds.). Oxford University Press, London.

Ryan, M. J., Pease, C. M., and Morris, M. R. (1992). A genetic polymorphism in the swordtail, *Xiphiphorus nigrensis*: Testing the prediction of equal fitness. *Am. Nat.* **139,** 21–31.

Ryder, N. B., and Westoff, C. F. (1971). "Reproduction in the United States 1965." Princeton University Press, New Jersey.

Sadowki, H., Ugarte, B., Kolvin, L., Kaplan, C., and Barnes, J. (1999). Early life family disadvantages and major depression in adulthood. *Br. J. Psychiatry* **174,** 112–120.

Schanberg, S. M., Evoniuk, G., and Kuhn, C. M. (1984). Tactile and nutritional aspects of maternal care: Specific regulators of neuroendocrine function and cellular development. *Proc. Soc. Exp. Biol. Med.* **175,** 135–146.

Schwabl, H. (1993). Yolk is a source of maternal testosterone for developing birds. *Proc. Natl. Acad. Sci. USA* **90,** 11446–11450.

Seckl, J. R. (1998). Physiologic programming of the fetus. *Clin. Perinatol.* **25,** 939–964.

Shahrohk, D., and Meaney, M. J. (2006). Oxytocin – dopamine interactions and individual differences in maternal behavior. *Abstr. - Soc. Neurosci.*

Shibata, D. M., and Rollo, C. D. (1988). Intraspecific variation in the growth rate of gastropods: Five hypotheses. *Mem. Entomol. Soc. Canada* **146,** 199–213.

Simerly, R. B. (2002). Wired for reproduction: Organization and development of sexually-dimorphic circuits in the mammalian forebrain. *Ann. Rev. Neurosci.* **25,** 507–536.

Simerly, R. B., Zee, M. C., Pendleton, J. W., Lubahn, D. B., and Korach, K. S. (1997). Estrogen-dependent sexual differentiation of dopmainergic neurons in the preoptic region of the mouse. *Proc. Natl. Acad. Sci.* **94,** 14077–14082.

Stearns, S. C. (1992). "The Evolution of Life Histories." Oxford University Press, Oxford.

Steinberg, L. (1988). Reciprocal relation between parent–child distance and pubertal maturation. *Dev. Psychol.* **24,** 122–128.

Stewart, J., and Cygan, D. (1980). Ovarian hormones act early in development to feminize open field behavior in the rat. *Horm. Behav.* **14,** 20–32.

Strobl, J. S. (1990). A role for DNA methylation in vertebrate gene expression. *Mol. Endocrinol.* **4,** 181–183.

Surbey, M. K. (1990). Family composition, stress, and the timing of human menarche. *In* "Monographs in Primatology, Socioendocrinology of Primate Reproduction" (T. E. Ziegler and F. B. Bercovitch, eds.), Vol. 13, pp. 11–32. Wiley-Liss, New York.

Szyf, M. (2001). Towards a pharmacology of DNA methylation. *Trends Pharmacol. Sci.* **22,** 350–354.

Szyf, M., Weaver, I. C., Champagne, F. A., Diorio, J., and Meaney, M. J. (2005). Maternal programming of steroid receptor expression and phenotype through DNA methylation in the rat. *Front. Neuroendocrinol.* **26,** 139–162.

Taborsky, M. (1994). Sneakers, satellites, and helpers: Parasitic and cooperative behaviour in fish reproduction. *Adv. Study Behav.* **23,** 1–100.

Tanner, J. M. (1962). "Growth and Adolescence," 2nd Edn., Blackwell Scientific, Oxford, England.

Tinbergen, N. (1972). Functional ethology and the human sciences. Proceedings of the Royal Society of London—Series B: Biological Sciences. **182,** 385–410.

Tollrian, R. (1995). Predator-induced morphological defenses: Costs, life history shifts, and maternal effects in Daphnia pulex. *Ecology* **76,** 1691–1705.

Tollrain, R., and Dodson, S. I. (1999). Inducible defenses in Caldocera: Constraints, costs and multipredator environments. *In* "The Ecology and Evolution of Inducible Defenses" (R. Tollrian and C. D. Harvell, eds.). Princeton University Press, Princeton.

Toran-Allerand, C. D. (1981). Gonadal steroids and brain development. *In vitro* veritas? *Trends Neurosci.* **4,** 118–121.

Toran-Allerand, C. D. (1995). Developmental interaction of estrogen with the neurotrophins and their receptors. *In* "Neurobiological Effects of Sex Steroids" (P. Micevych and R. P. Hammer, eds.). Cambridge University Press.

Tousignant, A., and Crews, D. (1995). Incubation temperature and gonadal sex affect growth and physiology in the leopard gecko (Eublepharis macularius), a lizard with temperature-dependent sex determination. *J. Morph.* **224,** 159–170.

Trivers, R. L. (1974). Parent-offspring conflict. *Am. Zool.* **14,** 249–264.

Turner, P. K., Runtz, M. G., and Galambos, N. L. (1999). Sexual abuse, pubertal timing, and subjective age in adolescent girls: A research note. *J. Reprod. Infant Psychol.* **17,** 111–118.

Udry, J. R. (1979). Age at menarche, at first intercourse, and at first pregnancy. *J. Biosoc. Sci.* **11,** 433–441.

Udry, J. R., and Cliquet, R. L. (1982). A cross-cultural examination of the relationship between ages at menarche, marriage, and first birth. *Demography* **19,** 53–63.

Vaughn, B., Egeland, B., Sroufe, L. A., and Waters, E. (1979). Individual differences in infant-mother attachment at twelve and eighteen months: Stability and change in families under stress. *Child Dev.* **50,** 971–975.

Veening, M. A., van Weissenbruch, M. M., Roord, J. J., and de Delemarre-van Waal, H. A. (2004). Pubertal development in children born small for gestational age. *J. Ped. Endocrinol.* **17,** 1497–1505.

Wagner, W. E., Jr. (1992). Deceptive or honest signaling of fighting ability? A test of alternative hypotheses for the function of changes in call dominant frequency by male cricket frogs. *Anim. Behav.* **44,** 449–462.

Weaver, I. C. G., Cervoni, N., D'Alessio, A. C., Champagne, F. A., Seckl, J. R., Szyf, M., and Meaney, M. J. (2004). Epigenetic programming through maternal behavior. *Nat. Neurosci.* **7,** 847–854.

Weaver, I. C., Champagne, F. A., Brown, S. E., Dymov, S., Sharma, S., Meaney, M. J., and Szyf, M. (2005). Reversal of maternal programming of stress responses in adult offspring through methyl supplementation: Altering epigenetic marking later in life. *J. Neurosci.* **25,** 11045–11054.

Weaver, I. C., D'Alessio, A. C., Brown, S. E., Hellstrom, I. C., Dymov, S., Sharma, S., Szyf, M., and Meaney, M. J. (2007). The transcription factor nerve growth factor-inducible protein a mediates epigenetic programming: Altering epigenetic marks by immediate-early genes. *J. Neurosci.* **27**(7), 1756–1768.

Wells, J. C. K. (2003). Parent-offspring conflict theory, signaling of need, and weight gain in early life. *Q. J. Biol.* **78,** 169–202.

West-Eberhard, M. (2003). "Developmental Plasticity and Evolution." Oxford University Press, Oxford, England.

Wierson, M., Long, P. J., and Forehand, R. L. (1993). Toward a new understanding of early menarche: The role of environmental stress in pubertal timing. *Adolescence* **28,** 913–924.

Whiting, J., and Whiting, B. B. (1978). Strategy for psychocultural research. *In* "The Making of Pychological Anthropology" (G. Spindler, ed.). University of California Press, Berkeley, CA.

Williams, L. A., Evans, S. F., and Newnham, P. J. (1997). Prospective cohort study of factors influencing the relative weights of the placenta and the newborn infant. *Br. Med. J.* **314,** 1864–1868.

Wilson, M., and Daly, M. (1997). Life expectancy, economic inequality, homicide, and reproductive timing in Chicago neighbourhoods. *Br. Med. J.* **314,** 1271–1274.

Worthman, C. M. (1999). Evolutionary perspectives on the onset of puberty. *In* "Evolutionary Medicine" (W. Trevethan, E. O. Smith, and J. J. McKenna, eds.), pp. 135–163. Oxford University Press, New York.

Yang, L. Y., and Clemens, L. G. (1996). Relation of intromission to the female's postejaculatory refractory period in rats. *Physiol. Behav.* **60,** 1505–1511.

Young, L. J., Muns, S., Wang, Z., and Insel, T. R. (1997). Changes in oxytocin receptor mRNA in rat brain during pregnancy and the effects of estrogen and interleukin-6. *J. Neuroendocrinol.* **9,** 859–865.

Zabin, L. S., Emerson, M. R., and Rowland, D. L. (2005). Childhood sexual abuse and early menarche: The direction of their relationship and its implications. *J. Adolesc. Health* **36,** 393–400.

8 Genomic Imprinting and the Evolution of Sex Differences in Mammalian Reproductive Strategies

E. B. Keverne
Sub-Department of Animal Behaviour, University of Cambridge, Madingley Cambridge, CB3 8AA, United Kingdom

ABSTRACT

Two major developments have occurred that have influenced the evolution of sexually dimorphic reproductive strategies of mammals. Viviparity and development of a placenta is one such development, especially in small-brained rodent lineages, where there has been a major impact of placental hormones on the maternal brain. In the Old World primate/hominoid lineages, the massive expansion of the brain through growth of the neocortex has radically changed how reproductive strategies are determined. Genomic imprinting has played a significant part in both of these developments.

Advances in Genetics, Vol. 59
Copyright 2007, Elsevier Inc. All rights reserved.
0065-2660/07 $35.00
DOI: 10.1016/S0065-2660(07)59008-5

Most of the imprinted genes investigated to date are expressed in the placenta and a subset are expressed in both placenta and hypothalamus. Based on phenotypes derived from targeted mutagenesis, a hypothesis is developed for the coadaptive evolution of placenta and hypothalamus, particularly in the context of neurohormonal regulation of maternalism. In small-brained mammals, maternalism places a severe restriction on sexual activity, which in the case of a female rodent is little more than several hours in a lifetime compared with the several weeks given over to maternalism. The consequent sparsity of oestrous, sexually receptive females imposes a rigorous competitive reproductive strategy in males, with the onus being on the male's ability to find oestrous females. This has resulted in a marked sex difference in the chemosensory system, particularly the VNO accessory olfactory system, for the engagement of male sexual behavior in response to oestrous females. Genomic imprinting, together with neonatal androgens, has also played a role in the developing accessory olfactory system and its role in detecting oestrous females.

With the evolutionary expansion of the neocortex seen in Old World primates and hominids, reproductive strategies are complex and embedded in the social structure and hierarchies which characterize primate societies. Reproductive strategies depend far more on intelligent behavioral determinants than they do on hormonal determinants. In females, sexual activity is not restricted to oestrous periods, indeed most of the sexual activity is not reproductive. Male Old World primates continue to mate for years after castration, but loss of dominance status leads to a loss of sexual interest within days. The genetic basis for the expansion of neocortical development is complex, but those parts of the brain which have expanded are undoubtedly under the influence of imprinted genes, as studies using parthenogenetic and androgenetic chimeras and allometric analysis of brains across comparative phylogenies have shown. Sex differences in behavior owe much to social structure, social learning, and the deployment of intelligent behavioral strategies. The epigenetic effects of social learning on brain development have become equally as important as the epigenetic effects of hormones on brain development and both contribute to sex differences in behavior in large-brained primates. © 2007, Elsevier Inc.

I. INTRODUCTION

There have been a number of important developments in the evolution of mammalian reproductive success, important among which has been viviparity and placentation, lactation, and extended parental care. Viviparity is not unique to mammals, but unlike that found in certain fish and reptiles, it goes beyond the simple protection of incubating yolky eggs and involves the evolutionary development of a unique mammalian tissue, trophectoderm (Rossant, 1995). The most advanced form of placentation is invasive haemochorial in which trophoblast cells invade the maternal blood supply of the uterine endometrium, disrupting endothelial cells and thereby achieving direct contact with maternal blood (Moffet and Loke, 2006). This type of placentation characterizes rodents and primates, which are the dominant land animals of today. What might placentation have to do with sexually dimorphic behaviors? The placenta is integral to lactation and maternal care because both events are dependent on priming by hormonal secretions of the placenta. Together with pregnancy, these evolutionary developments have preoccupied female behavioral time budgets, such that sexual behavior is a relatively rare event for most female mammals. Female rodents take weeks to reach puberty, spend only hours sexually receptive, weeks pregnant, and weeks lactating until the offspring are weaned. Hence, sexually receptive females are at a premium, while most adult males are permanently sexually active. A male has potential for siring more offspring than a female, providing he is successful at finding oestrous females and competing for them with other males. Thus, in the context of reproduction, males and females have developed strategies subserved by sexual dimorphisms in behavior which optimize the reproductive success of their own sex (Emlen and Oring, 1977).

The mechanisms for regulating these dimorphic strategies in female reproductive behavior are strongly dependent on the epigenetic effects of placental hormones on the brain, which in turn have required coadaptation between placental development and hypothalamic development. In this chapter, I will consider how the placenta influences female reproductive behavior and the importance of imprinted genes in this context. Since imprinted genes are autosomal and expressed according to parent-of-origin, any such genes that determine brain development in rodents might therefore be expected, *in utero*, to develop a brain that is similar in males and females. The bias of phenotype in this developing brain is likely to be female based since it is developing and evolving under the influence of the same imprinted genes that are developing the placenta which produces the hormones that in turn regulate the hypothalamus for maternalism. Masculine traits are likely to be determined postnatally as a consequence of the epigenetic effects of hormone secretions from the male gonad. Indeed, this appears to be the case since male rodents castrated at birth are demasculinized and have the potential for female patterns of sexual behavior, while female rodents given androgen early postnatally adopt potential for masculine behavior patterns (Baum, 1979). Moreover, exposure to postnatal androgen in males or females does not prevent the brain from expressing parental care, provided sufficient exposure to pups is given that brings about sensitization (Meyer and Rosenblatt, 1980). An important question therefore is what are male androgens modifying in the developing rodent brain that determines adult male sexual strategies in behavior. A second important question is in what way has brain evolution and the development of a large neocortex such as that seen in monkeys and humans impacted on these phylogenetically older reproductive strategies and sexually dimorphic behavior.

II. PLACENTAL REGULATION OF MATERNAL ENDOCRINE FUNCTION AND BEHAVIOR

The placenta, developed from the cell lineage of fetal trophectoderm, exerts considerable influence on maternal endocrine function. Progesterone is the steroid hormone that dominates pregnancy and is necessary to sustain pregnancy. High levels of progesterone during pregnancy are a function of the fetal placenta either directly (primates) or indirectly (rodents) by production of placental hormones that sustain the ovarian corpus luteum. Progesterone has a broad spectrum of effects by acting on many maternal tissues (Keverne, 2006). Notable among these is the brain, particularly the hypothalamus (Fig. 8.1). In the hypothalamic region of the brain, high levels of progesterone exert a negative feedback on the pulsatile release of gonadotropin-releasing hormone (GnRH) preventing ovulatory cycles and curtailing female sexual behavior. The GnRH neurons are not themselves steroid receptive cells but it is thought that negative feedback inhibition occurs via the action of steroids on γ-aminobutyric acid (GABA-ergic) interneurons (Everitt and Keverne, 1986). GnRH results in the release of luteinizing hormone (LH) from the anterior pituitary, which is essential for folliculogenesis and reactivation of the reproductive cycle. In most mammalian species, progesterone-negative feedback inhibits folliculogenesis and hence oestrogen production, thereby resulting in a complete suppression of female fertility and sexual receptivity. In the large-brained primates, where sexual receptivity has become substantially emancipated from endocrine determinants, high levels of progesterone in pregnancy still reduce sexual activity but

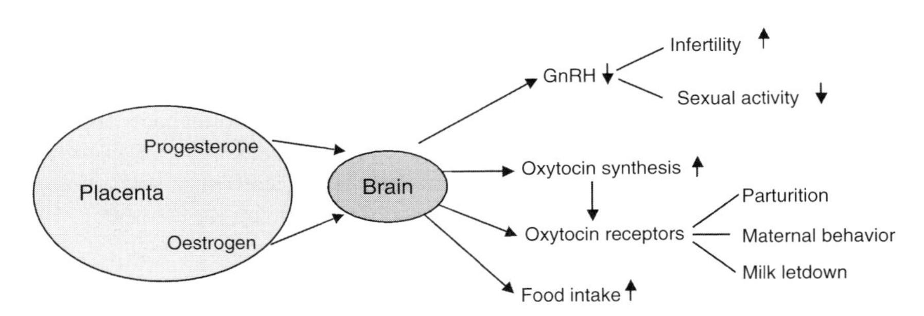

Figure 8.1. Placentally regulated steroids: multiple effects on maternalism.

mainly as a consequence of the peripheral actions of this hormone on the reproductive tract and sexual skin swellings (Keverne, 1984).

Placental progesterone is important not only in suppressing pregnant female sexual behavior but also for priming the brain for promotion of maternal behavior. Other placental hormones of importance in the context of maternal behavior are prolactin and oestrogen. In the late stages of pregnancy, prior to parturition and the onset of maternal behavior, notable changes occur in the circulating levels of hormones in the female which have much in common across most mammalian species studied. These changes involve a fall in progesterone levels and an increase in oestrogen and prolactin, maternal endocrine changes that are dependent on the placenta. Prolactin-like hormones produced by the trophectoderm cell lineages become the primary luteotropins in the latter third of rodent pregnancy and peak in the maternal circulation at parturition (Grattan, 2002). Oestrogen levels that increase toward the end of pregnancy are also directly dependent on the placenta. Placental trophoblast cells are the major site for conversion of progestins into androgen, which then serves as the precursor for aromatization to oestrogen by the maternal ovary. In the female rodent brain, placental lactogens (PL-I and PL-II) have been shown to promote maternal behavior by priming the brain possibly by an action on dopamine neurons (Mann and Bridges, 2001). Progesterone and oestrogen are steroids that readily permeate the blood–brain barrier and both of these hormones have effects on oxytocinergic neurons. Oxytocin is produced in the hypothalamic (paraventricular) parvocellular neurons to activate maternal behavior, while the magnocellular production of oxytocin is important for parturition and milk letdown. High levels of progesterone promote oxytocin synthesis but inhibit neural firing and hence oxytocin release (Kendrick and Keverne, 1992). Around the time of parturition, the falling levels of progesterone and increasing levels of oestrogen promote the synthesis of oxytocin receptors in the brain, uterus, and mammary gland (Broad *et al.*, 1999; Johnson *et al.*, 1989). Hence, the conceptus, via the hormones of its extraembryonic trophectoderm, capitalizes on the maternal neuroendocrine system to determine its own destiny by ensuring the synchronization of birth with milk letdown and maternal care. In order to ensure fetal control over the timing of parturition, the placental trophoblast cell lineage silences the firing of maternal oxytocinergic neurons by producing or inducing high levels of progesterone. Any prepartum oxytocin leakage from the posterior pituitary is taken care of by a proteolytic enzyme (oxytocinase) again produced by the placenta, rendering this oxytocin biologically inactive.

In addition to the production of prolactin under the regulatory control of the placental genome, the maternal pituitary also produces prolactin. This is under the secretory control of the maternal hypothalamus via the tuberoinfundibular dopamine neurons, the release of dopamine acting as a prolactin inhibitory factor on the pituitary lactotrophs. During pregnancy, these lactotrophs are primed for prolactin production by progesterone which induces dephosphorylation and inactivation of tyrosine hydroxylase, the precursor of dopamine (Arbogast and Voogt, 2002). Thus, the placenta by both a direct placental and indirect maternal action via the hypothalamus ensures that the mammary gland is well provided with milk in readiness for birth as well as providing a means of sustaining this production after the placenta is lost.

III. GENOMIC IMPRINTING: COADAPTIVE EVOLUTION OF BRAIN AND PLACENTA

The placental trophoblast is an extraordinary tissue capable of producing a vast range of endocrine secretions, which enable the fetus to regulate its own destiny. Most of these placental hormones function by acting on maternal receptors, an interaction that has required genomic coadaptation between mother and fetus. Hence, the functioning of two genomes (maternal and fetal) as part of a single phenotype (pregnant mother) provides a template for coadaptive selection pressures to operate. Early mortality often accounts for the majority of variance in viability fitness in many species (Wolf, 2000), thereby providing a substantial opportunity for selection on traits that are influenced by gene expression in the fetoplacental unit.

In 1984, two independent studies produced findings that mouse placental and embryonic development required both a maternal and paternal genomic contribution in order to survive (McGrath and Solter, 1984; Surani *et al.*, 1984). Diploidy for either maternal (parthenogenetic–Pg) or paternal (androgenetic–Ag) genomes could not sustain embryonic development due to impaired (Pg) or excessive (Ag) development of the placenta. These early findings were subsequently explained by the discovery of imprinted genes that were reciprocally imprinted (e.g., Igf2 and Igf2r) and which confer reciprocal functions on placental growth (Haig and Graham, 1991). Since these early studies, more than 100 imprinted genes have been identified and nearly all that have been investigated are expressed in the placenta. These imprinted genes are concerned with a balanced growth of the placenta, hormone production, vascular supply, and adequate transfer of nutrients to the fetus (Coan *et al.*, 2005).

Considering the relative sparsity of imprinted genes (110 discovered so far; MRC Mammalian Genetics Unit, 2006), their expression in certain tissues like the brain and placenta is greater than might be expected. Moreover, all of the genes so far identified that are expressed in both placenta and brain, especially the hypothalamus, are paternally expressed (Table 8.1). These distinct organ types, fetal placenta and maternal hypothalamus, function as one in the pregnant mother, although they are encoded by different genotypes. Detailed studies on the phenotype of maternal hypothalamic versus fetal-placental

Table 8.1. Imprinted Genes Expressed in Brain and Placenta

Gene	Parental imprint	Brain expression	Family/protein	References
Necdin	(M)	Hypo	Neuronal growth suppressor	Coan *et al.*, 2005; Watrin *et al.*, 2005
Dlk1	(M)	Hypo/pit	Negative regulator of notch 1	Coan *et al.*, 2005; Croteau *et al.*, 2005
Sgce	(M)	Hypo	Sarcoglycan family	Coan *et al.*, 2005; Yokoi *et al.*, 2005
Peg3	(M)	Hypo/pit	Bac transport—apoptosis	Coan *et al.*, 2005; Li *et al.*, 1999
Peg1	(M)	Hypo/pit	β-hydrolase fold family	Coan *et al.*, 2005; Lefebvre *et al.*, 1998
Magel2	(M)	Hypo	MAGE like protein	Coan *et al.*, 2005; Watrin *et al.*, 2005
Snrp	(M)	Hypo	Nuclear ribonuclear protein	Kaneko-Ishino *et al.*, 2003
Nnat	(M)	Hypo	Phosphorylates CREB	Kaneko-Ishino *et al.*, 2003

expression of such imprinted genes have been demonstrated for the maternally imprinted, paternally expressed genes, Peg1 (Lefebvre *et al.*, 1998; Mayer *et al.*, 2000) and Peg3 (Hiby *et al.*, 2001; Li *et al.*, 1999). Each of these imprinted genes influences growth of the placenta with mutations resulting in 25–30% reduction in placental growth and hence impairment of nutrient transfer and hormone production. Because imprinted genes are expressed according to parent-of-origin, a mother that carries a homozygous null mutation when mated with a wild-type father will produce offspring that express the wild-type normal allele, while a mutant father mated with a wild-type mother will produce mutant offspring that carry and express the mutant allele. In this way, the effects of the mutation on the maternal phenotype and the infant phenotype can be investigated independently. The maternal consequences of expressing this targeted deletion have much in common with lesions of the maternal hypothalamus, namely reduced food intake, severely impaired maternal care, inability to maintain body temperature under a cold challenge, and a severe impairment in milk letdown (Curley *et al.*, 2005). As a consequence, maternal weight gain and fat reserves during pregnancy are impaired and those pups that survive suffer reduced pre- and postnatal growth, even though litter size is smaller (Curley *et al.*, 2005). The similarity of the genetic lesions with neural lesions is not surprising since Peg3 regulates developmental apoptosis (Relaix *et al.*, 1998a) and the neurons in relevant hypothalamic nuclei that regulate these maternal functions are significantly reduced in number (Li *et al.*, 1999).

What is especially interesting about these paternally expressed alleles is that at the same time they are functioning in the placenta to hormonally regulate the maternal hypothalamus, the very same alleles are also actively engaged in developing the fetal hypothalamus and placenta (Fig. 8.2). Moreover, the effects of mutating one such gene (Peg3) either in the maternal hypothalamus or separately and independently in the placenta and fetal hypothalamus produce remarkably similar phenotypes (Fig. 8.3) (Curley *et al.*, 2004). The expression and functional effects of these imprinted genes in the maternal hypothalamus, fetal placenta, and fetal hypothalamus overlap in time (i.e., both genomes are in the pregnant mother, Fig. 8.2), a period when the fetal placenta is actively regulating the maternal mammalian hypothalamus via hormonal secretions. Such coordinated effects across genomes (maternal/fetal) have undoubtedly constrained the evolutionary development of the hypothalamus, but since imprinted genes are expressed according to their parent-of-origin, there is the potential for a very strong maternal effect on both the male and female hypothalamus.

It has been suggested that placental progesterone may influence differentiation of the hypothalamus since there is a significant difference in the expression of progesterone receptors in the medial preoptic nucleus of the female and male during fetal development (Wagner *et al.*, 1998). This region of the brain regulates sexual behavior in the male and maternal behavior in the

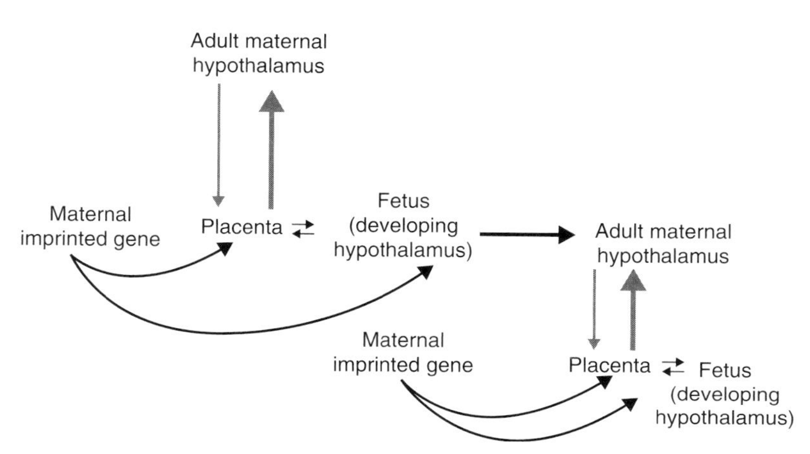

Figure 8.2. Flow chart illustrating the coadaptive effects of an imprinted gene (black arrows) on development of the fetal hypothalamus and fetal placenta across two generations. These structures are developing together and, at the same time, the fetal placenta interacts intimately with the maternal hypothalamus (gray arrows). The action of a gene important for the development of both structures is constrained in the brain by its simultaneous action in the placenta, the hormones of which also act simultaneously on maternal and fetal hypothalami.

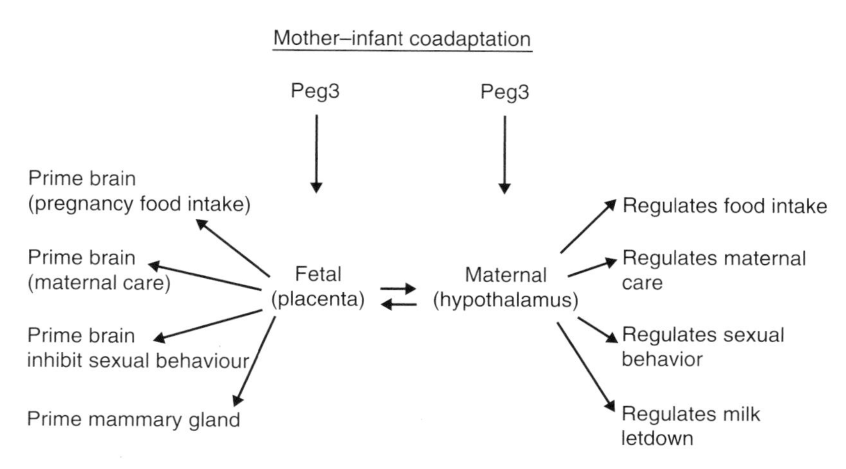

Figure 8.3. Mutations in Peg3 (a paternally expressed gene) independently in the fetal placenta with a wild-type mother or maternal hypothalamus with wild-type offspring produce remarkably similar phenotypes.

female. Moreover, a progesterone antagonist given neonatally influences the developmental size of this nucleus in both males and females (Quadros *et al.*, 2002). Progesterone receptors in the fetal sheep brain are expressed in the preoptic area of the hypothalamus and in the amygdala at the critical period for sexual differentiation (Roselli *et al.*, 2006). Moreover, progesterone metabolites (allopregnenalone) acting via the GABA(-A) receptors in the fetal hypothalamus increase the affinity for GABA binding (Crossley *et al.*, 2000). Interestingly, GABA-ergic neurons are targets for steroid hormones in the adult brain and have been implicated in sex differences in GnRH pulsatile release that leads to ovulation in the female, but are inhibitory to male sexual behavior via GABA-A receptors in the medial preoptic area (MPOA) without impairing motor function (McArthur *et al.*, 2006). Moreover, the GABA-A receptor itself has been reported to be expressed from an imprinted gene in the human brain (Lalande, 1997). Thus, the effects of imprinted genes on the hypothalamus and placenta, both genetic on development and epigenetic through placental hormone action, link these structures in a firmly coadapted embrace. There is undoubtedly a bias through the placenta to enhance hypothalamic parts of the brain toward maternalism, and even males when given sufficient exposure to pups will adopt maternal care. In the context of parental conflict theory of genomic imprinting (Haig, 1992), it is notable that all of the imprinted genes so far discovered that act in both placenta and hypothalamus are paternally expressed (Table 8.1). Could this have provided paternal constraints on the degree of maternalism to which a mammalian brain might evolve, and to what extent have restraints been relaxed in those few mammalian species which are monogamous with biparental care? Clearly, a major constraint on paternal care is produced by the postnatal epigenetic effects of male hormones secreted by the testes and under the developmental influence of the male SRY gene.

IV. GENOMIC IMPRINTING, HYPOTHALAMIC DEVELOPMENT, AND BEHAVIOR

The dual action of imprinted genes on the placenta and hypothalamus has both genetically and epigenetically ensured that the female reproductive strategy is strongly biased toward maternalism. Sexual behavior is a rare event in the life of a female rodent and is epitomized by a reflexive lordosis reactive response to the male. Male sexual behavior is very different, being a frequent behavioral preoccupation that involves proactively finding oestrous females and competing aggressively for them with other males. Since imprinted genes have such a notable impact on female reproductive strategy, and because they are expressed in both females and male progeny, the question arises as to how this same gene might influence male reproductive strategy. By determining maternal strategies that make females rarely sexually available, such imprinted genes would be difficult to establish in the population, unless they enabled males to overcome this apparent skew in the sex ratio brought about by the reduction in oestrous, sexually receptive females. Our work on the Peg3 mutant mouse has revealed a significant effect on male reproductive success. This same gene (Peg3) that impacts on maternalism has been shown to influence male sexual behavior both in terms of their latency to mate and their ability to discriminate oestrous females. Male rodent sexual behavior and male interest in oestrous female urine increase after sexual experience but not in Peg3 mutant males (Swaney *et al.*, 2007). These effects have been traced to defects in the development of the accessory olfactory projection pathway (Swaney *et al.*, 2007).

Olfaction is the primary sensory modality in rodents, and in the male this plays a significant role in all aspects of their sexual behavior. Most interestingly, the vomeronasal chemosensory system is particularly significant for gender-specific behavior. This has been illustrated in studies of targeted mutagenesis of the gene coding for the vomeronasal specific TRP2c ion channel, which cause these receptor neurons to fail in responding with action potentials to pheromonal cues from the female (Stowers *et al.*, 2002). The striking behavioral phenotypes observed in these mutant male mice deficient in the vomeronasal organ (VNO) TRP2c ion channel include a failure to display aggression to other males but to initiate sexual behavior with both males and females (Fig. 8.4). Since the main olfactory system remains functional, this finding suggests that

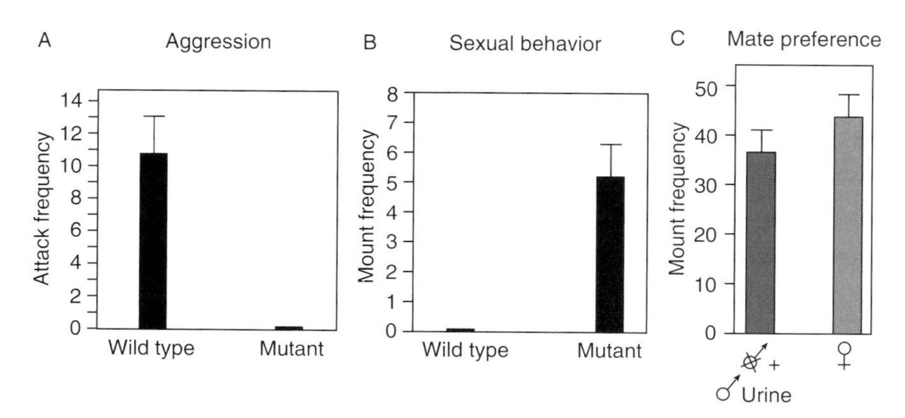

Figure 8.4. The effects of a targeted mutation to the TRP2c ion channel in vomeronasal neurons on male behavior. (A) Lack of aggressive behavior to intruder male in the mutant compared with wild type. (B) Mating behavior of mutant male and wild-type male with another male. (C) Mount frequency of mutant male with male and female presented simultaneously. The stimulus male is castrate to avoid aggression but male urine is applied to give male identity. [Modified after Stowers *et al.* (2002).]

VNO neurons provide an essential sensory activity for engaging sexual and aggressive gender-specific behavior (Leypold *et al.*, 2002; Stowers *et al.*, 2002). The TRP2c mutant males also fail to form normal dominant and subordinate relationships in their territorial marking behavior to intruder males and initiate ultrasonic courtship with both males and females. Thus, all the sexually dimorphic traits that constitute male reproductive strategy are impaired. The same mice have normal levels of testosterone, readily find food, and when they are introduced to the territory of other males, the TRP2c mutants are able to retaliate to the aggression of the resident male. Hence, without a functional VNO receptor system, male mice appear to experience problems in identifying gender cues but not in the engagement of sex-specific motor programs. This linkage between discriminating and engagement is an important distinction to make because it places the impairment on sensory processing by the VNO and how this might engage the gender-specific motor patterns of behavior, rather than an impairment in the behavior itself. While it is not surprising to find that VNO receptor neurons with a nonfunctional ion channel fail to signal the communicative nature of pheromones, it is surprising to find an essential role of the VNO pathway in ensuring the gender specificity of the mating, territorial marking and aggressive behavior, integral components of the reproductive strategy of males. TRP2c mutants appear not to be sexually demotivated, but the behavior occurs inappropriately to the sexual context in which the male is placed.

V. SEX DIFFERENCES IN THE RODENT VNO PATHWAY

The rodent vomeronasal system exhibits sexual dimorphisms at multiple levels along its projection pathway. The MeA, BNST, and MPOA are all larger in the male than female, while subsequent projections to the hypothalamic nuclei, particularly those concerned with female endocrine regulation and maternal care, tend to be larger in the female (Segovia and Guillamon, 1996) (Fig. 8.5). Lesions in discrete parts of this projection pathway enhance components of female typical behavior, whereas in the male such lesions inhibit mating behavior. In both sexes, these differences are sensitive to the regulatory effects of male gonadal steroids early in development (Segovia and Guillamon, 1996). This sexual dimorphism in the VNO pathway correlates with adult immediate early gene (IEG) expression in response to sexual stimuli with males showing enhancement in the number of IEG-expressing neurons in these regions when paired with the female and allowed to mate (Halem *et al.*, 1999). These central connections of the VNO pathway at the level of the amygdala allow for integration of other sensory stimuli, as shown by infusions of the peptide, β-endorphin, which is known to be inhibitory to rodent sexual behavior. At the level of the amygdala, such disruption of the VNO projection pathway impairs the

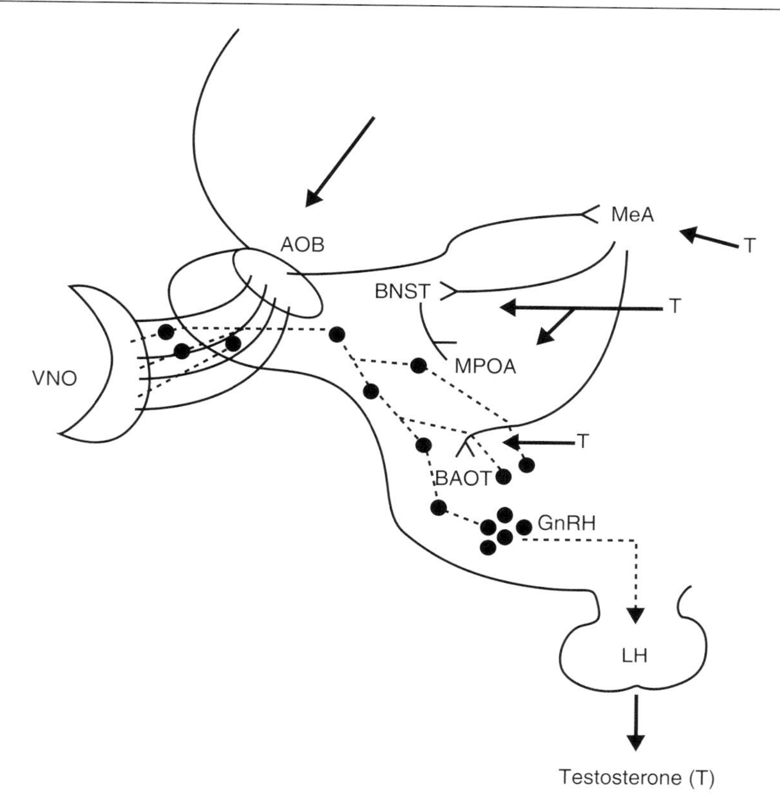

Figure 8.5. VNO central projection pathway (continuous line) and origin of GnRH neurons passing from the olfactory placode via the VNO to the hypothalamus (dotted line). The hormonal cascade from these GnRH neurons initiates the perinatal testosterone surge (T) in the male that masculinizes the VNO relays. Imprinted Peg3 together with P57 and P73 sculptures the differentiation of these relays by regulating apoptopic pathways.

integration of chemosensory stimuli; males fail to mate and show reduced olfactory investigation of the female. Further downstream in the VNO pathway at the level of the MPOA, the same neurochemical interference locks the male into chasing and increased investigation of the female, but again the male does not engage the mating motor response (Herbert, 1993; McGregor and Herbert, 1992). However, after the mating sequence has commenced, this same neurochemical interruption has no effect on sexual behavior. This suggests that the integration of sensory signals (pheromones and somatosensory information) occurs sequentially along the VNO projection pathway to provide for complete engagement of mating behavior (Fig. 8.5).

Interestingly, olfactory sex discrimination of male or female urine does still persist after VNO removal, but when physical access to the stimuli is permitted, the male mice with VNO removal are not able to form a preference between oestrous female and gonadally intact male urine (Pankevich *et al.*, 2006). This further illustrates the importance of the VNO connections for engagement of behavior relevant to gender-specific pheromonal cues which they can nevertheless detect via the main olfactory system. Since the TRP2c mutant VNO neurons have been silent throughout development, including times at which the action of male hormones are thought to determine sex-typical male behavior, a possible consideration for the dissociation of gender discrimination from gender-specific engagement of sexual behavior may relate to events in development. Evidence for developmental events of importance to VNO in the context of sexually dimorphic behavior seems to reside in the activity of intrinsic GABA-ergic neurons at the VNO pathway relays (Segovia and Guillamon, 1996).

Perinatally, GABA accelerates excitotoxic cell death induced by excitatory amino acids, whereas blockade of GABA protects against excitotoxicity (LoTurco *et al.*, 1995). P53 is a gene that acts as an important regulator of apoptotic cell death in the nervous system (Levine, 1997). In its absence, neurons survive excitotoxic insults, and even in animals with only one P53 allele, there is a reduction in cell death of developing neurons (Levine, 1997). The pruning process of neurons by the P53 gene activity is counterbalanced by the actions of a truncated form of the P73 gene lacking its transactivation domain (Pozniak *et al.*, 2000). Interesting in this context is the finding that P73 is involved in neurogenesis of the developing VNO projection pathway and counteracts the proapoptotic effects of P53 (Yang *et al.*, 2000). P73 rescues neurons from apoptosis following NGF withdrawal or P53 overexpression. P73 gene expression has an intense signal in VNO neurons and the AOB, but is also present in secondary and tertiary projections of the VNO pathway to the amygdala and MPOA (Yang *et al.*, 2000). Behaviorally, P73 null mutant male mice lack interest in sexually mature females and fail to show aggressive behavior to intruder males. These findings have been attributed not simply to the absence of a sensory signal but also to the developmental actions in the male brain consequent on the absence of P73 in downstream neurons (Yang *et al.*, 2000). The imprinted Peg3 gene also affects apoptosis, and there is a sex difference at the early postnatal period in the apoptopic protein, caspase 3, in all of the nuclear relays of the accessory olfactory pathway (Broad *et al.*, 2007). However, the interaction of Peg3, P73, and P53 is complex with Peg3 cooperating with Siah-1a in P53-mediated apoptosis (Relaix *et al.*, 1998a), while the NF-κB signal transduction pathway, which is also regulated by Peg3 (Relaix *et al.*, 1998b), is known to regulate the stability and activity of P73 by inducing its proteolytic pathway through a ubiquitin-dependent proteosome pathway (Kikuchi *et al.*, 2006).

VI. BRAIN EVOLUTION AND BEHAVIOR: A ROLE FOR GENOMIC IMPRINTING

In early shrewlike mammals and other small-brained rodents, the brain's regulation of behavior could be viewed as making an integrative contribution to physiological homeostasis. Such small brains may be considered as interfacing between internal bodily needs and the outside world in which the animal survives, serving to provide information about the habitat and social environment. Feeding is stimulated by hunger, sexual behavior is determined by the internal secretion of gonadal hormones, and maternal behavior is dependent on the hormones of pregnancy generated by the fetal placenta. The sensory cues that serve these motivational needs is dominated by olfaction in small-brained rodent-like mammals, partly because these mammals are mainly nocturnal with a poorly developed visual system and have a vocal system generating ultrasound and an auditory system tuned to this high-frequency sound which is only effective over short distances. In the context of reproductive strategies, the olfactory and vomeronasal neurons, like the hypothalamic reproductive neurons (GnRH), take their origin from the olfactory placode (Schwanzel-Fukada *et al.*, 1996; Yoshida *et al.*, 1995) (Fig. 8.5) and have many conserved developmental genes in common (KAL1, reelin, ephrin) (Gamble *et al.*, 2005; Maggi *et al.*, 2005).

Some mammals have developed complex sociobehavior, which reflects the different reproductive strategies of males and females. Reproductive success in males is determined through competition with other males to mate with as many females as possible, with emphasis on same sex aggressive social interactions. Females form strong social bonds with their infants and female–female relationships are affiliative among matrilineal kin who often assist with infant care. These differences are reflected in the evolutionary development of certain neuropeptides. The invertebrate nanopeptide representing the oxytocin–vasopressin complex is phylogenetically old and was coded for by a single gene which underwent duplication to give rise to the mammalian oxytocin and vasopressin (Gilligan *et al.*, 2003; Van Kesteren *et al.*, 1995). In male mammals, vasopressin is important for influencing male courtship and aggressive behavior, while oxytocin plays a dominant role in maternal behavior and female social affiliative interactions (Insel and Young, 2000). An important component of maternal care is the need to recognize offspring, which in small-brained mammals is dependent on olfaction. The hormones of pregnancy induce the synthesis of oxytocin receptors in the olfactory and vomeronasal projection pathways (accessory and main olfactory bulbs, medial amygdala, and medial preoptic area) as well as oxytocin and dopamine receptors in the nucleus accumbens, an area of the brain concerned with social and many other aspects of reward. Thus, if socially relevant behavior occurs in the same context as neutral odors, a conditioned association can develop which enhances the attractiveness of these second order cues.

The VNO inputs are particularly important in the context of male sexual behavior, enhancing the reward salience of female cues to the male by activating nucleus accumbens (Pankevich *et al.*, 2006).

Small-brained mammalian rodents clearly have a well-developed sense of smell that is integral to determining sex differences in reproductive behavioral strategies. Indeed, the largest gene family in the mouse genome, >1000 out of 25,000 genes (Zhang and Firestein, 2002), is given over to encoding receptors for olfactory molecules. In addition, some 300 genes code for vomeronasal receptors, which respond to nonvolatile odors (pheromones), such as those odors found in urine and amniotic fluid that are key to sociobehavior (Dulac and Axel, 1995; Herrada and Dulac, 1997). These receptors represent the first processing stage of the vomeronasal (or "accessory") olfactory system, which is the major pheromone chemosensory system in small-brained mammals (Brennan and Keverne, 2004; Keverne, 2002). This chemosensory system is distinct from the main olfactory system, which is present in all mammals and processes volatile airborne odors. However, it has been shown that some of these receptors do respond to peptide fractions (Liberles and Buck, 2006), a property normally associated with VNO receptors. The importance of the vomeronasal system for sociobehavior in rodents as reported earlier has been illustrated in studies with mice carrying mutations in the vomeronasal genes coding for pheromone detection (V1r gene family) (Del Punta *et al.*, 2002) and receptor transduction (e.g., V1r gene family and V2r TRP2 ion channel) (Stowers *et al.*, 2002). These mice suffer maternal and reproductive behavioral deficits, including impaired maternal aggression, impaired male sexual behavior, and failure to make gender recognition.

Although it appears from comparative genome studies that ancestral primates could process olfactory information via the vomeronasal system, this ability became vestigial ~23 million years ago in the ancestry of modern-day New World and Old World primates and apes (Liman and Innan, 2003; Zhang and Webb, 2003). Further evidence for functional degeneracy comes from comparative phylogenetic analysis of the genes that encode olfactory receptors in marmoset monkeys (Whinnett and Mundy, 2003). These studies have estimated from sequence disruptions that >30% of olfactory receptor genes are nonfunctional pseudogenes in nonhuman primates, rising to >60% in the human genome (Gilad *et al.*, 2003, 2004). Coupled with these genetic changes, there has also been a dramatic reduction in the relative size of the olfactory cortex, from 65% of total cortex in insectivorous mammals to <4% in Old World primates (Stephan *et al.*, 1982).

This decline in olfactory processing has been driven by the need of large-brained Old World primates to gather their social and foraging information from visual cues as they evolved from nocturnal to diurnal lifestyles. Arguably, the most significant visual change in Old World primates was the evolution of trichcromacy (Surridge and Mundy, 2002; Surridge *et al.*, 2003), which occurred at approximately the same time as the pseudogenization of the Old World primate vomeronasal genome (Webb *et al.*, 2004). This also coincided with the development of colorful sexual adornments that signal reproductive receptivity in females and dominance in males and a transition to vision as the dominant sensory system. Sexual behavior was no longer restricted by oestrus and most sexual interactions became nonreproductive, but served for sexual bonding. Postpartum maternal care extended mother–infant bonding beyond the period of suckling when infant mobility required the complex visual recognition of faces at a distance.

The visual cortex of Old World primates has also become especially enlarged, comprising up to 50% of the total neocortex (Van Essen *et al.*, 1992). Visual association areas have become increasingly complex, with several regions devoted to the differential processing of visual information, such as facial expression (Perrett *et al.*, 1992). Indicative of this dramatic shift in the regulation of sociobehaviors from the reliance on olfactory information in small-brained rodents, to visual information in Old World primates, is the negative correlation found between the size of an area of the brain central to relaying visual information (the lateral geniculate nucleus) and that which relays chemosensory information (the olfactory bulb) (Barton, 1998).

These important anatomical changes in the evolutionary development of the mammalian brain have been crucial in the reorganization of reproductive strategies involving sex differences in the brain and behavior. Especially important has been the pseudogenization of vomeronasal and olfactory receptor genes and the downregulation of gonadal and placental hormones in determining sexual and maternal care together with the upregulation of social determinants of behavior. Castrate male primates continue to show a sexual interest in females years after gonadectomy (Michael and Wilson, 1973) but lose sexual interest within days of losing dominance and social status (Keverne, 1992). Reproductive strategies are therefore very complex and embedded in social learning and social structure of the group in which primates live. Moreover, delaying the onset of puberty and extending the period of postnatal care has permitted extensive growth and enlargement of the neocortex. In Old World female primates, sexual behavior is not restricted to a few hours determined by oestrous, nor do females require the hormones of pregnancy to become maternal. Hence, female sexual activity is not primarily reproductive with most of it occurring outside the oestrus ovulatory period. Moreover, in the context of maternalism, nonpregnant females play an important role in infant care, which extends way beyond the infant's weaning period. Parenting and alloparenting are lifetime occupations for social living primates, an evolutionary development that has produced a profound impact on the brain. Brain growth is energetically costly, but postponing the greater part of this to the postpartum period has not only constrained the energetic demands of pregnancy on mother but has also ensured that the brain develops in an environment that facilitates the kind of social learning that is integral to adult reproductive success. A significant question to address is what has taken over from the vestigial, genetically redundant vomeronasal/olfactory system of rodents, and the determining role of gonadal/placental hormones in sex differences in behavior of primates. Reproductive strategies are complex in primates with large brains and depend more on intelligent behavioral determinants than they do on hormonal determinants. The kinds of decisions that are required for reproductive success depend on forward planning and complex decision-taking that requires cooperative behavior in complex primate societies. Hence, there has been evolution away from a determining effect of genes or hormones on behavior, while epigenetic effects of the environment (social and physical) are particularly effective during the extended postnatal period of brain development. Indeed, the postnatal social environment shapes brain development. The primate brain's reward system has been crucially important in this context. Higher order cortical, particularly prefrontal cortex, has replaced the vomeronasal olfactory input to ventral striatum which is important for behavioral reward in small-brained mammals with little neocortex (Curley and Keverne, 2005). These functional changes in the evolution of the mammalian brain and reward system are very clear from a comparison of small- with large-brained mammals. What part, if any, has genomic imprinting played in these neural developments and how might this be investigated? The remarkable similarities in gene number and gene sequences across phylogenies, from mouse to chimpanzee to human, suggest that it is not the evolution of new genes but the reorganization in expression of conserved genes that are primarily responsible for the many different mammalian phenotypes and phylogenies. Genomic imprinting, which among vertebrates is unique to placental mammals, is one way of regulating gene expression and gene dosage.

A significant approach to investigating genomic imprinting in the brain has been achieved by the construction of chimeras (Allen *et al.*, 1995; Keverne *et al.*, 1996a). Embryos constructed from a mixture of cells that are parthenogenetic/normal or androgenetic/normal survive, but survival requires that less than 50% of the cells to be chimeric. The precise locations in the brain to which these chimeric cells (i.e., those expressing exclusively maternal or paternal genes) participate in development can be determined by the presence of a genetic marker (β globin or LacZ). Using these techniques, a clear and distinct patterning in brain development emerges. At birth, cells that are disomic for paternal genome (i.e., both alleles are from the father) contribute substantially to those parts of the brain that are important for primary motivated behavior (the hypothalamus, septum, preoptic area, and BNST—bed nucleus of stria terminalis) and are excluded from the developing cortex and striatum (Keverne *et al.*, 1996a) (Fig. 8.6). By contrast, parthenogenetic cells (i.e., both alleles) inherited from the mother are excluded from these mediobasal forebrain areas, but

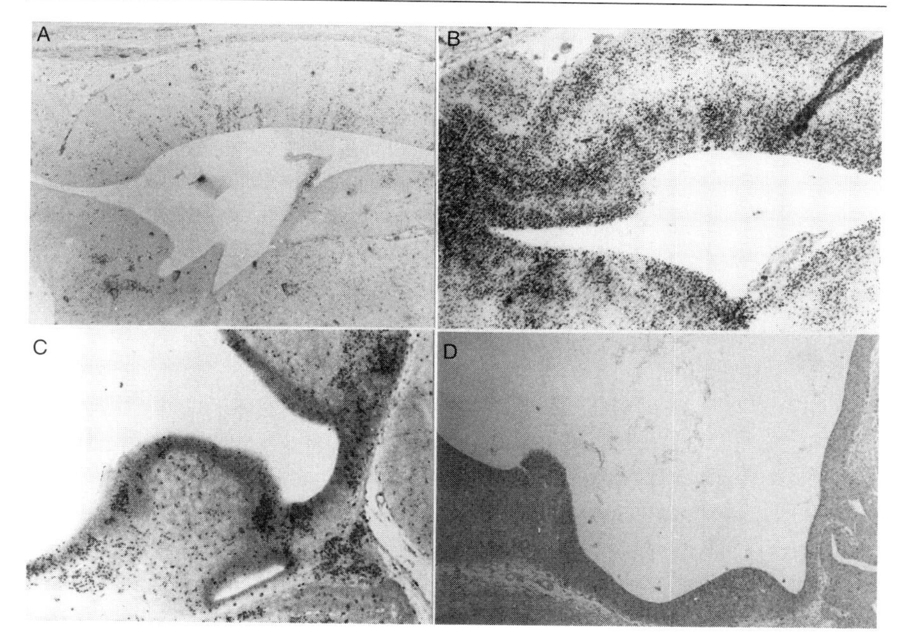

Figure 8.6. Parasagittal sections of the mouse brain (embryonic day 17) illustrating differential contribution of chimeric cells disomic for paternal (Ag) alleles (A and C) and chimeric cells disomic for maternal (Pg) cells (B and D). Relatively few Ag cells can be seen in the cortex (A) compared with Pg cells (B), Pg cells make a substantial contribution to cortex and striatum (B) but are virtually absent from the developing hypothalamus (D) compared with Ag cells (C).

selectively accumulate in those regions where androgenetic cells are excluded, especially the neocortex and striatum (Keverne *et al.*, 1996a) (Fig. 8.6). Furthermore, growth of the brain of parthenogenetic chimeras is enhanced by this increased gene dosage from maternally expressed genes, whereas the brains of androgenetic chimeras are smaller, both in absolute measurements and especially relative to body weight (Fig. 8.7) (Allen *et al.*, 1995).

It is not only surprising that parthenogenetic cells seem to proliferate at the expense of normal cells and to produce a larger telencephalon in chimeras, but this enlarged brain appears anatomically and functionally normal. This is surprising because a large number of genes have been silenced in these cells (i.e., all the imprinted genes that are paternally expressed) and others that are maternally expressed have been duplicated. Human neurological findings of brain function in Prader–Willi syndrome and Angelmans syndrome where disomies of restricted chromosal regions occur are congruent with these experimental findings (Lalande, 1997; Wagstaff *et al.*, 1992).

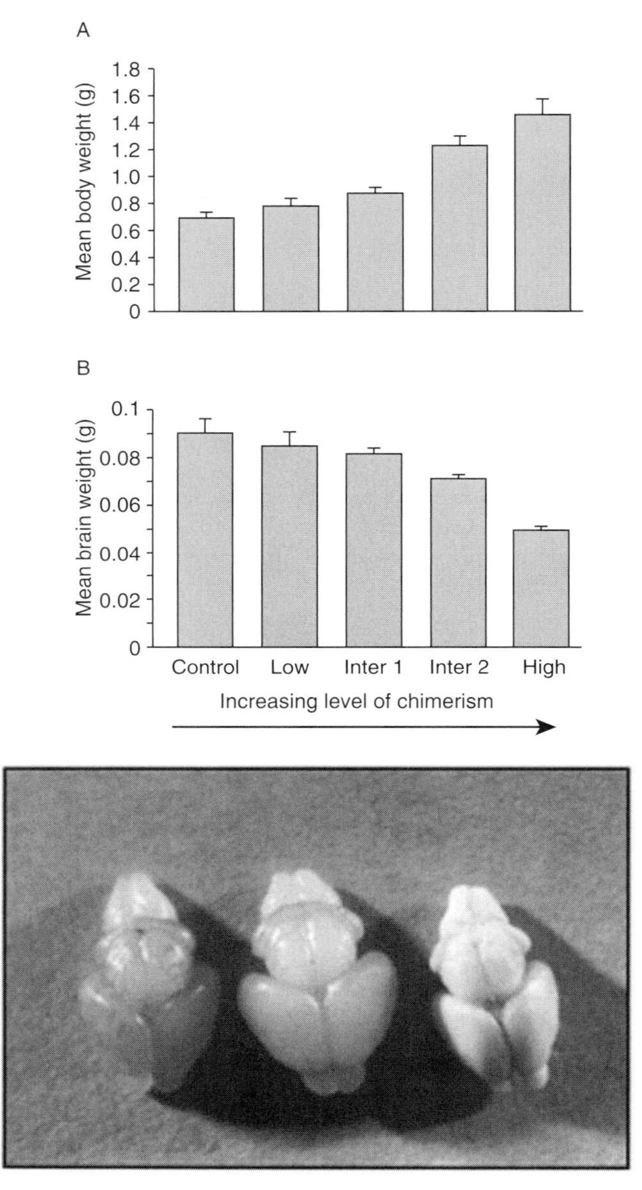

Figure 8.7. Androgenetic chimeras show graded increase in body weight at birth dependent on Ag contribution (A), but decreased brain weight at birth dependent on Ag contribution (B). The brain size at birth is represented in the picture from left to right by wild-type control, Pg chimera, Ag chimera. Pg cells can be seen to increase forebrain cortical size relative to wild type, while Ag chimera has smaller forebrain size at birth than both wild-type and Pg chimeras.

The distinct patterning in the distribution of parthenogenetic and androgenetic cells and their differential effects on brain growth suggest that genomic imprinting may have been important in forebrain evolution, a suggestion supported by comparative studies of the brain. Allometric scaling of the parts of the brain to which maternally or paternally expressed genes differentially contribute reveals that a remodeling of the brain has occurred during mammalian evolution (Keverne *et al.*, 1996b). On moving across phylogenies from insectivorous mammals to prosimian and then simian primates, it can be seen that the neocortex and striatum have increased significantly in size relative to the rest of the brain and body, while the hypothalamus, MPOA, and the septum have decreased in size (Keverne *et al.*, 1996b). Genomic imprinting may thus have facilitated a rapid, nonlinear expansion of the brain (especially the neocortex and striatum) relative to body size during its development over an evolutionary timescale. Of course this neocortical expansion is not specific to the development of sex differences in the brain, but it has been an important evolutionary development enabling behavior to be primarily brain generated dependent on experience within and across generations. Together with the brain's reward system, this has probably been enabling for the vast repertoire of sexual proclivities and preferences of human species.

VII. CONCLUSIONS

This chapter focuses on a relatively new area of mammalian gene regulation, genomic imprinting, that undoubtedly deserves further investigation in the context of sex differences in brain development and reproductive behavior. It is clear that imprinted genes play a major role in placental and hypothalamic development, and it is plausible to hypothesize coadaptive evolution of these structures, particularly in the context of maternalism. Maternalism has been integral to the female's reproductive strategy, a strategy which in turn imposes a major constraint on male sexual behavior and male reproductive strategies. It is therefore important to consider sexual differences in reproductive strategy as coadaptively evolving between the sexes and shaping the development of the brain which controls them. Looking back to reptilian ancestors, we find a brain that is not influenced by maternal care in these egg-laying vertebrates, the placenta has not evolved nor has genomic imprinting and sexual dimorphisms in behavior are not sexually distinct (Crews, 1998). Some species have temperature-dependent sex determination and show reproductive behavioral variation within sexes dependent on the incubation temperature. Other reptilian species which only have females (parthenogenomes) exhibit individuals alternating between female- and male-like sexual behavior. Hence, although sexually dimorphic behavior is observed in reptiles, this does not provide for distinct behavioral strategies between the sexes. In viviparous placental mammals where disproportionate investment by the female in energetics and time budgets are dedicated to reproductive success, a distinct strategy for reproduction is integral to being female. This female strategy in turn has secondary impact on the reproductive behavior which characterizes males. Male reproductive strategies are undoubtedly dependent on the epigenetic effects of gonadal hormones, but such effects occur postnatally in rodents after the basic structure of the hypothalamus has developed but its interconnections remain to be refined. The core sex differences in complexity of mammalian reproductive strategies have incorporated the genetics of maternalism. Although the sexes are distinct, they are not evolving as distinct species. Both sexes require a placenta, develop *in utero* in the female, and require maternal care. Such genetic coadaptivity across the sexes is perhaps best regulated by genes expressed according to parent-of-origin.

The mammalian brain has increased in size relative to body weight more than any other structure and has evolved to a level of complexity that is unique to mammals. Most of this complexity is in terms of neocortical expansion and its intraconnections, as well as in interconnections with other neural structures, especially the striatum. In general terms, genomic imprinting has contributed to these evolutionary developments, and this has been revealed from studies with chimeras in which the chimeric cells have been genetically modified in such a way as to increase or decrease their imprinted gene contribution. In mice, the size of the neocortex increases when gene dosage from maternally expressed genes is increased and decreases in size when it is reduced (Allen *et al.*, 1995; Keverne *et al.*, 1996). Moreover, those parts of the brain where these experimental chimeric cells proliferate, when examined across modern-day phylogenies, reveal a nonlinear logarithmic shift in size on moving from insectivores to prosimians to Old World monkeys and apes (Keverne *et al.*, 1996).

One fundamental consequence of this increase in mammalian neocortex has been the level of emancipation which has occurred in the regulation of behavior from hormonal determinants. The epigenetic effects of hormones on the limbic brain are presumably still manifest, but the increase in neocortical brain size has placed more emphasis on social learning and memory. The epigenetic effects from the bodily internal hormonal milieu have given way to the epigenetic effects of the social environment in which the neocortex develops.

We still know very little about individual imprinted genes that may be responsible for the evolutionary expansion of neocortex other than those imprinted genes that are affected by pathologies (e.g., Angelmans syndrome, Retts syndrome). It is known, however, that certain imprinted genes are concerned with the fundamentals of forebrain development. Ttl2 is differentially expressed between ventral and dorsal telencephalon (future neocortex) at a critical timepoint in the generation and migration of cortical neurons (McLaughlin *et al.*, 2006). Mest gene expression is directly induced by Fgf8 in neocortical explant cultures and is a candidate gene for setting up the cortical protomap (Sansom *et al.*, 2005). None of these genes are concerned with sexual differentiation, but they are concerned with developing the vast forebrain structure that has been integral to all rewarded behavior, including sexual and social.

References

Allen, N. D., Logan, K., Lally, G., Drage, D. J., Norris, M. L., and Keverne, E. B. (1995). Distribution of parthenogenetic cells in the mouse brain and their influence on brain development and behaviour. *Proc. Natl. Acad. Sci. USA* **92,** 10782–10786.

Arbogast, L. A., and Voogt, J. L. (2002). Progesterone induces dephosphorylation and inactivation of tyrosine hydroxylase in rat hypothalamic dopaminergic neurons. *Neuroendocrinology* **75,** 273–281.

Barton, R. (1998). Visual specialization and brain evolution in primates. *Proc. Biol. Sci.* **265,** 1933–1937.

Baum, M. J. (1979). Differentiation of coital behaviour in mammals. *Neurosci. Biobehav.* **3,** 265–284.

Brennan, P., and Keverne, E. B. (2004). Something in the air? New insights into mammalian pheromones. *Curr. Biol.* **14,** R81–R89.

Broad, K. C., Levy, F., Evans, G., Kimura, T., Keverne, E. B., and Kendrick, K. M. (1999). Previous maternal experience potentiates the effect of parturition on oxytocin receptor MRNA in the paraventricular nucleus. *Eur. J. Neurosci.* **11,** 3725–3737.

Broad, K. C., Curley, J. P., and Keverne, E. B. (2007). An increase in apoptosis during neonatal brain development underlies the behavioural deficits seen in mice lacking functional paternally expressed gene-3 (Peg3). *Eur. J. Neurosci.* In press.

Coan, P. M., Burton, G. J., and Ferguson-Smith, A. C. (2005). Imprinted genes in the placenta–a review. *Placenta* **26,** S10–S20.

Crews, D. (1998). On the organization of individual differences in sexual behavior. *Am. Zool.* **38,** 118–132.

Crossley, K. J., Walker, D. W., Beart, P. M., and Hirst, J. J. (2000). Characterisation of GABA(A) receptors in fetal, neonatal and adult ovine brain: Region and age related changes and the effects of allopregnanolone. *Neuropharmacology* **39,** 1514–1522.

Croteau, S., Roquis, D., Charron, M. C., Frappier, D., Yavin, D., Loredo-Osti, J. C., Hudson, T. J., and Naumova, A. K. (2005). Increased plasticity of genomic imprinting of DIK1 in brain is due to genetic and epigenetic factors. *Mamm. Genome* **16,** 127–135.

Curley, J. P., and Keverne, E. B. (2005). Genes, brains and mammalian social bonds. *Trends Ecol. Evol.* **20,** 561–567.

Curley, J. P., Barton, S., Surani, A., and Keverne, E. B. (2004). Coadaptation in mother and infant regulated by a paternally expressed gene. *Proc. R. Soc. Lond. B* **271,** 1303–1309.

Curley, J. P., Pinnock, S. B., Dicson, S. L., Thresher, R., Miyoshi, N., Surani, M. A., and Keverne, E. B. (2005). Increased body fat in mice with a target mutation of the paternally expressed imprinted gene Peg3. *FASEB J.* **19,** 1302–1304.

Del Punta, K., Leinders-Zufall, T., Rodriguez, I., Jukam, D., Wysocki, C. J., Ogawa, S., Zufall, F., and Mombaerts, P. (2002). Deficient pheromone responses in mice lacking a cluster of vomeronasal receptor genes. *Nature* **419,** 70–74.

Dulac, C., and Axel, R. (1995). A novel family of putative pheromone receptors in mammal with topographically organized and sexually dimorphic distribution. *Cell* **90,** 763–773.

Emlen, S. T., and Oring, L. W. (1977). Ecology, sexual selection and the evolution of mating system. *Science* **197,** 215–223.

Everitt, B. J., and Keverne, E. B. (1986). Reproduction. *In* "Neuroendocrinolgy" (S. L. Lightman and B. J. Everitt, eds.), pp. 427–537. Blackwell Scientific Publication, Oxford.

Gamble, J. A., Karunadasa, D. K., Pape, J. R., Skynner, M. J., Todman, M. G., Bicknell, R. J., Allen, J. P., and Herbison, A. E. (2005). *J. Neurosci.* **25,** 3142–3150.

Gilad, Y., Man, P., Paabo, S., and Lancet, D. (2003). Human specific loss of olfactory receptor genes. *Proc. Natl. Acad. Sci. USA* **100,** 3324–3327.

Gilad, Y., Wiebe, V., Przeworski, M., Lancet, D., and Pääbo, S. (2004). Loss of olfactory receptor genes coincides with the acquisition of full trichromaic vision in primates. *PLoS Biol.* **2,** E5.

Gilligan, P., Brenner, S., and Venkatesh, B. (2003). Neurone-specific expression and regulation of pufferfish isotocin and vasotocin genes in transgenic mice. *J. Neuroendocrinol.* **15,** 1027–1036.

Grattan, D. R. (2002). Behavioural significance of prolactin signalling in the central nervous system during pregnancy and lactation. *Reproduction* **123,** 497–506.

Haig, D. (1992). Genomic imprinting and the theory of parent-offspring conflict. *Semin. Dev. Biol.* **3,** 153–160.

Haig, D., and Graham, C. (1991). Genomic imprinting and the strange case of the insulin-like growth factor II receptor. *Cell* **64,** 1045–1046.

Halem, H. A., Cherry, J. A., and Baum, M. J. (1999). Vomeronaal neuroepithelium and forebrain Fos responses to male pheromones in male and female mice. *J. Neurobiol.* **390,** 251–260.

Herrada, V., and Dulac, C. (1997). A novel family of putative pheromone receptors in mammal with a topographically organized and sexually dimorphic distribution. *Cell* **90,** 763–773.

Herbert, J. (1993). Peptides in the limbic system: Neurochemical codes for co-ordinated adaptive responses to behavioural and physiological demand. *Prog. Neurobiol.* **41,** 723–791.

Hiby, S. E., Lough, M., Keverne, E. B., Surani, M. A., Loke, Y. U., and King, A. (2001). Paternal monoallelic expression of PEG3 in the human placenta. *Hum. Mol. Genet.* **10,** 1093–1100.

Insel, T., and Young, L. (2000). Neuropeptides and the evolution of social behavior. *Curr. Opin. Neurobiol.* **10,** 784–789.

Johnson, A. E., Ball, G. F., Coirini, H., Harbaugh, C. R., McEwen, B. S., and Insel, T. R. (1989). Time course of the estradiol-dependent induction of oxytocin receptor binding in the ventromedial hypothalamic nucleus of the rat. *Endocrinology* **125,** 1414–1419.

Kaneko-Ishino, T., Kohda, A., and Ishino, F. (2003). The regulation and biological significance of genomic imprinting in mammals. *J. Biochem.* **133,** 699–711.

Kendrick, K. M., and Keverne, E. B. (1992). Control of synthesis and release of oxytocin in the sheep brain. *In* "Annals of the New York Academy of Sciences, Volume 652: Oxytocin in Maternal, Sexual, and Social Behaviors" (C. A. Pederson, J. D. Caldwell, C. F. Jirikowski, and T. R. Insel, eds.), pp. 102–121. New York Academy of Science, New York.

Keverne, E. B. (1984). Reproductive behaviour. *In* "Reproduction in Mammals, Book 4, Reproductive Fitness" (C. R. Austin and R. V. Short, eds.), pp. 133–175. Cambridge University Press, Cambridge.

Keverne, E. B. (1992). Primate social relationships: Their determinants and consequences. *In* "Advances in the Study of Behaviour" (P. J. B. Slater, J. S. Rosenblatt, C. Beer, and Manfred Milinski, eds.), Vol. 21, pp. 1–37. Academic Press, San Diego.

Keverne, E. B., Fundele, R., Narashimha, M., Barton, S. C., and Surani, M. A. (1996a). Genomic imprinting and the differential roles of parental genomes in brain development. *Dev. Brain Res.* **92,** 91–100.

Keverne, E. B., Martel, F. L., and Nevison, C. M. (1996b). Primate brain evolution, genetic and functional considerations. *Proc. R. Soc. Lond. (Biol.)* **264,** 1–8.

Keverne, E. B. (2002). Pheromones, vomeronasal function, and gender-specific behavior. *Cell* **108,** 735–738.

Keverne, E. B. (2006). Trophoblast regulation of maternal endocrine function and behaviour. *In* "Biology and Pathology of Trophoblast" (A. Moffett, C. Loke, and A. McLaren, eds.), pp. 148–168. Oxford University Press, Oxford.

Kikuchi, H., Ozaki, T., Furuya, K., Hanamoto, T., Nakanishi, M., Yamamoto, H., Yoshida, K., Todo, S., and Nakagawara, A. (2006). NF-kappaB regulates the stability and activity of p73 by inducing its proteolytic degradation through a ubiquitin-dependent proteasome pathway. *Oncogene* **25,** 7608–7617.

Lalande, M. (1997). Parental imprinting and human disease. *Annu. Rev. Genet.* **30,** 173–195.

Lefebvre, L., Viville, S., Barton, S. C., Ishino, F., Keverne, E. B., and Surani, M. A. (1998). Abnormal maternal behaviour and growth retardation associated with loss of the imprinted gene *Mest*. *Nat. Genet.* **20,** 163–169.

Levine, A. J. (1997). P53, the cellular gatekeeper for growth and division. *Cell* **88,** 323–331.

Leypold, B. G., Leinders-Zufall, T., Kim, M. M., Zufall, F., and Axel, R. (2002). Altered sexual and sexual and social behaviours in trp2 mutant mice. *Proc. Natl. Acad. Sci. USA* **99,** 6376–6381.

Li, L. L., Keverne, E. B., Sparacio, S. A., Ishino, F., Barton, S. C., and Surani, M. A. (1999). Regulation of maternal behavior and offspring growth by paternally expressed *Peg3*. *Science* **284,** 330–396.

Liberles, S. D., and Buck, L. B. (2006). A second class of chemosensory receptors in the olfactory epithelium. *Nature* **442,** 645–650.

Liman, E. R., and Innan, H. (2003). Relaxed selective pressure on an essential component of pheromone transduction in primate evolution. *Proc. Natl. Acad. Sci. USA* **100,** 3328–3332.

LoTurco, J. J., Owens, D. F., Heath, M. J. S., Davies, M. B. E., and Kriegstein, A. R. (1995). GABA and glutamate depolarize cortical progenitor cells and inhibit DNA synthesis. *Neuron* **15,** 1287–1298.

Maggi, R., Cariboni, A., Zaninetti, R., Samara, A., Stossi, F., Pimpinelli, F., Giacobini, P., Consalez, G. G., Rugarli, E., and Piva, F. (2005). Factors involved in the migration of neuroendocrine hypothalamic neurons. *Arch. Ital. Biol.* **143,** 171–178.

Mann, P. E., and Bridges, R. S. (2001). Lactogenic hormone regulation of maternal behavior. *Prog. Brain Res.* **133,** 251–262.

Mayer, W., Hemberger, M., Frank, H. G., Grummer, R., Winterhager, E., Kaufmann, P., and Fundele, R. (2000). Expression of the imprinted genes MEST/Mest in human and murine placenta suggest a role in angiogenesis. *Dev. Dyn.* **217,** 1–10.

McArthur, S., Siddique, Z. L., Christian, H. C., Capone, G., Theogaraj, E., John, C. D., Smith, S. F., Morris, J. F., Buckingham, J. C., and Gillies, G. E. (2006). Perinatal glucocorticoid treatment disrupts the hypothalamolactotroph axis in adult female, but not male, rats. *Endocrinology* **147,** 1904–1915.

McGrath, J., and Solter, D. (1984). Completion of mouse embryogenesis requires both the maternal and paternal genomes. *Cell* **37,** 179–183.

McGregor, A., and Herbert, J. (1992). The effects of β-endorphin infusions into the amygdala on visual and olfactory sensory processing during sexual behaviour in the male rat. *Neuroscience* **46,** 173–179.

McLaughlin, D., Vidaki, M., Renierie, E., and Karagogeos, D. (2006). Expression pattern of the maternally imprinted gene Gtl2 in the forebrain during embryonic development and adulthood. *Gene Expr. Patterns* **6,** 394–399.

Meyer, A., and Rosenblatt, J. S. (1980). Hormonal interaction with stimulus and situational factors in the initiation of maternal behaviour in non-pregnant rats. *J. Comp. Physiol. Psychol.* **94,** 1040–1059.

Michael, R. P., and Wilson, M. I. (1973). Effects of castration and hormone replacement in fully adult male rhesus monkeys. *Endocrinology (Baltimore)* **95,** 150–159.

Moffet, A., and Loke, C. (2006). Immunology of placentation in eutherian mammals. *Nature Immunol.* **6,** 584–594.

MRC Mammalian Genetics Unit (2006). Mouse Imprinted Genes. http://www.mgu/har.ac.uk/research/imprinted/index.html

Pankevich, D. E., Cherry, J. A., and Baum, M. J. (2006). Effect of vomeronasal organ removal from male mice on their preference for and neural Fos responses to female urinary odors. *Behav. Neurosci.* **120,** 925–936.

Perrett, D., Hietanen, J. K., Oram, M. W., and Benson, P. J. (1992). Organization and function of cells responsive to faces in temporal cortex. *Phil. Trans. R. Soc. Lond. B Biol. Sci.* **335,** 23–30.

Pozniak, C. D., Radinvic, S., Yang, A., McKeon, F., Kaplan, D. R., and Miller, F. D. (2000). An anti-apoptotic role for the p53 family member, p73, during developmental neuron death. *Science* **289,** 304–306.

Quadros, P. S., Lopez, V., De Vries, G. J., Chung, W. C., and Wagner, C. K. (2002). Progesterone receptors and the sexual differentiation of the medial preoptic nucleus. *J. Neurobiol.* **51,** 24–32.

Relaix, F., Wei, X.-J., Wu, X., and Sassoon, D. A. (1998a). Peg3/ Pw1 is a potential cell death mediator and cooperates with Siah1a in p53-mediated apoptosis. *Proc. Natl. Acad. Sci. USA* **97,** 2105–2110.

Relaix, F., Wei, X. J., Wu, X., and Sassoon, D. A. (1998b). Peg3/Pw1 is an imprinted gene involved in the TNF-NFkappaB signal transduction pathway. *Nat. Genet.* **18,** 287–291.

Roselli, C. E., Resko, J. A., and Stormshak, F. (2006). Expression of steroid hormone receptors in the fetal sheep brain during the critical period for sexual differentiation. *Brain Res.* **1110**(1), 76–80.

Rossant, J. (1995). Development of the extra-embryonic lineages. *Semin. Dev. Biol.* **6,** 271–275.

Sansom, S. N., Hebert, J. M., Thammongkol, U., Smith, J., Nisbet, G., Surani, M. A., McConnell, S. K., and Livesey, F. J. (2005). Genomic characterisation of a Fgf-regulated gradient-based neocortical protomap. *Development* **132,** 3947–3961.

Schwanzel-Fukada, M., Crrossin, K. L., Pfaff, D. W., Bouloux, P. M., Hardelin, J. P., and Petit, C. (1996). Migration of luteinising hormone-releasing hormone (LHRH) neurons in early human embryos. *J. Comp. Neurol.* **366,** 547–557.

Segovia, A., and Guillamon, A. (1996). Searching for sex differences in the vomeronasal pathway. *Horm. Behav.* **30,** 618–626.

Stephan, H., Baron, G., and Frahm, H. D. (1982). Comparison of brain structure volumes in Insectivora and Primates. II. Accessory olfactory bulb (AOB). *J. Hirnforsch.* **23,** 575–591.

Stowers, L., Holy, T. E., Markus, M., Dulac, C., and Koentges, G. (2002). Electrophysiological characterization of chemosensory neurons from the mouse vomeronasal organ. *J. Neurosci.* **16,** 4625–4637.

Surani, M. A. H., Barton, S. C., and Norris, M. L. (1984). Roles of paternal and maternal genomes in mouse development. *Nature* **311,** 374–376.

Surridge, A., and Mundy, N. (2002). Trans-specific evolution of opsin alleles and the maintenance of trichromatic colour vision in Callitrichine primates. *Mol. Ecol.* **11,** 2157–2169.

Surridge, A., Osorio, D., and Mundy, N. I. (2003). Evolution and selection of thrichromatic vision in primates. *Trends Ecol. Evol.* **18,** 198–205.

Swaney, W. T., Curley, J. P., Champagne, F. A., and Keverne, E. B. (2007). Genomic imprinting mediates sexual experience–dependent olfactory learning in male mice. *Proc. Natl. Acad. Sci. USA* **104,** 6084–6089.

Van Essen, D., Anderson, C. H., and Felleman, D. J. (1992). Information processing in the primate visual system: An integrated systems perspective. *Science* **255,** 419–423.

Van Kesteren, R. E., Smit, A. B., De Lange, R. P., Kits, K. S., Van Golen, F. A., Van der Schors, R. C., De With, N. D., Burke, J. F., and Gaeaerts, W. P. (1995). Structural and functional evolution of the vasopressin/oxytocin superfamily: Vasopressin-related conopressin is the only ember present in Lymnaea, and is involved in the control of sexual behavior. *J. Neurosci.* **15,** 5989–5998.

Wagner, C. K., Nakayama, A. Y., and De Vries, C. J. (1998). Potential role of maternal progesterone in the sexual differentiation of the brain. *Endocrinology* **139,** 3658–3661.
Wagstaff, J., Knoll, J. M. H., Glatt, K. A., Shugart, Y., Sommer, A., and Lalande, M. (1992). Maternal but not paternal transmission of 15q11–13-linked nondeletion Angleman syndrome leads to phenotypic expression. *Nat. Genet.* **1,** 291–294.
Watrin, F., Le Meur, E., Roeckel, N., Ripoche, M-A., Dandolo, L., and Muscatelli, F. (2005). The Prader-Willi syndrome murine imprinting center is not involved in the spatio-temporal transcriptional regulation of the Necdin gene. *BMC Genetics* **6,** 2156–2161.
Webb, D., Cortés-Ortiz, L., and Zhang, J. (2004). Genetic evidence for the coexistence of pheromone perception and full trichromatic vision in howler monkey. *Mol. Biol. Evol.* **21,** 697–704.
Whinnett, A., and Mundy, N. (2003). Isolation of novel olfactory receptor genes in marmosets (*Callithrix*): Insights into pseudogene formation and evidence for functional degeneracy in non-human primates. *Gene* **304,** 87–96.
Wolf, J. B. (2000). Gene interactions for maternal effects. *Evolution* **54,** 1882–1898.
Yang, A., Walker, N., Bronson, R., Kaghad, M., Oosterwegel, M., Bonnin, J., Vagner, C., Bonnet, H., Dikkes, P., Sharpe, A., McKeon, F., and Caput, D. (2000). p73-deficient mice have neurological, pheromonal and inflammatory defects but lack spontaneous tumours. *Nature* **404,** 99–103.
Yokoi, F., Dang, M. T., Mitsui, S., and Li, Y. (2005). Exclusive paternal expression and novel alternatively spliced variants of epsilon-sarcoglycan mRNA in mouse brain. *FEBS Lett.* **579,** 4822–4828.
Yoshida, K., Tobet, S. A., Crandall, J. E., Jiminez, T. P., and Schwarting, G. A. (1995). The migration of luteinising hormone-releasing hormone neurons in the developing rat is associated with a transient, caudal projection of the vomeronasal nerve. *Neuroscience* **15,** 7769–7777.
Zhang, J., and Webb, D. (2003). Evolutionary deterioration of the vomeronasal pheromone transduction pathway in catarrhine primates. *Proc. Natl. Acad. Sci. USA* **100,** 8337–8341.
Zhang, X., and Firestein, S. (2002). The olfactory receptor gene superfamily of the mouse. *Nat. Neurosci.* **5,** 124–133.

9

Sex Differences in Brain and Behavior: Hormones Versus Genes

Sven Bocklandt and Eric Vilain

Laboratory of Sexual Medicine, Department of Urology, David Geffen School of Medicine at UCLA, Gonda Center, Los Angeles, California 90095

ABSTRACT

Sex determination is the commitment of an organism toward male or female development. Traditionally, in mammals, sex determination is considered equivalent to gonadal determination. Since the presence or the absence of the testes ultimately determines the phenotype of the external genitalia, sex determination is typically seen as equivalent to testis determination. But what exactly does sex determine? The endpoint of sex determination is almost invariably seen as the reproductive structures, which represent the most obvious phenotypic difference

Advances in Genetics, Vol. 59
Copyright 2007, Elsevier Inc. All rights reserved.
0065-2660/07 $35.00
DOI: 10.1016/S0065-2660(07)59009-7

between the sexes. One could argue that the most striking differences between males and females are not the anatomy of the genitals, but the size of the gametes—considerably larger in females than males. In fact, there could be many different endpoints to sex determination, leading to differences between the sexes: brain sexual differences, behavioral differences, and susceptibility to disease. The central dogma of sexual differentiation, stemming initially from the gonad-transfer experiments of Alfred Jost, is that sexual dimorphisms of all somatic tissues are dependent on the testicular secretion of the developing fetus.

In this chapter, we will take the example of sex differences in brain and behavior as an endpoint of sex determination. We will argue that genetic factors play a role in sexually dimorphic traits such as the number of dopaminergic cells in the mesencephalon, aggression, and sexual orientation, independently from gonadal hormones. © 2007, Elsevier Inc.

I. ROLE OF SRY IN SEX DETERMINATION

The primary event of sexual development in mammals is the development of the gonadal sex, from a bipotential and undifferentiated gonad into either testes or ovaries. This process, known as sex determination, is triggered by the sex-determining region Y chromosome (SRY) gene. Evidence that SRY was sex determining initially came from the microinjection of a 14.6-kb genomic DNA sequence containing the mouse SRY gene into chromosomally female embryos. The resulting transgenic mice developed phenotypically as males (Koopman *et al.*, 1991). SRY belongs to the *Sox* (SRY-box) family, whose members are characterized by a common high-mobility group (HMG) DNA-binding motif (Laudet *et al.*, 1993; Wegner, 1999). *Sox* genes have been documented in a wide range of developmental processes, including neurogenesis (*Sox2, 3,* and *10*) (Hargrave *et al.*, 1997; Rex *et al.*, 1997; Uwanogho *et al.*, 1995) and sex determination (*Sox9*) (Pevny and Lovell-Badge, 1997). Encoding a 204-amino acid protein, SRY is thought to bind and sharply bend DNA by means of its HMG box to regulate male-specific gene expression (Ferrari *et al.*, 1992; Harley *et al.*, 1992; King and Weiss, 1993; Nasrin *et al.*, 1991). The transient expression of SRY during a brief period in the developing mouse genital ridge, between 10.5 and 12.5 days post coitum, is what triggers testis development from a bipotential gonad (Koopman *et al.*, 1990). After this strictly regulated window of expression in mouse fetal gonads, SRY is reexpressed in the adult testis. However, while SRY RNA is expressed in the developing genital ridges as a linear transcript of about 5 kb (Hacker *et al.*, 1995; Jeske *et al.*, 1995), in the adult germ cells, SRY RNA is expressed as a circular transcript of about 1.3 kb, presumably untranslatable because it is not associated with ribosomes (Capel *et al.*, 1993). This pattern (linear in embryonic gonads and circular in adult ones) is opposite in the brain, as we will discuss later.

II. MALE AND FEMALE BRAINS ARE DIFFERENT

Many of the sex differences seen in the brain manifest themselves in a variety of ways, such as sizes of particular regions of the brain, number of nerve cells in specific regions, distribution of neurotransmitters, and even development of behavior (Goldstein *et al.*, 2001; Hutchinson, 1997; Segovia *et al.*, 1999). In humans, for instance, there are region-specific dimorphisms, with some structures larger in females (caudate nucleus, hippocampus, Broca's area, anterior commissure, and right parietal lobe) and some larger in males (hypothalamus, stria terminalis, and amygdala) (Goldstein *et al.*, 2001). In rodents, numerous sexually dimorphic nuclei have been studied, the following two perhaps most extensively. One is the sexually dimorphic nucleus of the preoptic area (SDN-POA), which is larger in males than females, and another is the anteroventral periventricular nucleus (AVPv), which is larger in females than males (Davis *et al.*, 1995; Gorski *et al.*, 1980; Simerly *et al.*, 1988). Sex differences in distributions of neurotransmitters have also been documented. For instance, neurochemical analysis of the AVPv showed that females harbor more cells secreting TH than males, while males synthesize more enkephalin than females (Simerly and Swanson, 1987; Simerly *et al.*, 1985). Most sex differences in the brain have been investigated in regions important for sexual function and reproduction, such as the hypothalamus. The influence of gonadal hormones on the sexual differentiation of these structures has been studied extensively, for instance in the case of the SDN-POA (Swaab and Fliers, 1985).

III. THE CENTRAL DOGMA OF SEXUAL DIFFERENTIATION

In the late 1940s, the classic gonad-transfer experiments of Alfred Jost led to the development of the central dogma of sexual differentiation in mammals and birds (Jost, 1947, 1970). According to this view, sexual dimorphisms of somatic tissues are dependent primarily on testicular secretions from the developing fetus. The presence of testes induces male development through the actions of two secreted testicular hormones, Müllerian inhibiting substance and testosterone. Absence of testicular hormones results in female development. Phoenix *et al.* (1959) were the first to successfully apply the concept of hormonal control of sexual differentiation directly to the brain. When testosterone was administered to pregnant female guinea pigs, profound effects on sex behaviors were observed, most notably in female offspring. Exposure to testosterone during early stages of development permanently masculinized and defeminized female guinea pig copulatory behavior patterns as adults, as female offspring were less likely to show lordosis and more likely to display masculine mounting behaviors compared to normal females. Subsequently, a great deal of experimental evidence continued to reveal that gonadal steroids are responsible for inducing brain sexual differentiation. Testosterone acts, directly or via local conversion into estradiol by aromatase, to promote the formation of neural circuits responsible for masculine phenotypes while preventing the formation of neural circuits responsible for feminine phenotypes (Arnold and Gorski, 1984; Goy and McEwen, 1980; Whalen, 1968).

IV. SEX HORMONES IN BRAIN SEXUAL DIFFERENTIATION

Over the past several decades, the roles of testosterone and estradiol have been intensely investigated in the brain, while sex differences in central nervous system (CNS) structures were characterized. The sites and modes of action of sex hormones have been delineated (Goy and McEwen, 1980; MacLusky and Naftolin, 1981; McEwen, 1981) as well as their masculinizing actions (Gorski, 1991). Testosterone and estrogen promote differentiation by affecting many cellular processes, such as protein synthesis, cell division and migration, neuronal growth and survival, as well as apoptosis, and synaptic remodeling (Nordeen *et al.*, 1985; Toran-Allerand, 1976). Manipulations of hormones during critical periods of development revealed the significance of gonadal hormones in the development of the sexually differentiated brain. When females are administered androgens or males deprived of androgens (such as in castration), structures previously found to be sexually dimorphic within the CNS could be sex reversed (Breedlove *et al.*, 1982; Dohler *et al.*, 1984; Gorski, 1978; Nordeen *et al.*, 1985). Dohler *et al.* (1984, 1986) observed that when administering estrogen to developing female rats, the SDN-POA was masculinized. Conversely, when estrogen action was blocked in developing male rats, masculinization in the SDN-POA was prevented (Dohler *et al.*, 1984, 1986). In fact, most sexual dimorphisms found in the brain could be explained by documented differences in hormone secretions (Arai and Matsumoto, 1978; Arnold and Gorski, 1984; Goy and McEwen, 1980; Suzuki *et al.*, 1983).

V. EXCEPTIONS TO THE DOGMA

Not all sexual dimorphisms found in the brain and other somatic tissues are solely attributable to the actions of gonadal sex hormones (Arnold, 1996). *In vitro* cultured mesencephalic dissociated cells developed into more dopaminergic neurons when the cultures were composed of XY cells than when they were composed of XX cells, regardless of the gonadal phenotype of the embryos from which the cells were harvested (Carruth *et al.*, 2002). This suggests that the signal for initiating this sexually dimorphic neuronal growth cannot be exclusively dependent on hormonal action and supports the role for a sex chromosome effect on the sexual dimorphism observed in DA neurons. When male and female embryonic mouse brains were analyzed for gene expressed, 54 genes were found to be differentially expressed between embryonic male and female brains before the development of gonadal and adrenal tissues, the main source of circulating steroid hormones (Dewing *et al.*, 2003). This suggests that sexual dimorphisms of embryonic brain gene expression is not dependent on gonad hormonal influence.

Another example arises from endocrine manipulations in zebra finch, whose neural song circuit is sexually dimorphic. The neural circuit for song is significantly larger in males than females, and males sing whereas females do not. However, when females are treated with fadrozole, an aromatase inhibitor, ovarian development is blocked allowing testicular development to progress. Yet the presence of testicular tissue in genetic females fails to masculinize the song circuit of the zebra finch (Wade and Arnold, 1996). This and other evidence make it unlikely that sexually dimorphic neural development is solely dependent on gonadal secretions. An alternative model to the hormone-dependent classic model where genetic factors would play a direct role in sex differences in the brain has been pioneered by Arnold and colleagues in the early 1990s. The role of sex chromosomes in brain sexual differentiation has been investigated in genetically modified rodent models where the testis-determining gene SRY was deleted from the Y chromosome and inserted back on an autosome, producing mice in which the complement of sex chromosomes of each cell (XX or XY) was independent from the gonadal sex (De Vries *et al.*, 2002). Specifically, XY^{-} mice, deleted for the male-determining gene SRY, develop as females and can be compared to XX female mice. Similarly, XX^{-SRY} mice carrying SRY as a transgene develop as males and can be compared to XY^{-SRY} who are males. Two-by-two comparison of animals with these four-core genotypes allows measurement of the effect sex chromosomes and of gonadal sex or a variety of brain phenotypes. For many sexually dimorphic traits that were observed, including the thickness of the parietal cortex, male copulatory behaviors, and social investigation behavior, males (i.e., testes-bearing animals) were masculine and females (i.e., ovaries-bearing animals) were feminine, suggesting the role of gonadal sex, and not chromosomal sex, for these traits.

However, for some dimorphic traits, differences between males and females were attributed to the chromosomal complement and not the gonadal sex. For example, XY^{-SRY} had a higher, more masculine density of vasopressin fibers in the lateral septum, suggesting an effect of sex chromosome for this sexual dimorphism.

When midbrain mesencephalic embryonic neurons were grown *in vitro*, XY cultures were shown to develop more dopaminegic neurons than XX cultures.

Finally, using the same four-core genotype model, sex chromosome complement was shown to influence two sexually dimorphic social behaviors: intruder-directed aggression and pup retrieval (Gatewood *et al.*, 2006).

VI. EVIDENCE FOR A DIRECT ROLE OF SRY IN THE BRAIN

In addition to the gonads, SRY expression has been discovered in both the adult and embryonic mouse brain. For instance, the hypothalamus and the mesencephalon (midbrain) were both positive for SRY expression in RT-PCR experiments (Lahr *et al.*, 1995; Mayer *et al.*, 1998). Interestingly, these are two areas that show functional sex differences (Reisert *et al.*, 1994; Vadasz *et al.*, 1988). A more comprehensive profile of SRY expression was described using mouse brains as early as embryonic day 11 (E11) through adulthood [postnatal day 90 (P90)] (Mayer *et al.*, 2000). For all embryonic stages studied, whole brains were obtained for analysis, while at postnatal stages, brain regions such as midbrain, diencephalon, and cortex, were isolated. SRY expression was seen at all stages and in all regions ascertained. Particular emphasis was placed on the detection of linear versus circular SRY transcripts in this study (Mayer *et al.*, 2000). During E11 through E19, SRY transcripts were found to be in the circular form, presumably an untranslatable form since they cannot bind to polysomes. In contrast, postnatal brain SRY transcripts were found to be in the linear (translatable) form suggesting that SRY in the brain is developmentally regulated. This switch in transcript form is directly opposite to the one seen in the gonads. All the expression studies of SRY have been performed by RT-PCR until recently, which raised two concerns: (1) since SRY is a single exon gene expressed at low levels, the RT-PCR data are often difficult to interpret and (2) RT-PCR studies do not provide specific anatomical localization of expression. Using *in situ* hybridization and immunohistochemistry, it was shown that, in rats, SRY is expressed in specific regions of the brain (the substantia nigra and the mammillary bodies) and maintains the number and the motor function of tyrosine hydroxylase-expressing neurons in the substantia nigra (Dewing *et al.*, 2006). SRY was also shown to control *in vivo* the expression of tyrosine hydroxylase, a key enzyme in dopamine synthesis, and the consequent effects on motor behavior when SRY is downregulated in the brain. For instance, reduction of SRY in the substantia nigra resulted in a 50% reduction of the number of TH-positive neurons, and akinesia symptoms, with a significant increase of time to initiate movement.

Why do females, who do not carry SRY, show no signs of akinesia? One hypothesis is that SRY compensates for a factor present in females, but lacking in males. A possible candidate is estradiol, known to influence TH. For instance, short-term injection of estradiol benzoate in rats increased TH mRNA levels in dopaminergic neurons of the SNc, but long-term administration did not (Serova *et al.*, 2004). Consistently, in primates, short-term ovariectomy (10 days) decreased the number of TH-positive cells in the SNc, a reversible effect, whereas long-term ovariectomy (30 days) results in a permanent loss of the SNc dopamine cells (Leranth *et al.*, 2000). The actual transcriptional regulation of TH by estrogen is complex and poorly known, and it seems to depend on the type of estrogen receptors: estrogen increases TH activity with ERα but decreases it with ERβ (Maharjan *et al.*, 2005). In addition, there are sex differences in the response of estrogen, as effects seen in females on striatal dopamine release are not seen in males (Castner *et al.*, 1993), and estrogen functions as a neuroprotectant against metamphetamine in females, but not in males (Dluzen and McDermott, 2004).

Rodent models, such as the mouse four-core genotype, and the modification of SRY expression in the rat substantia nigra are examples of direct effects of specific genetic factors on brain sexual differences *in vivo*, independently from gonadal hormonal influence, shifting the classic paradigm of hormonal influence on brain sexual differences to a genetically controlled model (Fig. 9.1).

VII. SEXUAL ORIENTATION IS A SEXUALLY DIMORPHIC TRAIT

Of all behavioral differences between males and females, partner choice is the most pronounced. With very few exceptions in the animal kingdom, males generally choose females as sexual partners, and females choose males. Although sexual selection has been shown to be a driving force of evolution, little is known about the molecular basis of partner preference. In the remainder of this chapter, we will discuss the current state of scientific knowledge on the biological basis of sexual orientation.

Sexual orientation—that is being sexually attracted to predominantly males or females—is one of the most critical biological behavioral traits and yet one of the least studied and understood.

Human sexual orientation is a complex phenotype. Although most males report primarily heterosexual attractions, a significant minority (~2–6%) of males report predominantly homosexual attractions (Diamond, 1993; Laumann *et al.*, 1994; Wellings *et al.*, 1994). The Kinsey Scale is widely used to measure sexual orientation as a continuum from exclusive heterosexuality (Kinsey 0) through various degrees of bisexuality to exclusive homosexuality (Kinsey 6) (Kinsey *et al.*, 1948, 1953). The combination of sexual fantasy, attraction, and behavior is largely bimodal in men (Diamond, 1993; Hamer *et al.*, 1993) when measured

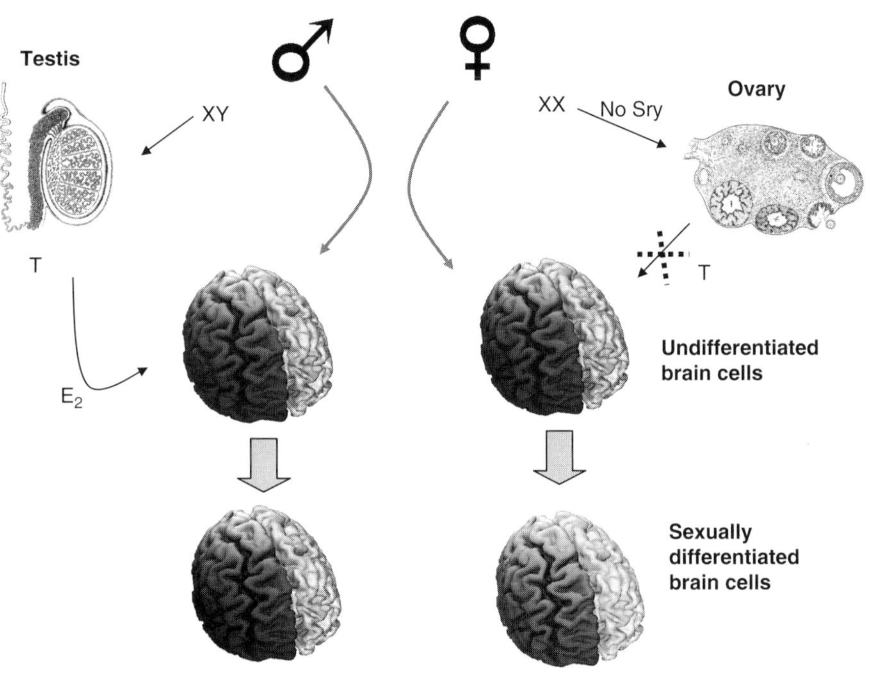

Figure 9.1. Traditional and alternative paradigm of brain sexual differentiation. Classic theory (black arrow): In XY mammals, Sry triggers the differentiation of the testis, which secretes testosterone, which in turn masculinizes the brain, directly or via aromatization to estradiol. In XX mammals, Sry is absent, the gonads develop into ovaries, which do not secrete testosterone, and the brain develops as feminized. Alternative theory (orange arrow): In addition to hormonal effects, the genes on the X and Y chromosome directly influence sex differences in the brain. (See Color Insert.)

using this scale. Furthermore, studies suggest that very few men (even those who openly identify as bisexual) show comparable physical attraction to both men and women (Rieger *et al.*, 2005).

The distribution is more complex in women, in which the fraction of women that show exclusive same-sex attraction is lower than men (1–3%), but many more women than men report erotic fantasies toward both sexes (Hu *et al.*, 1995).

Male sexual orientation is closely linked to childhood gender-typical behavior (Bailey and Zucker, 1995; Bell *et al.*, 1981; Green, 1987) and to adult gender-related behavior as well, including occupational and recreational interests (Bailey, 1996; Lippa, 2002). A longitudinal study on gender-nonconforming boys showed that about three out of four report a homosexual or bisexual orientation in adulthood, but without adulthood gender identity disorder (transsexualism) (Association, 2000; Green, 1987). Six follow-up reports of boys with gender identity showed similar results (Zucker and Bradley, 1995).

Multiple lines of evidence suggest that biological factors play a role in explaining individual differences in sexual orientation. For example, several differences have been reported between heterosexual and homosexual men in neuroanatomic structures. The strong link between adult sexual orientation and childhood gender-related traits expressed at a very early age suggests that such biological influences act early in development, possibly prenatally.

VIII. HOMOSEXUAL BRAINS ARE DIFFERENT

Neuropsychological studies have reported differences in performance on tasks that show sex differences, such as spatial processing (Rahman and Wilson, 2003), which may indicate differences in relevant neural structures.

Neuroanatomical differences have been reported for three brain regions based on sexual orientation in males: the arginine vasopressin neuronal population of the suprachiasmatic nucleus (Swaab *et al.*, 1997), the third interstitial nucleus of the anterior hypothalamus (LeVay, 1991), and the anterior commissure (Allen and Gorski, 1992). In all instances of significant neuroanatomical differences, gay men were reported to be skewed in the female direction.

The most interesting anatomical finding is that the third interstitial nuclei of the human anterior hypothalamus, which is significantly smaller in females, is also smaller in homosexual males (LeVay, 1991). Byne *et al.* (2001) followed up on this finding by reporting a trend for INAH3 to occupy a smaller volume in homosexual men than in heterosexual men, with no significant difference in the number of neurons within the nucleus.

A comparable size difference was found in sheep (Roselli *et al.*, 2004). Sheep are a unique animal model in which to study sexual partner preference since variations in sexual attraction occur spontaneously in domestic ram populations. Most domestic rams are sexually attracted to and active with estrous ewes but as many as 8–10% of rams exhibit a sexual partner preference for other males. These male oriented rams mount and ejaculate like other males; only their choice of sexual partner is different. They are thus ideal animal models in that their coital behavior is masculinized, but not their sexual partner preference. An analogue of the SDN [ovine SDN (oSDN)], a hypothalamic nucleus thought to be involved in mate selection, was identified in the sheep brain (Roselli *et al.*, 2004). Furthermore, the oSDN was found to be larger in female-oriented rams than in male-oriented rams (MORs), and similar in size in MORs and ewes. It is hypothesized that the oSDN corresponds with human INAH3 mentioned above. The observation of a similar size difference in a comparable hypothalamic nucleus in sheep and humans suggests that sheep are indeed a good animal model for human sexual orientation and that neuroanatomical pathways are conserved between mammalian species.

IX. THE ROLE OF PRENATAL ANDROGEN EXPOSURE ON SEXUAL ORIENTATION: MYTH OR REALITY?

The most influential theory about the biological regulation of homosexuality implicates sex-atypical hormone levels during gestation. This theory is based on animal research which we discussed earlier. Since homosexual males show sex-atypical partner preference, they were considered incompletely masculinized, and the neurohormonal theory for sexual orientation was born: homosexual men were exposed to androgen levels that were too low for a complete masculinization of the brain, resulting in female-typical partner preference.

A more detailed look at the animal experiments, however, reveals a number of caveats in this theory. The offspring of the treated females typically have masculinized genitalia, which is not the case for gay males. Furthermore, the sex atypical behavior is only seen in offspring where the normal adult hormone levels are blocked either chemically or through removal of the gonads, and only after additional injections with sex hormones (Phoenix *et al.*, 1959). The treatment necessary to change the sexual behavior of the offspring goes far beyond any naturally occurring variation in androgen levels. This may make any real comparison between this data and the occurrence of homosexuality in humans impossible. Finally, the behavior studied in rodents does not clearly compare to human sexuality. Increased occurrence of lordosis in male rats, for instance, does not compare to male homosexuality—only a consistent choice of same sex partners by the rodents would be comparable, and this type of behavior has never been induced through hormone treatment.

Case studies in humans with various genetic defects in the androgen pathway show only limited support for the hypothesis. Although a wide range of masculinization of the external genitals exists in humans with disorders of sex development, there is no evidence that the sexual orientation differs from what is typical for the gender identity these patients identify with. The only known exception to date is for women with congenital adrenal hyperplasia (CAH). Here, abnormal activity of the embryonic adrenal glands leads to an overexposure of females to androgens, sometimes to levels that far exceed those found in typical male fetuses. Only in these cases of very high androgenization has an increase in same sex activity and self-identification been shown (Hines *et al.*, 2004). Other studies on CAH women found no such increase (Stikkelbroeck *et al.*, 2003). It is noteworthy that these same studies also find that CAH women have a more masculine core gender identity and childhood play, suggesting an overall masculinization of the brain, not limited to sexual orientation.

Two studies looked at candidate genes involved in the testosterone pathway in homosexual and heterosexual men. Both the androgen receptor (Macke *et al.*, 1993) and the aromatase gene (CYP19) (DuPree *et al.*, 2004) were studied and no role for variations in these genes in establishing differences in sexual orientation was found.

Since animal studies show that maternal stress during pregnancy influences brain masculinization of the offspring through an aberrant hormone exposure, several studies have looked retrospectively at stressful events that occurred while mothers were pregnant of their gay or lesbian children. The results of those studies are inconsistent (Bailey *et al.*, 1991; Ellis and Cole-Harding, 2001), and no convincing evidence for any effect of environmental conditions during pregnancy has been found.

X. INDIRECT HORMONAL MEASURES

Because direct measurements of prenatal androgen levels and their effect on adult sexual orientation are hard to achieve, other anthropometric characteristics have been studied as assumed indirect measures of prenatal androgen exposure. The most attention was paid to a finding that the ratio of the length of the second and fourth finger (2D:4D ratio) was significantly different in lesbian women compared with heterosexual women (Williams *et al.*, 2000). A larger study that controlled for ethnicity did not find these differences (Lippa, 2003). Several smaller follow-up studies have been published that show mixed results both in women and men (McFadden *et al.*, 2005), although several confirm the more male-typical 2D:4D ratio in lesbian women (Kraemer *et al.*, 2006). It is at this point, however, unclear if this finger length ratio is a measure of any specific prenatal condition in the first place. This variation was suggested as being caused by differences in prenatal androgen exposure, mostly because this 2D:4D ratio is a sexually dimorphic trait and "because all nongonadal somatic sex differences in humans appear to be the result of fetal androgens that masculinize males, the sex difference in the 2D:4D ratio probably reflects the prenatal influence of androgen on males" (Williams *et al.*, 2000)—a short-sighted claim that is obviously incorrect (Arnold, 2004). Racial differences in 2D:4D far exceed sex differences (Loehlin *et al.*, 2006), in that for instance females of African origin have a more masculine ratio than typical Caucasian males. Even studies on the extreme prenatal androgen levels seen in women with CAH are contradicting depending on whether bone (Buck *et al.*, 2003) or soft tissue of the fingers (Brown *et al.*, 2002; Okten *et al.*, 2002) is measured.

Similar arguments were employed to link other sexually dimorphic anthropometric characteristics to sexual orientation, even though for most it has not been shown that physiological occurring variations in androgen levels reflect on differences in any of these traits. Some other characteristics for which sexual orientation effects were published include soft sounds emitted by the inner ear (otoacoustic emissions) and brain waves in response to sounds (auditory evoked potentials) (McFadden and Champlin, 2000; McFadden and Pasanen, 1999). For sexually dimorphic traits like fingerprint asymmetry, age of puberty, height, and weight, a correlation with sexual orientation has not consistently been found (summarized in Mustanski *et al.*, 2002).

XI. THE GENETICS OF SEXUAL ORIENTATION

Although there is no convincing evidence linking differences in sexual orientation to variations in prenatal androgens, there is abundant evidence for a strong genetic component.

Family studies (Bailey and Bell, 1993; Bailey and Benishay, 1993; Bailey and Pillard, 1991; Bailey *et al.*, 1995, 1999; Pattatucci and Hamer, 1995; Pillard, 1990; Pillard and Weinrich, 1986) showed an increased rate of homosexuality in siblings of gays and lesbians and in the maternal uncles of gay men (a median rate of 9% for brothers of gay men; Bailey and Pillard, 1995).

However, family studies are not able to distinguish between genetic and environmental effects, something that twin studies are able to do. Concordance rates of homosexuality in twins vary between studies, depending on ascertainment methods (Bailey *et al.*, 2000; Bailey and Pillard, 1991; Kendler *et al.*, 2000; Kirk *et al.*, 2000), but all studies show strong evidence of a substantial genetic component in the development of sexual orientation.

There has only been very limited research at the level of molecular genetics. Hamer *et al.* (1993) reported maternal loading of male homosexuality. They found a pattern of male homosexuals at the mother's side of the family, more often than at the father's side. Other studies reported evidence of increased maternal transmission of male homosexuality (Rice *et al.*, 1999b), while some found no increase relative to paternal transmission (Bailey *et al.*, 1999; McKnight and Malcolm, 2000). Maternal transmission suggested X-linked inheritance, and the X chromosome has accumulated genes that influence sex, reproduction, and cognition (Graves *et al.*, 2002). Thus, a linkage scan of the X chromosome was performed which mapped male homosexuality to the X chromosome region Xq28 (Hamer *et al.*, 1993), producing evidence of significant linkage, based on Lander and Kruglyak criteria (Lander and Kruglyak, 1995). Another study, from the same laboratory but with a new sample, reported a significant replication of these findings (Hu *et al.*, 1995). An independent group produced inconclusive results regarding linkage to Xq28 ($p = 0.04$; discussed in Sanders and Dawood, 2003), but did not publish the findings in a peer-reviewed journal. All three of these studies excluded families showing evidence for nonmaternal transmission. A fourth study from another independent group found no support for linkage (Rice *et al.*, 1999a). A meta-analysis of the results across all four studies (Fig. 9.2) gives an estimated level of Xq28 allele sharing between gay brothers of 64% instead of the expected 50% (Hamer, 1999) and produced a statistically suggestive multiple scan probability (MSP) value of 0.00003 (Sanders and Dawood, 2003). However, almost 10 years after the initial findings, the exact gene involved has not been identified.

A recent study found a role for the X chromosome in male sexual orientation in a very different way. Because male cells contain only one X chromosome and female cells contain two, each cell in a female embryo randomly inactivates one X chromosome early in development, thus creating dosage compensation. Since the choice of which X chromosome to inactivate is made randomly in most tissues, and the inactive chromosome remains inactive in all daughter cells resulting from cell divisions, a tissue sample of a female typically contains cells that have one X chromosome inactivated and cells that have the other inactivated (Brown and Robinson, 2000). By comparing mothers of gay sons with women without gay sons, it was shown that the number of women with extreme skewing of X-inactivation was significantly higher in mothers of gay men ($13/97 = 13\%$) compared to controls ($4/103 = 4\%$), and increased in mothers with two or more gay sons ($10/44 = 23\%$) (Bocklandt *et al.*, 2006).

Whether the unusual X-inactivation influenced the sexual orientation of the sons directly or whether it is simply the consequence of a mechanism influencing sexual orientation is unclear. The most likely explanation of this finding would be that the sequence variation that influences sexual orientation in the children increases or decreases the growth or survival of the white blood

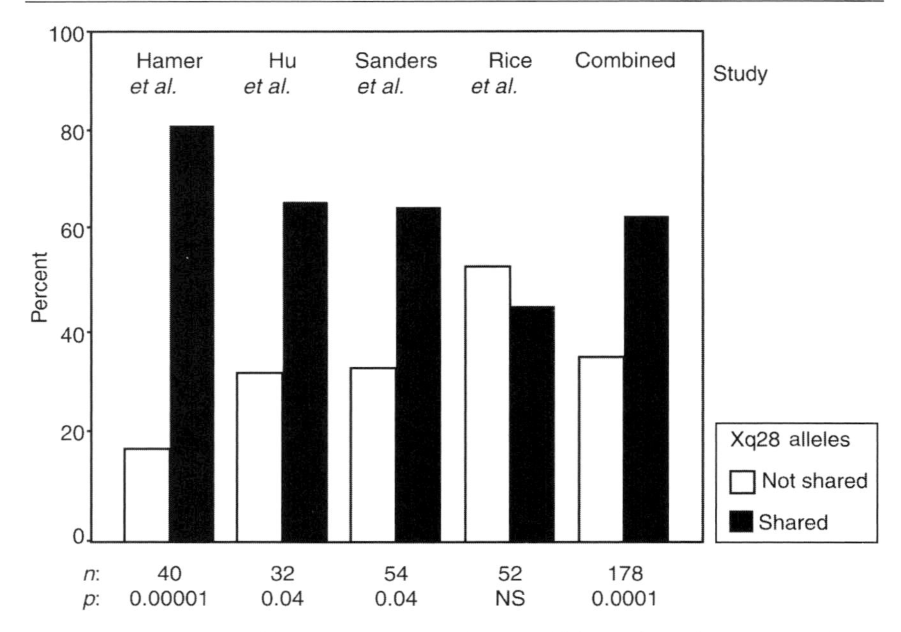

Figure 9.2. Comparison of four studies of linkage between alleles on the X chromosome and male homosexuality. Meta-analysis (combined data, last set of columns) shows overall estimated allele sharing of 64% ($p = 0.0001$). Sample size, n. Statistical significance, p. Adapted from (Hamer, 1999).

cells or stem cells in the mother, thus leading to a selection of cells that inactivate one allele or the other. Arguably, the effect of the X-chromosome gene(s) or mechanisms that influence sexual orientation in the sons is visible in the blood of their mothers. The unusual X chromosome methylation pattern in this sample of mothers of homosexual men supports a role for the X chromosome in regulating male sexual orientation.

Given the complexity of sexual orientation, it is likely to involve numerous genes, many of which are expected to be autosomal rather than sex-linked. Indeed, the modest levels of linkage that have been reported for the X chromosome can account for at most only a fraction of the overall heritability of male sexual orientation as deduced from twin studies. A genome-wide linkage scan to aid in the identification of genes contributing to variation in sexual orientation (Fig. 9.3) was published (Mustanski *et al.*, 2005). Unlike previous studies that focused solely on the X chromosome, and thus excluded families showing evidence of nonmaternal transmission, this study did not use transmission pattern as an exclusion criteria. Lod scores were calculated separately for maternal, paternal, and combined transmission. The maximum lod score was

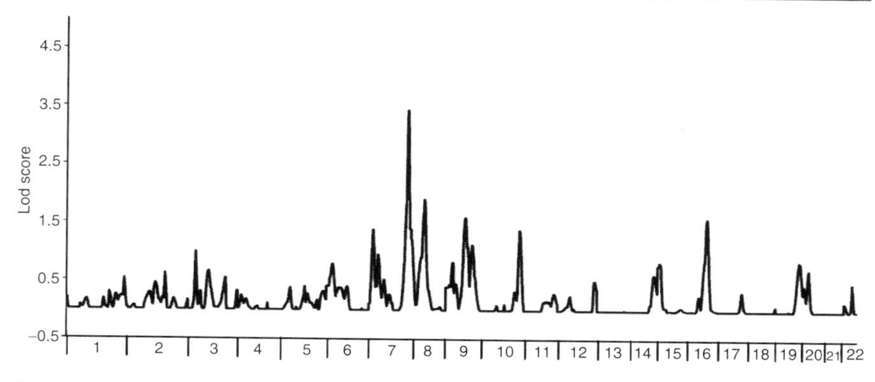

Figure 9.3. Genome scan results. The x-axis is the chromosome location, and the y-axis is the mlod score. Results are shown for combined maternal and paternal meioses. Adapted from Mustanski *et al.* (2005).

3.45 at a position near D7S798 on 7q36 with approximately equivalent maternal and paternal contributions. The second highest mlod score of 1.96 was located near D8S505 on 8p12, again with equal maternal and paternal contributions. A maternal origin effect was found near marker D10S217, located at 10q26, with a mlod score of 1.81 for maternal meioses and no paternal contribution. This suggests the presence of a maternally expressed, paternally silenced imprinted gene for sexual orientation in 10q26.

The relatively small sample size of this linkage scan lead to rather low mlod scores. A much larger linkage scan on 1000 homosexual male sib pairs is currently underway (Sanders, personal communication).

The presence of a possible imprinted gene on chromosome 10 is particularly interesting. Previously reported evidence of maternal loading of sexual orientation transmission was initially used to implicate the X chromosome in human sexual orientation, but it could just as well indicate epigenetic factors acting on autosomal genes. A role for imprinted genes in human sexual orientation was hypothesized earlier (Bocklandt and Hamer, 2003).

Interestingly, epigenetic regulation of sexual orientation could explain some of the discordance between identical twins. Although identical twins share all of their DNA sequence, there might be differences in DNA methylation that lead to variations in gene expression during a critical period of development (Fraga *et al.*, 2005). Differences in sexual orientation that are now attributed to each twin's unique environment could very well be caused by the "epigenetic environment." This hypothesis creates an entirely new research paradigm.

One biological influence on male sexual orientation that has been replicated over a dozen times, and is the best established in sexual orientation research, is the fraternal birth order effect. In men, sexual orientation is correlated with the number of older brothers, each additional older brother increasing the odds of homosexuality by ~33% (Blanchard, 1997; Jones and Blanchard, 1998). The odds to be gay are about twice as high for the fourth born son than for the first born. It was shown that only biological older brothers, and not the number of older brothers one grows up with, influence sexual orientation (Bogaert, 2006). Furthermore, this effect seems to be true only for right-handed homosexual men (Blanchard *et al.*, 2006).

The best hypothesis explaining this effect postulates an immunization of the mother by successive male pregnancies (Blanchard and Bogaert, 1996; Blanchard and Klassen, 1997). The mother would produce an immune response against male-specific antigens expressed by the fetus. This immune response would block the antigen function, preventing it from influencing brain masculinization. To date, no molecular studies of these Y-linked histocompatibility antigens and their role in sexual orientation have been undertaken, making them and their transcription regulators prime candidates for further genetic research.

Family studies, twin and adoption studies, linkage studies, and data on X chromosome inactivation all provide evidence for a genetic role of human sexual orientation, and the birth order effect found for homosexual males supports that sexual orientation is established before birth. The overwhelming dominance of heterosexual behavior in the animal kingdom points at a tight molecular regulation of this trait.

We hypothesize that one central neuronal pathway establishes sexual attraction to either males or females, usually toward the opposite sex. However, a variety of genetic and nongenetic biological effects might intersect this pathway. Hence, there might be several subgroups of gay men and women, each with their own specific biological origin.

References

Allen, L. S., and Gorski, R. A. (1992). Sexual orientation and the size of the anterior commissure in the human brain. *Proc. Natl. Acad. Sci. USA* **89,** 7199–7202.

Arai, Y., and Matsumoto, A. (1978). Synapse formation of the hypothalamic arcuate nucleus during post-natal development in the female rat and its modification by neonatal estrogen treatment. *Psychoneuroendocrinology* **3**(1), 31–45.

Arnold, A. P. (1996). Genetically triggered sexual differentiation of brain and behavior. *Horm. Behav.* **30**(4), 495–505.

Arnold, A. P. (2004). Sex chromosomes and brain gender. *Nat. Rev. Neurosci.* **5**(9), 701–708.

Arnold, A. P., and Gorski, R. A. (1984). Gonadal steroid induction of structural sex differences in the central nervous system. *Annu. Rev. Neurosci.* **7,** 413–442.

Association, A. P. (2000). Guidelines for psychotherapy with lesbian, gay, and bisexual clients. *Am. Psychol.* **55,** 1440–1451.

Bailey, J. M. (1996). Gender identity. *In* "The Lives of Lesbians, Gays, and Bisexuals: Children to Adults" (S. Williams, C. Ritch, and K. M. Cohen, eds.), pp. 71–93. Harcourt Brace College Publishers, Orlando, FL, US. xviii, 493 see book.

Bailey, J. M., and Bell, A. P. (1993). Familiality of female and male homosexuality. *Behav. Genet.* **23**(4), 313–322.

Bailey, J. M., and Benishay, D. S. (1993). Familial aggregation of female sexual orientation. *Am. J. Psychiatry* **150**(2), 272–277.

Bailey, J. M., and Pillard, R. C. (1991). A genetic study of male sexual orientation. *Arch. Gen. Psychiatry* **48**(12), 1089–1096.

Bailey, J. M., and Pillard, R. C. (1995). Genetics of human sexual orientation. *Annu. Rev. Sex Res.* **6**, 126–150.

Bailey, J. M., and Zucker, K. J. (1995). Childhood sex-typed behavior and sexual orientation: A conceptual analysis and quantitative review. *Dev. Psychol.* **31**, 43–55.

Bailey, J. M., Willerman, L., and Parks, C. (1991). A test of the maternal stress theory of human male homosexuality. *Arch. Sex. Behav.* **20**(3), 277–293.

Bailey, J. M., Bobrow, D., Wolfe, M., and Mikach, S. (1995). Sexual orientation of adult sons of gay fathers. *Dev. Psychol.* **31**, 124–129.

Bailey, J. M., Pillard, R. C., Dawood, K., Miller, M. B., Farrer, L. A., Trivedi, S., and Murphy, R. L. (1999). A family history study of male sexual orientation using three independent samples. *Behav. Genet.* **29**(2), 79–86.

Bailey, J. M., Dunne, M. P., and Martin, N. G. (2000). Genetic and environmental influences on sexual orientation and its correlates in an Australian twin sample. *J. Pers. Soc. Psychol.* **78**, 524–536.

Bell, A. P., Weinberg, M. S., and Hammersmith, S. K. (1981). "Sexual Preference: Its Development in Men and Women." Indiana University Press, Bloomington.

Blanchard, R. (1997). Birth order and sibling sex ration in homosexual versus heterosexual males and females. *Annu. Rev. Sex. Res.* **8**, 27–67.

Blanchard, R., and Bogaert, A. F. (1996). Homosexuality in men and number of older brothers. *Am. J. Psychiatry* **153**, 27–31.

Blanchard, R., and Klassen, P. (1997). H-Y antigen and homosexuality in men. *J. Theor. Biol.* **185**, 373–378.

Blanchard, R., Cantor, J. M., Bogaert, A. F., Breedlove, S. M., and Ellis, L. (2006). Interaction of fraternal birth order and handedness in the development of male homosexuality. *Horm. Behav.* **49**(3), 405–414.

Bocklandt, S., and Hamer, D. H. (2003). Beyond hormones: A novel hypothesis for the biological basis of male sexual orientation. *J. Endocrinol. Invest.* **26**(Suppl. 3), 8–12.

Bocklandt, S., Horvath, S., Vilain, E., and Hamer, D. H. (2006). Extreme skewing of X chromosome inactivation in mothers of homosexual men. *Hum. Genet.* **118**(6), 691–694.

Bogaert, A. F. (2006). Biological versus nonbiological older brothers and men's sexual orientation. *Proc. Natl. Acad. Sci. USA* **103**(28), 10771–10774.

Breedlove, S. M., Jacobson, C. D., Gorski, R. A., and Arnold, A. P. (1982). Masculinization of the female rat spinal cord following a single neonatal injection of testosterone propionate but not estradiol benzoate. *Brain Res.* **237**(1), 173–181.

Brown, C. J., and Robinson, W. P. (2000). The causes and consequences of random and non-random X chromosome inactivation in humans. *Clin. Genet.* **58**(5), 353–363.

Brown, W. M., Hines, M., Fane, B. A., and Breedlove, S. M. (2002). Masculinized finger length patterns in human males and females with congenital adrenal hyperplasia. *Horm. Behav.* **42**(4), 380–386.

Buck, J. J., Williams, R. M., Hughes, I. A., and Acerini, C. L. (2003). In-utero androgen exposure and 2nd to 4th digit length ratio-comparisons between healthy controls and females with classical congenital adrenal hyperplasia. *Hum. Reprod.* **18**(5), 976–979.

Byne, W., Tobet, S., Mattiace, L., Lasco, M. S., Kemether, E., Edgar, M. A., Morgello, S., Buchsbaum, M. S., and Jones, L. B. (2001). The interstitial nuclei of the human anterior hypothalamus: An investigation of variation within sex, sexual orientation and HIV status. *Horm. Behav.* **40**, 86–92.

Capel, B., Swain, A., Nicolis, S., Hacker, A., Walter, M., Koopman, P., Goodfellow, P., and Lovell-Badge, R. (1993). Circular transcripts of the testis-determining gene Sry in adult mouse testis. *Cell* **73**(5), 1019–1030.

Carruth, L. L., Reisert, I., and Arnold, A. P. (2002). Sex chromosome genes directly affect brain sexual differentiation. *Nat. Neurosci.* **5**(10), 933–934.

Castner, S. A., Xiao, L., and Becker, J. B. (1993). Sex differences in striatal dopamine: *In vivo* microdialysis and behavioral studies. *Brain Res.* **610**(1), 127–134.

Davis, E. C., Shryne, J. E., and Gorski, R. A. (1995). A revised critical period for the sexual differentiation of the sexually dimorphic nucleus of the preoptic area in the rat. *Neuroendocrinology* **62**(6), 579–585.

De Vries, G. J., Rissman, E. F., Simerly, R. B., Yang, L. Y., Scordalakes, E. M., Auger, C. J., Swain, A., Lovell-Badge, R., Burgoyne, P. S., and Arnold, A. P. (2002). A model system for study of sex chromosome effects on sexually dimorphic neural and behavioral traits. *J. Neurosci.* **22**(20), 9005–9014.

Dewing, P., Shi, T., Horvath, S., and Vilain, E. (2003). Sexually dimorphic gene expression in mouse brain precedes gonadal differentiation. *Brain Res. Mol. Brain Res.* **118**(1–2), 82–90.

Dewing, P., Chiang, C. W., Sinchak, K., Sim, H., Fernagut, P. O., Kelly, S., Chesselet, M. F., Micevych, P. E., Albrecht, K. H., Harley, V. R., and Vilain, E. (2006). Direct regulation of adult brain function by the male-specific factor SRY. *Curr. Biol.* **16**(4), 415–420.

Diamond, M. (1993). Homosexuality and bisexuality in different populations. *Arch. Sex. Behav.* **22**(4), 291–310.

Dluzen, D. E., and McDermott, J. L. (2004). Developmental and genetic influences upon gender differences in methamphetamine-induced nigrostriatal dopaminergic neurotoxicity. *Ann. NY Acad. Sci.* **1025**, 205–220.

Dohler, K. D., Coquelin, A., Davis, F., Hines, M., Shryne, J. E., and Gorski, R. A. (1984). Pre- and postnatal influence of testosterone propionate and diethylstilbestrol on differentiation of the sexually dimorphic nucleus of the preoptic area in male and female rats. *Brain Res.* **302**(2), 291–295.

Dohler, K. D., Coquelin, A., Davis, F., Hines, M., Shryne, J. E., Sickmoller, P. M., Jarzab, B., and Gorski, R. A. (1986). Pre- and postnatal influence of an estrogen antagonist and an androgen antagonist on differentiation of the sexually dimorphic nucleus of the preoptic area in male and female rats. *Neuroendocrinology* **42**(5), 443–448.

DuPree, M. G., Mustanski, B. S., Bocklandt, S., Nievergelt, C., and Hamer, D. H. (2004). A candidate gene study of CYP19 (aromatase) and male sexual orientation. *Behav. Genet.* **34**(3), 243–250.

Ellis, L., and Cole-Harding, S. (2001). The effects of prenatal stress, and of prenatal alcohol and nicotine exposure, on human sexual orientation. *Physiol. Behav.* **74**(1–2), 213–226.

Ferrari, S., Harley, V. R., Pontiggia, A., Goodfellow, P. N., Lovell-Badge, R., and Bianchi, M. E. (1992). SRY, like HMG1, recognizes sharp angles in DNA. *EMBO J.* **11**(12), 4497–4506.

Fraga, M. F., Ballestar, E., Paz, M. F., Ropero, S., Setien, F., Ballestar, M. L., Heine-Suner, D., Cigudosa, J. C., Urioste, M., Benitez, J., Boix-Chornet, M., Sanchez-Aguilera, A., *et al.* (2005). Epigenetic differences arise during the lifetime of monozygotic twins. *Proc. Natl. Acad. Sci. USA* **102**(30), 10604–10609.

Gatewood, J. D., Wills, A., Shetty, S., Xu, J., Arnold, A. P., Burgoyne, P. S., and Rissman, E. F. (2006). Sex chromosome complement and gonadal sex influence aggressive and parental behaviors in mice. *J. Neurosci.* **26**(8), 2335–2342.

Goldstein, J. M., Seidman, L. J., Horton, N. J., Makris, N., Kennedy, D. N., Caviness, V. S., Jr., Faraone, S. V., and Tsuang, M. T. (2001). Normal sexual dimorphism of the adult human brain assessed by *in vivo* magnetic resonance imaging. *Cereb. Cortex* **11**(6), 490–497.

Gorski, R. A. (1978). Sexual differentiation of the brain. *Hosp. Pract.* **13**(10), 55–62.

Gorski, R. A. (1991). Sexual differentiation of the endocrine brain and its control. *In* "Brain Endocrinology" (M. Motta, ed.), pp. 71–103. Raven Press, New York.

Gorski, R. A., Harlan, R. E., Jacobson, C. D., Shryne, J. E., and Southam, A. M. (1980). Evidence for the existence of a sexually dimorphic nucleus in the preoptic area of the rat. *J. Comp. Neurol.* **193**(2), 529–539.

Goy, R. W., and McEwen, B. (1980). "Sexual Differentiation of the Brain." MIT Press, Cambridge, MA.

Graves, J. A., Gecz, J., and Hameister, H. (2002). Evolution of the human X–a smart and sexy chromosome that controls speciation and development. *Cytogenet. Genome Res.* **99**(1–4), 141–145.

Green, R. (1987). "The "Sissy Boy Syndrome" and the Development of Homosexuality." Yale University Press, New Haven.

Hacker, A., Capel, B., Goodfellow, P., and Lovell-Badge, R. (1995). Expression of Sry, the mouse sex determining gene. *Development* **121**(6), 1603–1614.

Hamer, D. (1999). Genetics and male sexual orientation. *Science* **285**, 803.

Hamer, D., Hu, S., Magnuson, V., Hu, N., and Pattatucci, A. M. (1993). A linkage between DNA markers on the X chromosome and male sexual orientation. *Science* **261**(5119), 321–327.

Hargrave, M., Wright, E., Kun, J., Emery, J., Cooper, L., and Koopman, P. (1997). Expression of the Sox11 gene in mouse embryos suggests roles in neuronal maturation and epithelio-mesenchymal induction. *Dev.Dyn.* **210**(2), 79–86.

Harley, V. R., Jackson, D. I., Hextall, P. J., Hawkins, J. R., Berkovitz, G. D., Sockanathan, S., Lovell-Badge, R., and Goodfellow, P. N. (1992). DNA binding activity of recombinant SRY from normal males and XY females. *Science* **255**(5043), 453–456.

Hines, M., Brook, C., and Conway, G. S. (2004). Androgen and psychosexual development: Core gender identity, sexual orientation and recalled childhood gender role behavior in women and men with congenital adrenal hyperplasia (CAH). *J. Sex. Res.* **41**(1), 75–81.

Hu, S., Pattatucci, A., Patterson, C., Li, L., Fulker, D., Cherny, S., Kruglyak, L., and Hamer, D. (1995). Linkage between sexual orientation and chromosome Xq28 in males but not females. *Nat. Genet.* **11**, 248–256.

Hutchinson, J. B. (1997). Gender-specific steroid in neural differentiation. *Cell. Mol. Neurobiol.* **17**, 603–626.

Jeske, Y. W., Bowles, J., Greenfield, A., and Koopman, P. (1995). Expression of a linear Sry transcript in the mouse genital ridge. *Nat. Genet.* **10**(4), 480–482.

Jones, M. B., and Blanchard, R. (1998). Birth order and male homosexuality: Extension of Slater's index. *Hum. Biol.* **70**(4), 775–787.

Jost, A. (1947). Reserches sur la differenciation sexuelle de l'embryon de lapin. *Arch. Anat. Microsc.* **36**, 271–315.

Jost, A. (1970). Hormonal factors in the sex differentiation of the mammalian foetus. *Philos. Trans. R. Soc. Lond. B. Biol. Sci.* **259**(828), 119–130.

Kendler, K. S., Thornton, L. M., Gilman, S. E., and Kessler, R. C. (2000). Sexual Orientation in a U.S. National Sample of Twin and Nontwin Sibling Pairs. *The Am. J. Psychiatry* **157**, 1843–1846.

King, C. Y., and Weiss, M. A. (1993). The SRY high-mobility-group box recognizes DNA by partial intercalation in the minor groove: A topological mechanism of sequence specificity. *Proc. Natl. Acad. Sci. USA* **90**(24), 11990–11994.

Kinsey, A. C., Pomeroy, W. B., and Martin, C. E. (1948). "Sexual Behavior in the Human Male." Indiana University Press, Bloomington.

Kinsey, A. C., Pomeroy, W. B., Martin, C. E., and Gebhard, P. H. (1953). "Sexual Behavior in the Human Female." WB Saunders, Philadelphia.

Kirk, K. M., Bailey, J. M., Dunne, M. P., and Martin, N. G. (2000). Measurement models for sexual orientation in a community twin sample. *Behav. Genet.* **30**(4), 345–356.

Koopman, P., Munsterberg, A., Capel, B., Vivian, N., and Lovell-Badge, R. (1990). Expression of a candidate sex-determining gene during mouse testis differentiation. *Nature* **348**(6300), 450–452.

Koopman, P., Gubbay, J., Vivian, N., Goodfellow, P., and Lovell-Badge, R. (1991). Male development of chromosomally female mice transgenic for Sry. *Nature* **351**(6322), 117–121.

Kraemer, B., Noll, T., Delsignore, A., Milos, G., Schnyder, U., and Hepp, U. (2006). Finger length ratio (2D:4D) and dimensions of sexual orientation. *Neuropsychobiology* **53**(4), 210–214.

Lahr, G., Maxson, S. C., Mayer, A., Just, W., Pilgrim, C., and Reisert, I. (1995). Transcription of the Y chromosomal gene, Sry, in adult mouse brain. *Brain Res. Mol. Brain Res.* **33**(1), 179–182.

Lander, E., and Kruglyak, L. (1995). Genetic dissection of complex traits: Guidelines for interpreting and reporting linkage results. *Nat. Genet.* **11**(3), 241–247.

Laudet, V., Stehelin, D., and Clevers, H. (1993). Ancestry and diversity of the HMG box superfamily. *Nucleic Acids Res.* **21**(10), 2493–2501.

Laumann, E. O., Gagnon, J. H., Michael, R. T., and Michaels, S. (1994). "The Social Organization of Sexuality: Sexual Practices in the United States." University of Chicago Press, Chicago.

Leranth, C., Roth, R. H., Elsworth, J. D., Naftolin, F., Horvath, T. L., and Redmond, D. E., Jr. (2000). Estrogen is essential for maintaining nigrostriatal dopamine neurons in primates: Implications for Parkinson's disease and memory. *J. Neurosci.* **20**(23), 8604–8609.

LeVay, S. (1991). A difference in hypothalamic structure between heterosexual and homosexual men. *Science* **253**, 1034–1037.

Lippa, R. A. (2002). Gender-related traits of heterosexual and homosexual men and women. *Arch. Sex. Behav.* **31**(1), 83–98.

Lippa, R. A. (2003). Are 2D:4D finger-length ratios related to sexual orientation? Yes for men, no for women. *J. Pers. Soc. Psychol.* **85**(1), 179–188.

Loehlin, J. C., McFadden, D., Medland, S. E., and Martin, N. G. (2006). Population differences in finger-length ratios: Ethnicity or latitude? *Arch. Sex. Behav.* **35**(6), 739–742.

Macke, J. P., Hu, N., Hu, S., Bailey, M., King, V. L., Brown, T., Hamer, D., and Nathans, J. (1993). Sequence variation in the androgen receptor gene is not a common determinant of male sexual orientation. *Am. J. Hum. Genet.* **53**(4), 844–852.

MacLusky, N. J., and Naftolin, F. (1981). Sexual differentiation of the central nervous system. *Science* **211**(4488), 1294–1302.

Maharjan, S., Serova, L., and Sabban, E. L. (2005). Transcriptional regulation of tyrosine hydroxylase by estrogen: Opposite effects with estrogen receptors alpha and beta and interactions with cyclic AMP. *J. Neurochem.* **93**(6), 1502–1514.

Mayer, A., Lahr, G., Swaab, D. F., Pilgrim, C., and Reisert, I. (1998). The Y-chromosomal genes SRY and ZFY are transcribed in adult human brain. *Neurogenetics* **1**(4), 281–288.

Mayer, A., Mosler, G., Just, W., Pilgrim, C., and Reisert, I. (2000). Developmental profile of Sry transcripts in mouse brain. *Neurogenetics* **3**(1), 25–30.

McEwen, B. S. (1981). Sexual differentiation of the brain. *Nature* **291**(5817), 610.

McFadden, D., and Champlin, C. A. (2000). Comparison of auditory evoked potentials in heterosexual, homosexual, and bisexual males and females. *J. Assoc. Res. Otolaryngol.* **1**(1), 89–99.

McFadden, D., and Pasanen, E. G. (1999). Spontaneous otoacoustic emissions in heterosexuals, homosexuals, and bisexuals. *J. Acoust. Soc. Am.* **105**(4), 2403–2413.

McFadden, D., Loehlin, J. C., Breedlove, S. M., Lippa, R. A., Manning, J. T., and Rahman, Q. (2005). A reanalysis of five studies on sexual orientation and the relative length of the 2nd and 4th fingers (the 2D:4D ratio). *Arch. Sex. Behav.* **34**(3), 341–356.

McKnight, J., and Malcolm, J. (2000). Is male homosexuality maternally linked? *Psych. Evol. Gender* **2**, 229–239.

Mustanski, B. S., Chivers, M. L., and Bailey, J. M. (2002). A critical review of recent biological research on human sexual orientation. *Annu. Rev. Sex. Res.* **13**, 89–140.

Mustanski, B. S., Dupree, M. G., Nievergelt, C. M., Bocklandt, S., Schork, N. J., and Hamer, D. H. (2005). A genomewide scan of male sexual orientation. *Hum. Genet.* **116**(4), 272–278.

Nasrin, N., Buggs, C., Kong, X. F., Carnazza, J., Goebl, M., and Alexander-Bridges, M. (1991). DNA-binding properties of the product of the testis-determining gene and a related protein. *Nature* **354**(6351), 317–320.

Nordeen, E. J., Nordeen, K. W., Sengelaub, D. R., and Arnold, A. P. (1985). Androgens prevent normally occurring cell death in a sexually dimorphic spinal nucleus. *Science* **229**(4714), 671–673.

Okten, A., Kalyoncu, M., and Yaris, N. (2002). The ratio of second- and fourth-digit lengths and congenital adrenal hyperplasia due to 21-hydroxylase deficiency. *Early. Hum. Dev.* **70**(1–2), 47–54.

Pattatucci, A. M. L., and Hamer, D. H. (1995). Development and familiality of sexual orientation in females. *Behav. Genet.* **25**(5), 407–420.

Pevny, L. H., and Lovell-Badge, R. (1997). Sox genes find their feet. *Curr. Opin. Genet. Dev.* **7**(3), 338–344.

Phoenix, C. H., Goy, R. W., Gerall, A. A., and Young, W. C. (1959). Organizing action of prenatally administered testosterone propionate on the tissues mediating mating behavior in the female guinea pig. *Endocrinology* **65,** 369–382.

Pillard, R. C. (1990). The Kinsey Scale: Is it familial? *In* "Homosexuality/Heterosexuality: Concepts of Sexual Orientation. The Kinsey Institute Series" (D. P. McWhirter, S. A. Sanders, and J. M. Reinisch, eds.), pp. 88–100. Oxford University Press, New York.

Pillard, R. C., and Weinrich, J. D. (1986). Evidence of familial nature of male homosexuality. *Arch. Gen. Psychiatry* **43,** 808–812.

Rahman, Q., and Wilson, G. D. (2003). Large sexual-orientation-related differences in performance on mental rotation and judgment of line orientation tasks. *Neuropsychology* **17**(1), 25–31.

Reisert, I., Kuppers, E., and Pilgrim, C. (1994). Sexual differentiation of central catecholamine systems. *In* "Phylogeny and Development of Catecholamine Systems in the CNS of Vertebrates" (W. Smeets and A. Reiner, eds.), pp. 453–462. Cambridge University Press, Cambridge.

Rex, M., Uwanogho, D. A., Orme, A., Scotting, P. J., and Sharpe, P. T. (1997). cSox21 exhibits a complex and dynamic pattern of transcription during embryonic development of the chick central nervous system. *Mech. Dev.* **66**(1–2), 39–53.

Rice, G., Anderson, C., Risch, N., and Ebers, G. (1999a). Male homosexuality: Absence of linkage to microsatellite markers at Xq28. *Science* **284,** 665–667.

Rice, G., Risch, N., and Ebers, G. (1999b). Response: Genetics and male sexual orientation. *Science* **285,** 803.

Rieger, G., Chivers, M. L., and Bailey, J. M. (2005). Sexual arousal patterns of bisexual men. *Psychol. Sci.* **16**(8), 579–584.

Roselli, C. E., Larkin, K., Resko, J. A., Stellflug, J. N., and Stormshak, F. (2004). The volume of a sexually dimorphic nucleus in the ovine medial preoptic area/anterior hypothalamus varies with sexual partner preference. *Endocrinology* **145**(2), 478–483.

Sanders, A. R., and Dawood, K. (2003). "Nature Encyclopedia of Life Sciences." Nature Publishing Group, London.

Segovia, S., Guillamon, A., del Cerro, M. C., Ortega, E., Perez-Laso, C., Rodriguez-Zafra, M., and Beyer, C. (1999). The development of brain sex differences: A multisignaling process. *Behav. Brain Res.* **105**(1), 69–80.

Serova, L. I., Maharjan, S., Huang, A., Sun, D., Kaley, G., and Sabban, E. L. (2004). Response of tyrosine hydroxylase and GTP cyclohydrolase I gene expression to estrogen in brain catecholaminergic regions varies with mode of administration. *Brain Res.* **1015**(1–2), 1–8.

Simerly, R. B., and Swanson, L. W. (1987). The distribution of neurotransmitter-specific cells and fibers in the anteroventral periventricular nucleus: Implications for the control of gonadotropin secretion in the rat. *Brain Res.* **400**(1), 11–34.

Simerly, R. B., Swanson, L. W., Handa, R. J., and Gorski, R. A. (1985). Influence of perinatal androgen on the sexually dimorphic distribution of tyrosine hydroxylase-immunoreactive cells and fibers in the anteroventral periventricular nucleus of the rat. *Neuroendocrinology* **40**(6), 501–510.

Simerly, R. B., McCall, L. D., and Watson, S. J. (1988). Distribution of opioid peptides in the preoptic region: Immunohistochemical evidence for a steroid-sensitive enkephalin sexual dimorphism. *J. Comp. Neurol.* **276**(3), 442–459.

Stikkelbroeck, N. M., Beerendonk, C. C., Willemsen, W. N., Schreuders-Bais, C. A., Feitz, W. F., Rieu, P. N., Hermus, A. R., and Otten, B. J. (2003). The long term outcome of feminizing genital surgery for congenital adrenal hyperplasia: Anatomical, functional and cosmetic outcomes, psychosexual development, and satisfaction in adult female patients. *J. Pediatr. Adolesc. Gynecol.* **16**(5), 289–296.

Suzuki, Y., Ishii, H., and Arai, Y. (1983). Prenatal exposure of male mice to androgen increases neuron number in the hypogastric ganglion. *Brain Res.* **312**(1), 151–154.

Swaab, D. F., and Fliers, E. (1985). A sexually dimorphic nucleus in the human brain. *Science* **228** (4703), 1112–1115.

Swaab, D. F., Zhou, J. N., Fodor, M., and Hofman, M. A. (1997). Sexual differentiation of the human hypothalamus: Differences according to sex, sexual orientation, and transsexuality. *In* "Sexual Oreientation: Toward Biological Understanding Ellis, Lee" (Ebertz Linda, ed.), pp. 129–150, *xxii*, 276 see book. Praeger. Publishers/Greenwood Publishing Group, Inc., Westport, CT, US.

Toran-Allerand, C. D. (1976). Sex steroids and the development of the newborn mouse hypothalamus and preoptic area *in vitro*: Implications for sexual differentiation. *Brain Res.* **106**(2), 407–412.

Uwanogho, D., Rex, M., Cartwright, E. J., Pearl, G., Healy, C., Scotting, P. J., and Sharpe, P. T. (1995). Embryonic expression of the chicken Sox2, Sox3 and Sox11 genes suggests an interactive role in neuronal development. *Mech. Dev.* **49**(1–2), 23–36.

Vadasz, C., Kobor, G., Kabai, P., Sziraki, I., Vadasz, I., and Lajtha, A. (1988). Perinatal anti-androgen treatment and genotype affect the mesotelencephalic dopamine system and behavior in mice. *Horm. Behav.* **22**(4), 528–539.

Wade, J., and Arnold, A. P. (1996). Functional testicular tissue does not masculinize development of the zebra finch song system. *Proc. Natl. Acad. Sci. USA* **93**(11), 5264–5268.

Wegner, M. (1999). From head to toes: The multiple facets of Sox proteins. *Nucleic Acids. Res.* **27**(6), 1409–1420.

Wellings, K., Field, J., Johnson, A., and Wadsworth, J. (1994). "Sexual Behaviour in Britain." Penguin Books, New York.

Whalen, R. (1968). Differentiation of the neural mechanisms which control gonadotrophins secretion and sexual behavior. *In* "Reproduction and Sexual Behavior" (M. Diamond, ed.), pp. 303–340. Indiana University Press, Bloomington.

Williams, T. J., Pepitone, M. E., Christensen, S. E., Cooke, B. M., Huberman, A. D., Breedlove, N. J., Breedlove, T. J., Jordan, C. L., and Breedlove, S. M. (2000). Finger-length ratios and sexual orientation. *Nature* **404**, 455–456.

Zucker, K. J., and Bradley, S. J. (1995). "Gender Identity Disorder and Psychosexual Problems in Children and Adolescents." The Guilford Press, New York.

Index

Q

R

S

W

X

Y

Z

A

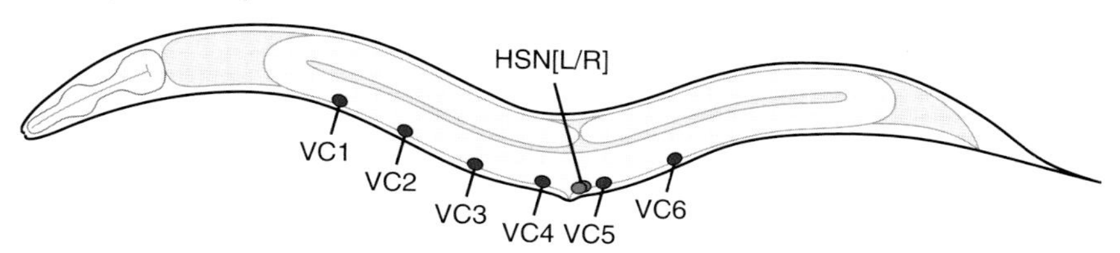

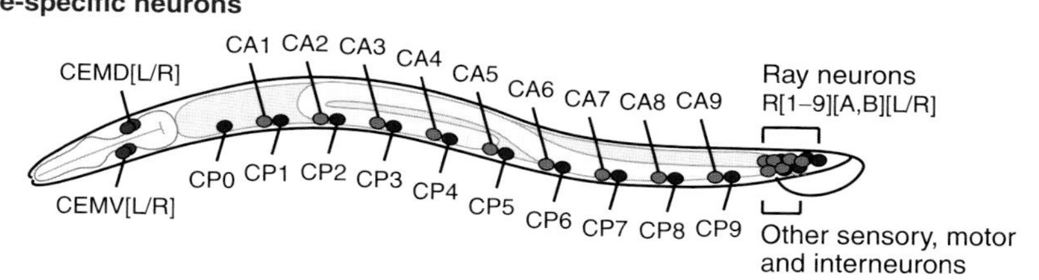

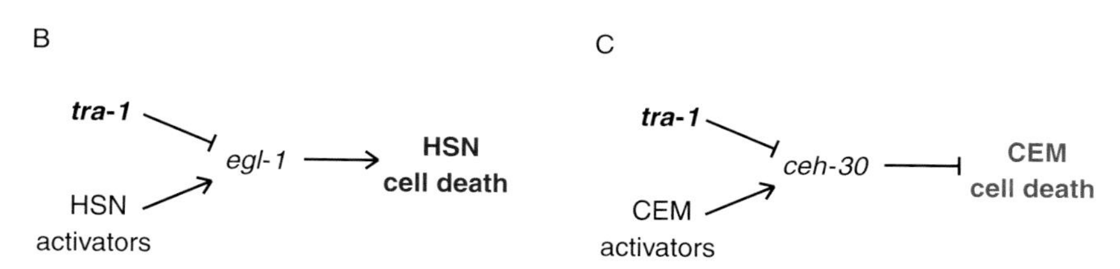

Chapter 1, Figure 1.2. Sexual dimorphism in the *C. elegans* nervous system. (A) The adult hermaphrodite (above) has eight sex-specific neurons: six VC neurons in the ventral cord and two HSN motor neurons. These neurons regulate hermaphrodite egg-laying behavior. The adult male (below) has 89 sex-specific neurons. Four of these, the CEMs, are sensory neurons in the head and are thought to have a role in detecting hermaphrodite pheromone cues (see text). The male ventral cord contains the CA and CP motor neurons; both of these are implicated in specific steps of male mating behavior (Loer and Kenyon, 1993; Schindelman *et al.*, 2006). The male tail contains a variety of sensory, motor, and interneurons; the largest class of these is the RnA and RnB ray sensory neurons. (B) Sexual dimorphism in the presence of the HSN is regulated by differential programmed cell death. In XX animals, *tra-1* is active and represses *egl-1* expression in the HSNs. In X0 animals, unknown activators promote *egl-1* expression in the HSNs, leading to cell death (Conradt and Horvitz, 1999). (C) Differential cell death also underlies the sex difference in the CEM neurons. In XX animals, *tra-1* prevents the expression of the survival factor *ceh-30* in the CEM neurons, leading to their death. In X0 animals, *ceh-30* prevents HSN death; whether this occurs through regulation of *egl-1* is unknown (see text for details).

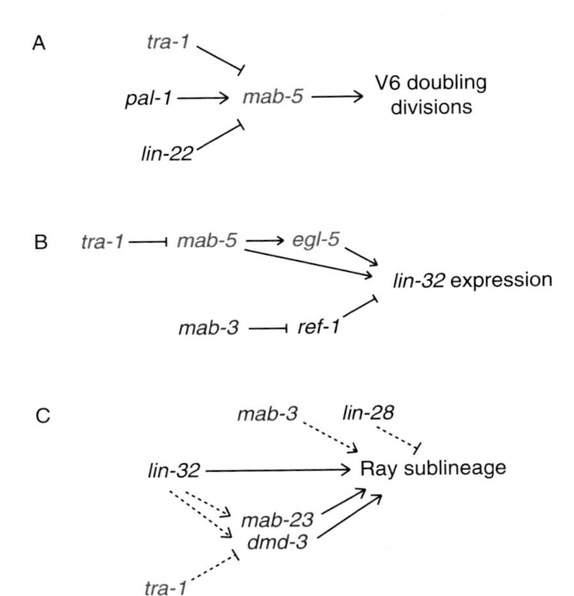

Chapter 1, Figure 1.4. Sexually dimorphic steps in ray development. (A) The male-specific lineage alterations in V6 require the Hox gene *mab-5*. *mab-5* expression requires the *caudal* orthologue *pal-1* (Hunter *et al.*, 1999); its anterior boundary is limited by the bHLH gene *lin-22* (Kenyon, 1986; Wrischnik and Kenyon, 1997). The male-specific maintenance of *mab-5* in the V6 lineage is likely to require *tra-1*, though the mechanism is not known. (B) Expression of *lin-32*, the ray proneural gene, requires the Hox genes *mab-5* and *egl-5* (Yi *et al.*, 2000; Zhao, 1995). In addition, the DM-domain gene *mab-3* promotes *lin-32* expression by inhibiting the action of the bHLH gene *ref-1* (Ross *et al.*, 2005). (C) The mechanisms that generate male-specific cell types (ray cells) from the ray sublineage are not well understood. *mab-3* is necessary for *lin-32* to efficiently drive ray production (Yi *et al.*, 2000), and *mab-23* and *dmd-3* have a role in the differentiation of the RnA neuron (Lints and Emmons, 2002; R. Lints and D.S. Portman, unpublished data). Whether *mab-23* and *dmd-3* act downstream of or in parallel with *lin-32* is not known, nor is whether *tra-1* may block *dmd-3* expression in the hermaphrodite seam. *lin-28* may act to confer temporal specificity to the ray sublineage, preventing a ray from being made in the L2 postdeirid lineage (Ambros and Horvitz, 1984).

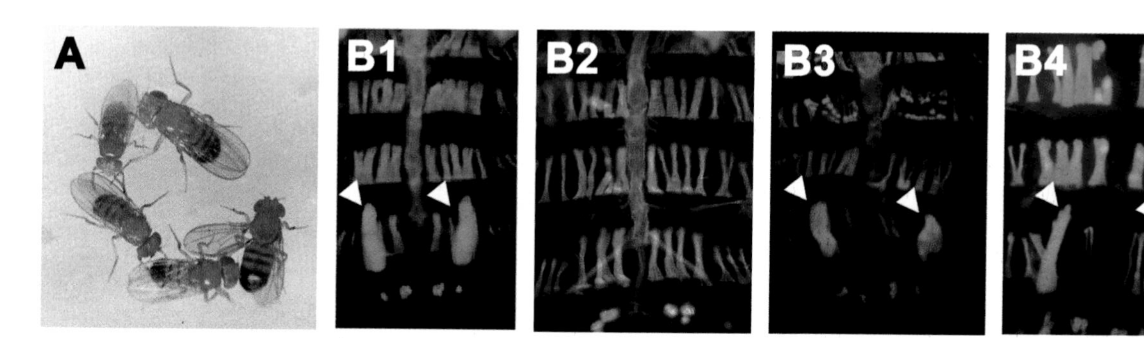

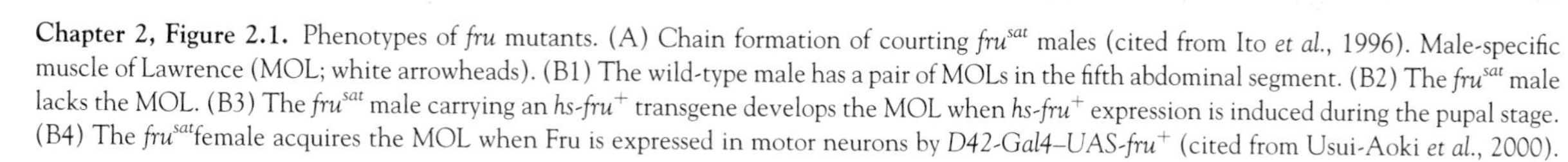

Chapter 2, Figure 2.1. Phenotypes of *fru* mutants. (A) Chain formation of courting *fru*^{sat} males (cited from Ito *et al.*, 1996). Male-specific muscle of Lawrence (MOL; white arrowheads). (B1) The wild-type male has a pair of MOLs in the fifth abdominal segment. (B2) The *fru*^{sat} male lacks the MOL. (B3) The *fru*^{sat} male carrying an *hs-fru*$^{+}$ transgene develops the MOL when *hs-fru*$^{+}$ expression is induced during the pupal stage. (B4) The *fru*^{sat}female acquires the MOL when Fru is expressed in motor neurons by *D42-Gal4–UAS-fru*$^{+}$ (cited from Usui-Aoki *et al.*, 2000).

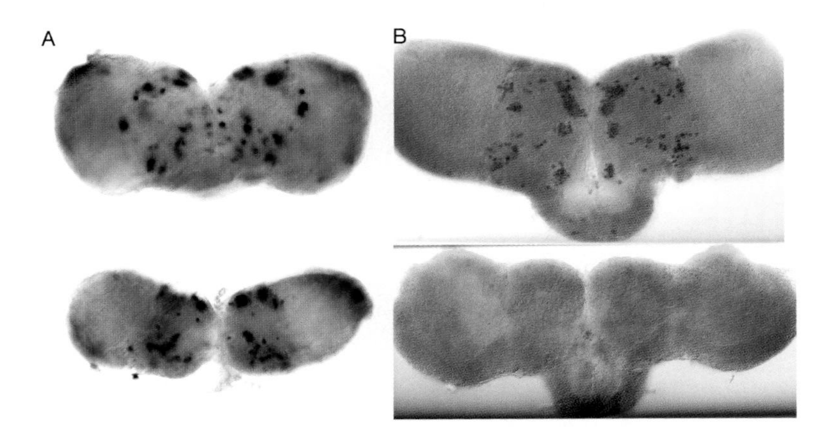

Chapter 2, Figure 2.2. Expression of *fru* gene products in pupal brain (cited from Usui-Aoki *et al.*, 2000). (A) *fru* mRNA expression as detected by *in situ* hybridization in wild-type male (upper) and female (lower). (B) Fru expression as detected by immunostaining with anti-Fru antibody in wild-type male (upper) and female (lower).

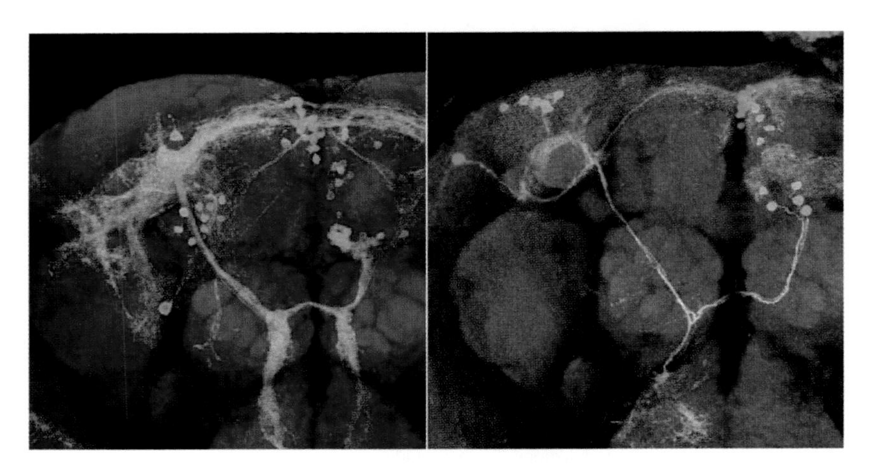

Chapter 2, Figure 2.3. Sexual dimorphism in mAL neurons that express *fru* (cited from Kimura *et al.*, 2005). The mAL cluster is composed of 30 neurons per hemisphere in males (left) and 5 neurons in females (right). mAL neurons in males have both contralateral and ipsilateral neurites (left), whereas those in females extend only contralateral neurites (right). Dendritic branches in the subesophageal ganglion (lowest part of the neurons) are horsetail-like in males (left) or forked in females (right).

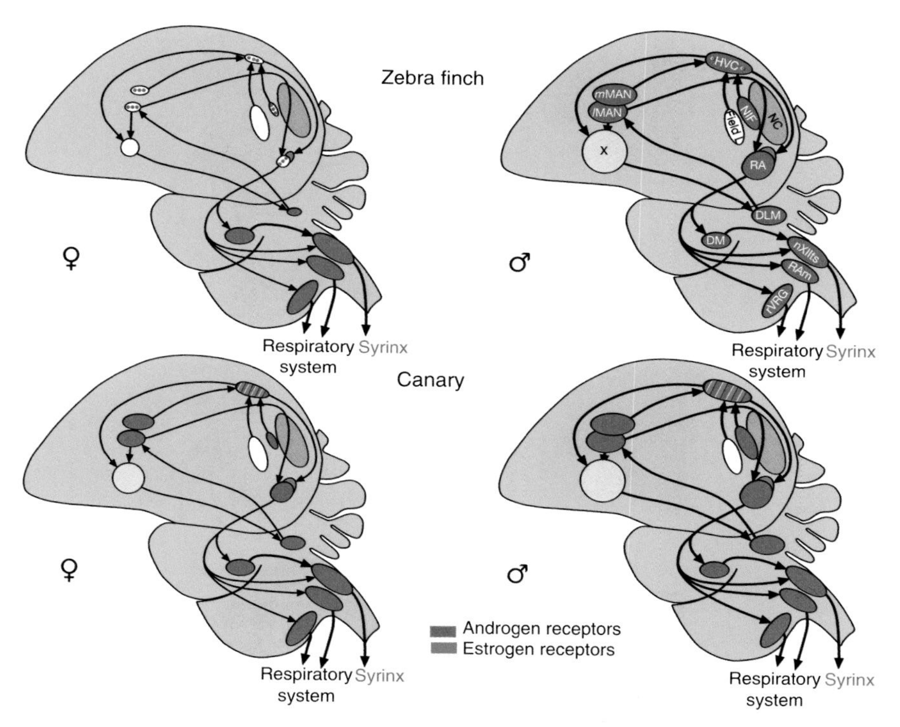

Chapter 3, Figure 3.1. The vocal control system of songbirds and degrees of sexual dimorphisms. This schematic diagram of a composite view of parasagittal sections of a songbird brain gives approximate positions of vocal nuclei and brain regions and their content of androgen receptors (red, rose) and estrogen receptors (green). All structures are bilateral—for reasons of clarity, only those in one-half of the brain are depicted. Further, some thalamic brain areas that appear important for coordination of the left and right vocal control network are omitted. The volumes of vocal control areas of adult zebra finches are highly sexually dimorphic, while those of canaries are to a lesser extent, indicated in the relative size of the areas. Despite these sex differences, all areas and connections as well as the sex hormone receptors are present in the female vocal control system and syrinx. The HVC of canaries contains higher amounts of estrogen receptors compared to the zebra finch. Area X and NC (rose) contains androgen receptors in only some animals. Androgen receptors are found in the caudal nidopallium including the caudomedial nidopallium. Abbreviations: DLM, nucleus dorsolateralis anterior, pars medialis; DM, dorsomedial nucleus of the midbrain nucleus intercollicularis; HVC, acronym used as a proper name; formerly known as high vocal center; Field L; lMAN, lateral magnocellular nucleus of the anterior nidopallium; mMAN, medial magnocellular nucleus of the anterior nidopallium; NC, caudal nidopallium; NIF, nucleus interface of the nidopallium; nXIIts, tracheosyringeal portion of the nucleus hypoglossus; RA, robust nucleus of the arcopallium; RAm, nucleus retroambigualis; rVRG, rostroventral respiratory group; Area X.

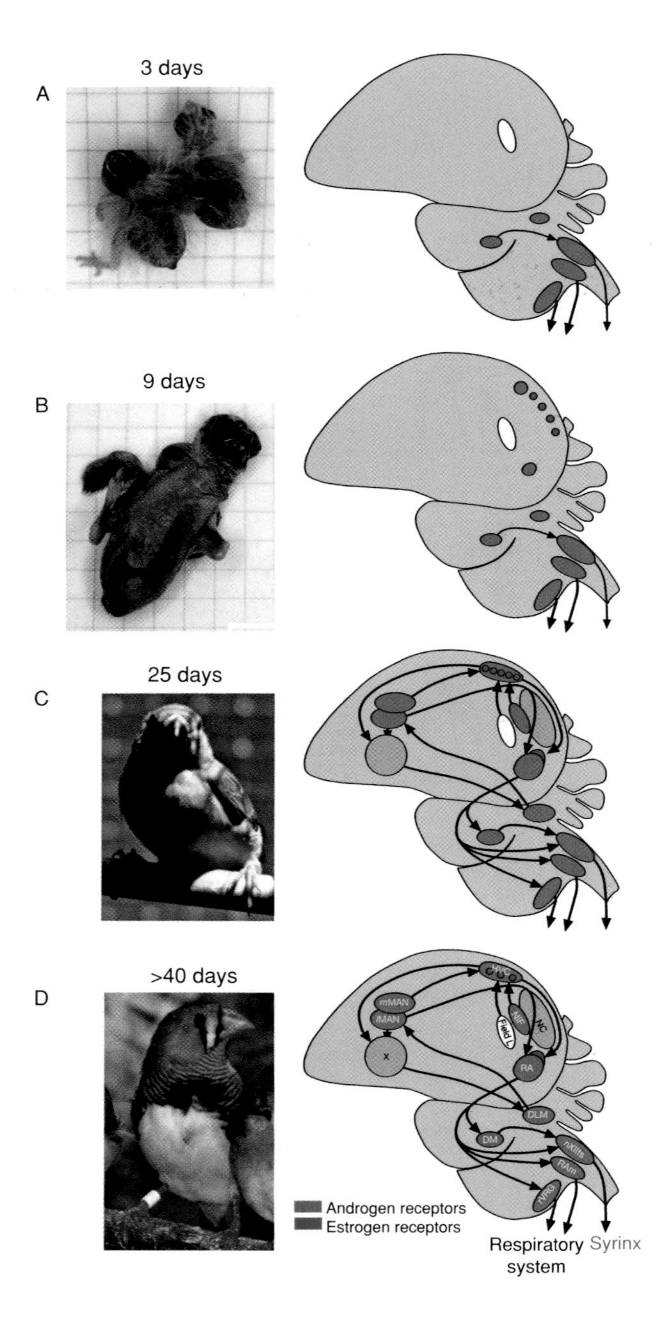

A
3 days
B
9 days
C
25 days
D
>40 days
HVC
mMAN
lMAN
NIF
Field L
NC
X
RA
DLM
DM
nXIIts
RAm
rVRG
Androgen receptors
Estrogen receptors
Respiratory system
Syrinx

Chapter 3, Figure 3.3. The development of the vocal control system of the zebra finch. Zebra finch chicks start to vocalize around posthatching (P) day 3–5 (A). At this time, forebrain vocal areas are not detectable. At P9 or later (B), HVC and RA are detectable and contain androgen receptors, and estrogen receptors are expressed along the lateral ventricle. Chicks produce regularly begging calls and continue to do so throughout ontogeny. At about P25 (C), all vocal areas are differentiated (although in part sexually dimorphic) and contain androgen (red) or estrogen (green) receptors in both sexes. Further, all areas are connected as in the adults except HVC and RA and juveniles are still not singing. After the formation of the HVC to RA synapse between P28 and P35 (D), juveniles start to sing and forebrain vocal areas continue to differentiate. In zebra finches, most HVC neurons of the females do not form this synapse and the forebrain vocal areas become highly sexually dimorphic till adulthood (see Fig. 1). In relation, only male zebra finches sing. In females of other species such as canaries (Fig. 1), larger numbers of HVC neurons develop the HVC to RA synapse and females sing in a species-typical manner. Further, the timing of these developmental steps differs between species.

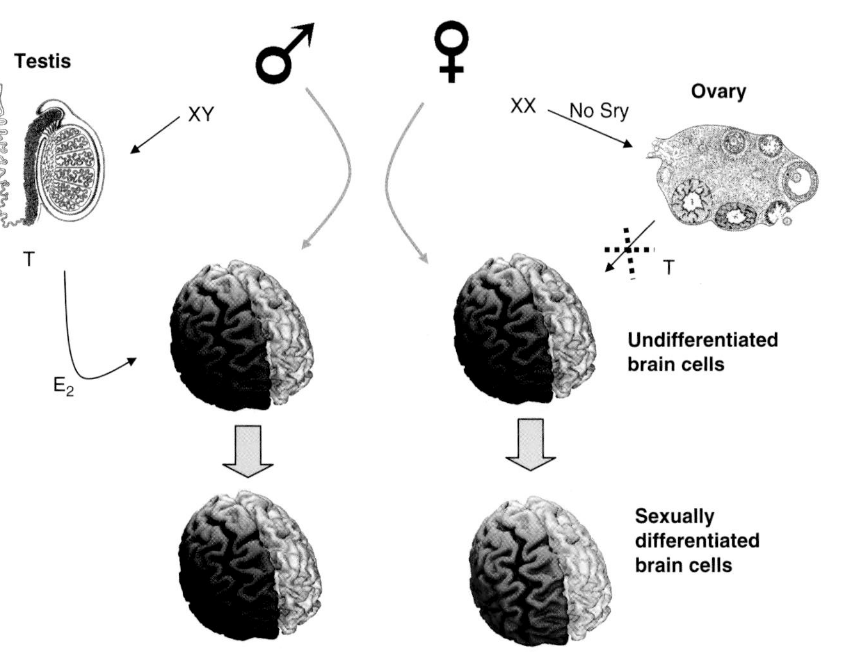

Chapter 9, Figure 9.1. Traditional and alternative paradigm of brain sexual differentiation. Classic theory (black arrow): In XY mammals, Sry triggers the differentiation of the testis, which secretes testosterone, which in turn masculinizes the brain, directly or via aromatization to estradiol. In XX mammals, Sry is absent, the gonads develop into ovaries, which do not secrete testosterone, and the brain develops as feminized. Alternative theory (orange arrow): In addition to hormonal effects, the genes on the X and Y chromosome directly influence sex differences in the brain.